Microbial Biosorption of Metals

Pavel Kotrba • Martina Mackova • Tomas Macek
Editors

Microbial Biosorption of Metals

Springer

Editors
Dr. Pavel Kotrba
Department of Biochemistry
and Microbiology
Faculty of Food and Biochemical
Technology
Institute of Chemical Technology
Technicka 3, 16628 Prague
Czech Republic
pavel.kotrba@vscht.cz

Dr. Martina Mackova
Department of Biochemistry
and Microbiology
Faculty of Food and Biochemical
Technology
Institute of Chemical Technology
Technicka 3, 16628 Prague
Czech Republic
martina.mackova@vscht.cz

Dr. Tomas Macek
IOCB & ICT Joint Laboratory
Institute of Organic Chemistry
and Biochemistry
Academy of Sciences of the Czech Republic
Flemingovo n. 2, 16610 Prague
Czech Republic
tom.macek@uochb.cas.cz

ISBN 978-94-007-0442-8 e-ISBN 978-94-007-0443-5
DOI 10.1007/978-94-007-0443-5
Springer Dordrecht Heidelberg London New York

Cover design: deblik, Berlin

Printed on acid-free paper

Springer is part of Springer Science+Business Media (www.springer.com)

Preface

The word *biosorption* unites a biological entity with a physico-chemical process of sorption. Indeed, the biosorption of metal ions is a metabolism-independent metal accumulation event, which takes place at the cell wall by polysaccharides, associated molecules, and functional groups. It is an ubiquitous property of living or dead biomass and derived products, and is undoubtedly an important process in the environment. Since the early 80s of the previous century, the biosorption with biosorbents formulated from non-living biomass has also become recognized as a promising biotechnology for heavy metal removal from liquid waste streams. When we examine the ISI Web of Science, we can see that the number of journal papers published with *biosorption* and *metal* in their subject matter is at nearly 2700. Also the continuing increase in research published on biosorption can be seen, especially during the last decade. While there were 96 metal biosorption articles in 2000, the figure nearly doubled in 2005 to 178 articles. In 2009, the number of articles jumped to 393. These studies inspected biosorption from different angles—from (micro)biology and (bio)chemistry to process engineering points of view—and significantly contributed to elucidation of the biosorption phenomenon and its biotechnological potential. This book attempts to collect review articles which do justice to the multidisciplinary nature of biosorption studies. We are well aware of the fact that a single volume could not cover all the particular aspects one could think about in connection with biosorption. However, we do believe that it provides a solid summation of the present state of the biosorption art. At this point, it is our great pleasure to thank the team of authors whose fine contributions made this book possible.

Prague, Czech Republic

Dr. Pavel Kotrba
Dr. Martina Mackova
Dr. Tomas Macek

Contents

Contributors

Yves Andrès Ecole des Mines de Nantes, GEPEA UMR CNRS 6144, BP 20722, 44307 Nantes Cedex 3, France
e-mail: yves.andres@emn.fr

Antonio Ballester Departamento de Ciencia de los Materiales e Ingeniería Metalúrgica, Facultad de Ciencias Químicas, Universidad Complutense, 28040 Madrid, Spain
e-mail: ambape@quim.ucm.es

María Luisa Blázquez Departamento de Ciencia de los Materiales e Ingeniería Metalúrgica, Facultad de Ciencias Químicas, Universidad Complutense, 28040 Madrid, Spain
e-mail: mlblazquez@quim.ucm.es

Pavel Dostálek Department of Fermentation Chemistry and Bioengineering, Institute of Chemical Technology Prague, Technicka 3, 166 28 Prague, Czech Republic
e-mail: pavel.dostalek@vscht.cz

Jan Fišer Department of Biochemistry and Microbiology, Institute of Chemical Technology Prague, Technicka 3, 166 28 Prague, Czech Republic
e-mail: jan.fiser@vscht.cz

Felisa González Departamento de Ciencia de los Materiales e Ingeniería Metalúrgica, Facultad de Ciencias Químicas, Universidad Complutense, 28040 Madrid, Spain
e-mail: fgonzalezg@quim.ucm.es

Camino García-Balboa Departamento de Ciencia de los Materiales e Ingeniería Metalúrgica, Facultad de Ciencias Químicas, Universidad Complutense, 28040 Madrid, Spain
e-mail: abipsoferro1@yahoo.es

Claire Gérente Ecole des Mines de Nantes, GEPEA UMR CNRS 6144, BP 20722, 44307 Nantes Cedex 3, France
e-mail: claire.gerente@emn.fr

Katerina Horska Department of Nanobiotechnology, Institute of Systems Biology and Ecology, Academy of Sciences, Na Sadkach 7, 370 05 Ceske Budejovice, Czech Republic
e-mail: horska@usbe.cas.cz

Pavel Kotrba Department of Biochemistry and Microbiology, Institute of Chemical Technology Prague, Technicka 3, 166 28 Prague, Czech Republic
e-mail: pavel.kotrba@vscht.cz

Kouichi Kuroda Division of Applied Life Sciences, Graduate School of Agriculture, Kyoto University, Sakyo-ku, Kyoto 606-8502, Japan
e-mail: k_kuro@kais.kyoto-u.ac.jp

Tomas Macek IOCB & ICT Joint Laboratory, Institute of Organic Chemistry and Biochemistry, Academy of Sciences of the Czech Republic, Flemingovo n. 2, 166 10 Prague 6, Czech Republic
e-mail: tom.macek@uochb.cas.cz

Department of Biochemistry and Microbiology, Institute of Chemical Technology Prague, Technicka 3, 166 28 Prague, Czech Republic

Martina Mackova Department of Biochemistry and Microbiology, Faculty of Food and Biochemical Technology, Institute of Chemical Technology Prague, Technicka 3, 166 28 Prague 6, Czech Republic
e-mail: martina.mackova@vscht.cz

Jesús Ángel Muñoz Departamento de Ciencia de los Materiales e Ingeniería Metalúrgica, Facultad de Ciencias Químicas, Universidad Complutense, 28040 Madrid, Spain
e-mail: jamunoz@quim.ucm.es

Ghinwa Naja Science Department, Everglades Foundation, 18001 Old Cutler Road, Palmetto Bay, FL 33157, USA
e-mail: mnaja@evergladesfoundation.org

Francesca Pagnanelli Department of Chemistry, Sapienza University of Rome, P.le A. Moro 5, 00185 Rome, Italy
e-mail: francesca.pagnanelli@uniroma1.it

Esther Romera Departamento de Ciencia de los Materiales e Ingeniería Metalúrgica, Facultad de Ciencias Químicas, Universidad Complutense, 28040 Madrid, Spain
e-mail: eromerag@hotmail.com

Lubomír Rulíšek Gilead Sciences & IOCB Research Center, Institute of Organic Chemistry and Biochemistry, Academy of Sciences of the Czech Republic, Flemigovo náměstí. 2, 166 10 Prague 6, Czech Republic
e-mail: lubos@uochb.cas.cz

Tomas Ruml Department of Biochemistry and Microbiology, Institute of Chemical Technology Prague, Technicka 3, 166 28 Prague, Czech Republic
e-mail: tomas.ruml@vscht.cz

Ivo Safarik Department of Nanobiotechnology, Institute of Systems Biology and Ecology, Academy of Sciences, Na Sadkach 7, 370 05 Ceske Budejovice, Czech Republic
e-mail: ivosaf@yahoo.com

Regional Centre of Advanced Technologies and Materials, Palacky University, Slechtitelu 11, 783 71 Olomouc, Czech Republic

Mirka Safarikova Department of Nanobiotechnology, Institute of Systems Biology and Ecology, Academy of Sciences, Na Sadkach 7, 370 05 Ceske Budejovice, Czech Republic
e-mail: mirkasaf@usbe.cas.cz

Asha Srinivasan Faculty of Engineering and Applied Science, University of Regina, Regina S4S0A2, Canada
e-mail: sriniash@uregina.ca

Mitsuyoshi Ueda Division of Applied Life Sciences, Graduate School of Agriculture, Kyoto University, Sakyo-ku, Kyoto 606-8502, Japan
e-mail: miueda@kais.kyoto-u.ac.jp

Kuppusamy Vijayaraghavan Singapore-Delft Water Alliance, National University of Singapore, 1 Engineering Drive 2, Singapore 117576, Singapore
e-mail: cvekv@nus.edu.sg

Thiruvenkatachari Viraraghavan Faculty of Engineering and Applied Science, University of Regina, Regina S4S0A2, Canada
e-mail: t.viraraghavan@uregina.ca

Bohumil Volesky Department of Chemical Engineering, McGill University, 3610 University Street, Montreal, QC H3A 2B2, Canada
e-mail: bohumil.volesky@mcgill.ca

Sung Wook Won Division of Environmental and Chemical Engineering, Research Institute of Industrial Technology, Chonbuk National University, Jeonjubuk 561-756, South Korea
e-mail: sungukw@chonbuk.ac.kr

Yeoung-Sang Yun Division of Environmental and Chemical Engineering, Research Institute of Industrial Technology, Chonbuk National University, Jeonjubuk 561-756, South Korea
e-mail: ysyun@chonbuk.ac.kr

Chapter 1
Microbial Biosorption of Metals—General Introduction

Pavel Kotrba

Abstract Discharge of waste contaminated with heavy metals and related elements is known to have an adverse effect on the environment and solving this problem has for long been presented as a challenge. Nowadays, continuing demand for and increasing value of high-tech metals and rare earth elements makes efficient recycling technologies of utmost importance. In solving these tasks, the biosorption—sequestration of heavy metals, radionuclides and rare earth elements usually by non-living biomass—can be a part of the solution.

Keywords Decontamination • Bacterial biosorbent • Fungal biosorbent • Algal biosorbent • Mechanism of biosorption • Modeling of biosorption

1.1 Brief View on Conventional Waste Stream Treatments

Industrialization has long been accepted as a hallmark of civilization. The Boulton and Watt steam engine, the synonym of industrial revolution, propelled huge changes in mining, metallurgical technology, manufacturing, transport and agriculture. Since then, progressive metallurgy and the use of metals and chemicals in numerous industries have resulted in a generation of large quantities of liquid effluent loaded with high levels of heavy metals, often as bioavailable mobile and thus toxic ionic species (Calderón et al. 2003; Peakall and Burger 2003; Gadd 1992a). Due to their elemental non-degradable nature, heavy metals always and regardless of their chemical form, pose serious ecological risk, when released into the environment. Not only is there the demand for cleanup of contaminated waste water to meet regulatory agency limits, but there is also increasing value of some metals which place a call for efficient and low-cost effluent treatment and metal recuperation technologies.

P. Kotrba (✉)
Department of Biochemistry and Microbiology, Institute of Chemical Technology Prague, Technicka 3, 166 28 Prague, Czech Republic
e-mail: pavel.kotrba@vscht.cz

P. Kotrba et al. (eds.), *Microbial Biosorption of Metals*,
DOI 10.1007/978-94-007-0443-5_1,

Conventional procedures for heavy metal removal from aqueous industrial effluents involve precipitation, ion exchange, electrochemical methods and reverse osmosis. Another promising approach is solvent extraction. Conversion of metal ions to insoluble forms by chemical precipitation is the most common method, reducing the metal content of solution to the levels of mg l^{-1}. The cheapest precipitation technique relies on alkalization of the metal solution (usually with lime) to achieve formation of insoluble metal species, namely of hydroxides. Chemical precipitation could also be achieved by the addition of other coagulants, such as of potash alum, sodium bisuphite, sulphide or iron salts. Though it is cost effective; such precipitation lacks the specificity, produces large volumes of high water content sludge and has low performance at low metal concentrations. Although adsorption using activated carbon is generally expensive (and not suitable for many metal species), it is an efficient method for the removal of metallic mercury following chemical reduction (e.g., with hydrazine) of mercuric ions in heavily contaminated process waters.

Ion exchange employing manmade synthetic organic resins is the most common method. It becomes the method of choice especially for its capacity to reduce the metal contents to μg l^{-1} levels in relatively large volumes of effluent, some possibility to formulate metal-selective resin and well established procedures for metal recovery from and reuse of the ion exchanger. This method is, however, relatively expensive, which therefore makes the processing of concentrated metal solutions cost intensive. Precautions should also be taken to prevent the poisoning of ion exchanger by organics and solids in solution.

Electro-winning, employing electro-deposition of metals on anodes is popularly used for the recuperation of metals in mining and metallurgical operations as well as in electrical industries and electronics. Electrodialysis involves the use of ion selective semi-permeable membranes fitted between the charged electrodes attracting respective ions (in the case of metal cations, the anode compartment is smaller to concentrate the metal in). The main disadvantage of electrodialysis operation is clogging of the membrane by metal hydroxides formed during the process. Like electrodialysis, reverse osmosis and ultrafiltration employ semi-permeable membranes which allow water to pass, while solutes, including heavy metals, remain contained in retentate. The advantages of membrane-based processes involve some selectivity of metal separation and tolerance to changes in pH. One disadvantage of membrane-based approaches is that they are cost intensive.

Reactive two-phase extraction complexing extractants specifically (or preferentially) dissolved in organic solvents has been suggested as another technological alternative (Schwuger et al. 2001). This approach may provide a viable method for the selective recuperation of metals, e.g., of platinum group metals from spent catalysts (Marinho et al. 2010). Suitable extractants for platinum involve organophosphorus compounds, aliphatic amines and ammonium quaternary salts. The main disadvantages of this process are the difficult recovery of metals from organic phase and the toxicity of extractants.

1.2 Bio-based Methods for Waste Water Treatment and Environment Restoration

The natural capacity of microrganisms, fungi, algae and plants to take up heavy metal ions and radionuclides and, in some cases, to promote their conversion to less toxic forms has sparked the interest of (micro)biologists, biotechnologists and environmental engineers for several decades. Consequently, various concepts for "bio-removal" of metals from waste streams and bioremediations of contaminated environment are being proposed, some of which were brought to pilot or industrial scale (Bargar et al. 2008; Macek et al. 2008; Muyzer and Stams 2008; Singh et al. 2008; Chaney et al. 2007; Sheoran and Sheoran 2006; Volesky 2004; Lloyd et al. 2003; Ruml and Kotrba 2003; Baker et al. 2000; Gadd 1992b; see also Chap. 2). The "bio" prefix refers to the involvement of biological entity, which is living organisms, dead cells and tissues, cellular components or products. The ultimate goal of these efforts is to provide an economical and eco-friendly technology, efficiently working also at metal levels below 10 mg l^{-1}. These are the features that living as well as dead biomass could be challenged for. There are generally three routes to follow considering "bio-removal" of metallic species from solutions. The first two rely on properties of living cells and involve active metal uptake—bioaccumulation (i.e., plasma membrane mediated transport of metal ion into cellular compartment) and eventual chemical conversion of mobile metal to insoluble forms. The later may occur in the cytoplasm, at the cell surface or in the solution by precipitation of metal ion with metabolites, via redox reactions or by their combination. The effectiveness of the process will depend on the (bio)chemistry of particular metal and on metabolic activity of eligible organism, which is in turn affected by the presence of metal ions. To this point, the use of metallotolerant species or physical separations of the production of metal-precipitating metabolite from metal precipitation in contaminated solution produce viable methods. For their importance in the treatment of industrial liquid streams as well as of the environmental pollution are some of these approaches discussed in Chap. 9. Several of them are to various extents dependent on or involve the metabolism-independent metal uptake event at the cell wall by polysaccharides, associated molecules, and functional groups. This metal sequestration capacity is commonly known as biosorption, which itself represents the third potent way of "bio-removal" of metals from solution.

Biosorption is a general property of living and dead biomass to rapidly bind and abiotically concentrate inorganic or organic compounds from even very diluted aqueous solutions. As a specific term, biosorption is used to depict a method that utilizes materials of biological origin—biosorbents formulated from non-living biomass—for the removal of target substances from aqueous solutions. Biosorption "traditionally" covers sequestration of heavy metals as well as rare earth elements and radionuclides or metalloids, but the research and applications extended to the removal of organics, namely dyes (Kaushik and Malik 2009; see also some examples with magnetic biocomposite biosorbents in Chap. 13), and biosorption is being proposed for the recovery of high-value proteins, steroids, pharmaceuticals and drugs (Volesky 2007).

Decades of biosorption research provided a solid understanding of the mechanism underlying microbial biosorption of heavy metals and related elements. It involves such physico-chemical processes as adsorption, ion-exchange, chelatation, complexation and microprecipitation. These depend on the type and ionic form of metal, the type of metal binding site available from microbial biomass, as well as on various external environmental factors (see Chap. 3). Accumulated knowledge resulted in the development of suitable modelling approaches comprehensively described in Chap. 4. When properly used, these models explain the equilibrium biosorption data, the kinetics in batch reactors and the dynamics in biosorption columns both for single and multimetal systems and provide a powerful tool for the design and development of the actual biosorption process. It should be noted here that it was due to a poor understanding of mechanisms and kinetics of AlgaSORB™ and AMT-Bioclaim™ processes commercialized in the early 1990s that hindered the adequate assessment of process performance and limitations, and thus the expected widespread industrial application of biosorption.

Biosorbents are derived from raw biomass selected for its superior metal-sequestering capacity. Investigated biomass types are of such diverse origins as bacterial, cyanobacterial, fungal (including filamentous fungi and unicellular yeasts), algal, plant or even animal (chitosan). This book covers development in major areas exploiting bacterial biomass (Chap. 5), fungal biomass (Chap. 6) and algal biomass (including macroalgae; Chap. 7) for the biosorption of heavy metal and radionuclides as well as for the sequestration of precious and rare earth elements (Chap. 8). It is noteworthy to add that the potential of plant-based biosorbents formulated from agricultural waste is attaining growing attention (Demirbas 2008; Sud et al. 2008; a few examples with magnetic biocomposite biosorbents are given in Chap. 13). The cheapest microbial biomass could be procured from selected fermentation industries as waste by-product or could be harvested from its natural habitat when it is produced in sufficient quantity there (e.g., marine macroalgae). Independent propagation of biomass under specific conditions optimizing its metallosorption properties is another option. Some efforts have been also devoted to modifications of yeast (Chap. 10) and bacterial (Chap. 11) cell walls through their genetic engineering, resulting in a surface display of particular amino acid sequences providing additional (even selective) metal-binding sites.

When derived from dead raw biomass featuring high metal uptake, the biosorbent for its practical application should exhibit some additional characteristics improving its stability and favoring hydrodynamic process conditions. To this end, biosorbent particle size, density and porosity, hardness, resistance to a broad range of physical and chemical conditions could be tailored by an appropriate immobilization method. Conventional strategies of biosorbent formulation from different types of microbial biomass are described in respective chapters as well as in Chap. 12. Chapter 13 further sets the biosorbent design forward to "smart materials", the magnetically responsive biocomposites improving biosorbents applicability by enabling their selective magnetic separation even from solutions containing suspended solids.

1.3 Future Thrusts in Biosorption

Compared with conventional or some biological methods for removing metal ions from industrial effluents, the biosorption process offers the advantages of low operating cost, minimization of the use of chemicals, no requirements for nutrients or disposal of biological or inorganic sludge, high efficiency at low metal concentrations, and no metal toxicity issues. The operation of biosorption shares many common features with ion-exchange technology and, despite shorter life cycle and less selectivity options, biosorbents could be considered direct competitors of ionex resins. The high cost of the ion-exchange process limits its application. Not all industries producing metal bearing effluents have financial resources for such sophisticated treatment and most opt only for basic decontamination techniques to meet regulation limits. The accumulated knowledge already provides a solid basis for the commercial exploitation of biosorption processes. Huge markets already exist (Volesky 2007) and they may even grow with progressively stricter legislation worldwide and increase demand on metal resources. Future efforts to improve selectivity and shelf life of biosorbents, further information on biosorption mechanisms and reliability and performance of biosorption models as well as more pilot scale demonstrations should bring convincing marketing arguments for large-scale applications. Biosorption also has the potential to find an industrial application in the future separation technologies with renewable biosorbents complementing conventional methods in hybrid or integrated installations.

References

Baker AJM, McGrath SP, Reeves RD, Smith JAC (2000) Metal hyperaccumulator plants: a review of the ecology and physiology of a biological source for phytoremediation of metal-polluted soil. In: Terry N, Bañuelos GS (eds) Phytoremediation of contaminated soil and water. CRC Press, Boca Raton, pp 85–107

Bargar JR, Bernier-Latmani R, Giammar DE, Tebo BM (2008) Biogenic uraninite nanoparticles and their importance for uranium remediation. Elements 4:407–412

Calderón J, Ortiz-Pérez D, Yáñez L, Díaz-Barriga F (2003) Human exposure to metals. Pathways of exposure, biomarkers of effect, and host factors. Ecotoxicol Environ Saf 56:193–103

Chaney RL, Angle JS, Broadhurst CL, Peters CA, Tappero RV, Sparks DL (2007) Improved understanding of hyperaccumulation yields commercial phytoextraction and phytomining technologies. J Environ Qual 36:1429–1443

Demirbas A (2008) Heavy metal adsorption onto agro-based waste materials: a review. J Hazard Mater 157:220–229

Gadd GM (1992a) Metals and microorganisms: a problem of definition. FEMS Microbiol Lett 79:197–203

Gadd GM (1992b) Microbial control of heavy metal pollution. In: Fry JC, Gadd GM, Herbert RA, Jones CW, Watson-Craik IA (eds) Microbial control of pollution. Cambridge University Press, Cambridge, pp 59–87
Kaushik P, Malik A (2009) Fungal dye decolourization: recent advances and future potential. Environ Int 35:127–141
Lloyd JR, Lovley DR, Macaskie LE (2003) Biotechnological application of metal-reducing microorganisms. Adv Appl Microbiol 53:85–128
Macek T, Kotrba P, Svatoš A, Demnerová K, Nováková M, Macková M (2008) Novel roles for GM plants in environmental protection. Trends Biotechnol 26:146–152
Marinho RS, Afonso JC, da Cunha JW (2010) Recovery of platinum from spent catalysts by liquid–liquid extraction in chloride medium. J Hazard Mater 179:488–494
Muyzer G, Stams AJ (2008) The ecology and biotechnology of sulphate-reducing bacteria. Nat Rev Microbiol 6:441–454
Peakall D, Burger J (2003) Methodologies for assessing exposure to metals: speciation, bioavailability of metals, and ecological host factors. Ecotoxicol Environ Saf 56:110–121
Ruml T, Kotrba P (2003) Microbial control of metal pollution: an overview. In: Fingerman M, Nagabhushanam R (eds) Recent advances in marine biotechnology, vol 8. Science Publishers Inc., Enfield, pp 81–153
Schwuger MJ, Subklew G, Woller N (2001) New alternatives for waste water remediation with complexing surfactants. Colloid Surface A 186:229–242
Sheoran AS, Sheoran V (2006) Heavy metal removal mechanism of acid mine drainage in wetlands: a critical review. Mineral Eng 19:105–116
Singh S, Kang SH, Mulchandani A, Chen W (2008) Bioremediation: environmental clean-up through pathway engineering. Curr Opin Biotechnol 19:437–444
Sud D, Mahajan G, Kaur MP (2008) Agricultural waste material as potential adsorbent for sequestering heavy metal ions from aqueous solutions—a review. Biores Technol 99:6017–6027
Volesky B (2004) Sorption and biosorption. BV-Sorbex Inc., St. Lambert
Volesky B (2007) Biosorption and me. Water Res 4017–4029

Chapter 2
Potential of Biosorption Technology

Tomas Macek and Martina Mackova

Abstract Heavy metal removal from inorganic effluent can be achieved by conventional treatment such as chemical precipitation, ion exchane or flotation, however each treatment has its limitation. Recently, sorption, namely biosorption has become one of the alternative treatments. Basically, sorption is a mass transfer process by which a substance is transferred from the liquid phase to the surface of a solid, and substance becomes bound by physical and/or chemical interactions. Due to large surface area, high sorption capacity and surface reactivity of sorbents, sorption can be utilized as low-cost alternative to conventional processes. For example, materials locally available in large quantities such as natural materials, living or dead biomass, agricultural waste or industrial byproducta can be used as biosorbents with quite little processing. This chapter discusses the significance of the heavy metal removal from waste streams and provides brief oveview of the potential of biosorbents and biosorption technology. Considered are various aspects of utilization of microbial and plant derived biomass in connection with biosorption and the possibility of exploiting such material for heavy metal removal form solutions.

Keywords Heavy metals • Biomass • Biosorption • Biosorbent • Bioavailability

2.1 Significance of Metal Recovery—Industrial and Environmental View

Heavy metal pollution is one of the most important environmental problems today. Various industries produce and discharge wastes containing different heavy metals into the environment, such as mining and smelting of metalliferous ores, surface

T. Macek (✉)
IOCB & ICT Joint Laboratory, Institute of Organic Chemistry and Biochemistry, Academy of Sciences of the Czech Republic, Flemingovo n. 2, 166 10 Prague 6, Czech Republic
e-mail: tom.macek@uochb.cas.cz

P. Kotrba et al. (eds.), *Microbial Biosorption of Metals*,
DOI 10.1007/978-94-007-0443-5_2, © Springer Science+Business Media B.V. 2011

finishing industry, energy and fuel production, fertilizer and pesticide industry and application, metallurgy, iron and steel, electroplating, electrolysis, electro-osmosis, leatherworking, photography, electric appliance manufacturing, metal surface treating, aerospace and atomic energy installation etc. (Wang and Chen 2009). Among these, the following four appear as the main priority targets, particularly in the industrialized world (Volesky 2007):

1. acid mine drainage (AMD)—associated with mining operations;
2. electroplating industry waste solutions (growth industry);
3. coal-based power generation (throughput of enormous quantities of coal);
4. nuclear power generation (uranium mining/processing and special waste generation).

Three kinds of heavy metals are of concern, including toxic metals (such as Hg, Cr, Pb, Zn, Cu, Ni, Cd, As, Co, Sn, etc.), precious metals (such as Pd, Pt, Ag, Au, Ru etc.) and radionuclides (such as U, Th, Ra, Am, etc.) (Wang and Chen 2009, 2006).

Methods for removing metal ions from aqueous solution mainly consist of physical, chemical and biological technologies. Conventional methods for removing metal ions from aqueous solution involve chemical precipitation, chemical and electro coagulation, filtration, ion exchange, electrochemical treatment, membrane technologies, adsorption on activated carbon, zeolite, evaporation etc. However, chemical precipitation and electrochemical treatment are ineffective, and also produce large quantity of sludge required to treat with great difficulty. Ion exchange, membrane technologies and activated carbon adsorption process are extremely expensive when treating large amount of water and wastewater containing heavy metal in low concentration, they cannot be used at large scale. The advantages and disadvantages of the conventional metal removal technologies were summarized by Volesky (2001).

The development and implementation of cost-effective process for removal/recovery of metals is essential to improve the competitiveness of industrial processing operations. Disadvantages, together with the need for more economical and effective methods for the recovery of metals from wastewaters, have resulted in the development of alternative separation technologies (Volesky and Naja 2007).

In recent years there has been a trend toward the implementation of passive treatment schemes. These take advantage of naturally occurring geochemical and biological processes to improve water quality with minimal operation and maintenance requirements. Biological removal includes the use of microorganisms (fungi, algae, bacteria), plants (live or dead) and biopolymers and may provide a suitable means for heavy metals treatment from wastewater.

Microoorganisms react with metals and minerals in natural and synthetic environments, altering their physical and chemical state, with metals and minerals also able to affect microbial growth, activity and survival. In addition, many minerals are biogenic in origin, and the formation of such biominerals is of global geological and industrial significance. Microbes can somehow interact with all elements found in the periodic table (including actinides, lanthanides, radionuclides). The elements can be accumulated by or be associated with microbial biomass depending on the context and environment. Microbes possess transport systems for essential metals; inessential metal species can also be taken up. Microbes are also capable of mediating metal and mineral bioprecipitation, e.g. by metabolite production, by changing the physico-chemical microenvironmental conditions around the biomass, and also by the indirect release of metal-precipitating substances from other activities, e.g. phosphate from organic decomposition or phosphate mineral solubilization. Microbial cell walls, outer layers, and exopolymers can sorb, bind or entrap many soluble and insoluble metal species as well as e.g. clay minerals, colloids, oxides, etc. which also have significant metal-sorption properties. Redox transformations are also widespread in microbial metabolism, some also mediated by the chemical activity of structural components.

Metals exhibit a range of toxicities towards microbes, and while toxic effects can arise from natural geochemical events, toxic effects on microbial communities are more commonly associated with anthropogenic contamination or redistribution of toxic metals in aquatic and terrestrial ecosystems. Such contamination can arise from aerial and aquatic sources, as well as agricultural and industrial activities, and domestic and industrial wastes. In some cases, microbial activity can result in remobilization of metals from waste materials and transfer into aquatic systems (Gadd 2010, 2009; Violante et al. 2008). It is commonly accepted that toxic metals, their chemical derivatives, metalloids and organometals can have significant effects on microbial populations and, under toxic conditions, almost every index of microbial activity can be affected (Giller et al. 2009).

There is a number of mechanisms involved in detoxification and transformation of metals depending on the organism and the cellular environment; mechanisms may be dependent on and/or independent of metabolism. A variety of mechanisms may be involved in transport phenomena contributing to decreased uptake and/or efflux. A variety of specific or non-specific mechanisms may also effect redox transformations, intracellular chelation and intracellular precipitation. Biomineral formation (biomineralization) may be biologically induced, i.e. caused by physico-chemical environmental changes mediated by the microbes, or biologically controlled. The mechanism by which microorganisms remove metals from solutions are: (1) extracellular accumulation/precipitation; (2) cell-surface sorption or complexation; and (3) intracellular accumulation (Muraleedharan et al. 1991). Among these mechanisms, extracellular accumulation/precipitation may be facilitated by using viable microorganisms, cell-surface sorption or complexation which can occur with alive or dead microorganisms, while intracellular accumulation requires microbial activity (Asku et al. 1991). Although living and dead cells are both capable of metal accumulation, there are differences in the mechanisms involved, given on the extent of metabolic dependence (Gadd and White 1990).

The major mechanisms of microbial metal transformations between soluble and insoluble metal species include chemolithotrophic leaching, chemoorganotrophic leaching, rock and mineral bioweathering and biodeterioration, biocorrosion, redox mobilization, methylation, complexation (with microbial products such as extracellular polymers (EPS) and metallothionein like proteins) in case of soluble metal species while for latter case we speak about biosorption, accumulation, biomineral formation, redox immobilization, metal sorption to biogenic minerals and formation of metalloid nanoparticles (Roane et al. 2005). The relative balance between such processes depends on the environment and associated physico-chemical conditions and the microbe(s) involved as well as relationships with plants, animals and anthropogenic activities. Chemical equilibria between soluble and insoluble phases are influenced by abiotic components, including dead biota and their decomposition products, as well as other physico-chemical components of the environmental matrix, e.g. pH, water, inorganic and organic ions, molecules, compounds, colloids and particulates. Solubilization can occur by chemolithotrophic (autotrophic) and chemo organotrophic (heterotrophic) leaching; siderophores, including phytosiderophores released by plants, and other complexing agents; redox reactions; methylation and demethylation; and biodegradation of organo-radionuclide complexes. Immobilization can occur by biosorption to cell walls, exopolymers, other structural components and derived/excreted products; precipitation can be a result of metabolite release (e.g. sulfide, oxalate) or reduction; transport, accumulation, intracellular deposition, localization and sequestration; and adsorption and entrapment of colloids and particulates. The overall system is also affected by reciprocal interactions between biotic and abiotic components of the ecosystem such as abiotic influence on microbial diversity, numbers and metabolic activity; ingestion of particulates and colloids (including bacteria) by phagotrophs; and biotic modification of physico-chemical parameters including redox potential, pH, O_2, CO_2, other gases and metabolites, temperature, and nutrient depletion. An important role play also plants and their metabolites in extraction influencing the composition of bacterial composition in soil (Uhlík et al. 2009; Macek et al. 2009). Plant biomass itself can exhibit improved metal accumulation capacity (Kotrba et al. 2009).

The combined effects of above mentioned parameteres influence so called speciation of the metals. At high pH metals are predominantly found as insoluble mineral phosphates and carbonates while at low pH they are more commonly found as free ionic species or as soluble organometals. Also redox potential of an environment influences metal speciation. Redox potential is established by oxidation/reduction reactions in the environment (reactions that are relatively slow), particularly in soils, but also metabolic activities of microorganisms play essential roles in establishing redox potential as well.

In contrast to metal speciation, metal bioavailability is determined by the solubility of metal species present and the sorption of metal species by solid surfaces including soil minerals, organic matter and colloidal materials. Organic matter is a significant source of metal complexation. Living organisms, organic debris and humus sorb metals, reducing metal solubility and bioavailability. Organic matter consists of humic and nonhumic material. Nonhumic substances include amino acids, carbohydrates, organic acids, fats etc. Humics consist of high molecular weight compounds altered from their original structures. Anionic functional groups bind cation metals, sequestering metal activity. Some organic complexing agents form soluble complexes with metals while others form insoluble structures. In latter case toxic metal concentrations in water phase may be reduced to nontoxic levels.

Metal bioavailability generally increases with decreasing pH. This is due to the presence of phosporic, sulfuric and carbonic acids which solubilize organic and particulate bound metals. For example solubility can increase in surface layers where plant exudates, microbial activity, moisture and leaching lower the pH (Roane et al. 2005).

2.2 Biosorption—A Suitable Approach for Heavy Metal Removal

Numerous strategies have potential applicability in the removal of metals, however only few field-based studies were performed. Biohydrometallurgy is a recent technical area that is based on specific interactions between microorganisms and minerals to extract metals from raw materials. The technological breakthroughs must allow the integration of innovative biotechnology-based processes for recovery and/or removal of metals from primary materials such as ores and concentrates, secondary materials such as mining wastes, metallurgical slags, and combustion/power plant ashes. The investigated biotechnologies, covering all the aspects of the application of biohydrometallurgy, have included bioleaching, biooxidation, biosorption, bioreduction, bioaccumulation, bioprecipitation, bioflotation, bioflocculation, and biosensors. They should give consideration for eco-design and a reduced impact on environment.

One such important and widely studied alternative is biosorption, where certain types of biomass are able to bind and concentrate metals from even very dilute aqueous solutions. Microbial biomass provides a metal sink, by biosorption to cell walls, pigments and extracellular polysaccharides, intracellular accumulation, or precipitation of metal compounds in and/or around cells, hyphae or other structures. All microbial materials can be effective biosorbents for metals except for mobile alkali metal cations like Na^+ and K^+, and this can be an important passive process in living and dead organisms (Gadd 1993).

A biosorption-based process offers a number of advantages when compared to the conventional methods used. However, for all practical priority reasons, the met al biosorption studies are focusing on mainly anthropogenic point sources of metal releases into the environment (Volesky 2007). The process of biosorption has many attractive features including the selective removal of metals over a broad range of pH and temperature, its rapid kinetics of adsorption and desorption and low capital and operation cost. Biosorbent an easily be produced using inexpensive growth media or obtained as a by-product from industry. It is desirable to develop biosorbents with a wide range of metal affinities that can remove a variety of metal cations. Alternatively a mixture of non-living biomass consisting of more than one type of microorganisms can be employed as biosorbents (Ahluwalia and Goyal 2007).

Based upon the metal binding capacities of various biological materials, biosorption can separate heavy metals from various waste material including wastewater (Vilar et al. 2007; Pavasant et al. 2006). Biosorption can be characterized as the removal of heavy metals using a passive binding process with nonliving microorganisms including bacteria, fungi, and yeasts (Parvathi et al. 2007), and other biomass types that are capable of efficiently collecting heavy metals. Obviously, some of the advantages biosorption has over conventional treatment methods include low cost, high efficiency for dilute concentration solutions, a minimal amount of chemical and/or biological sludge, no additional nutrients required and the possibility of biosorbent regeneration and metal recovery (Vilar et al. 2007). The sorption of heavy metals onto these biomaterials is attributed to their constituents, which are mainly proteins, carbohydrates and phenolic compounds, since they contain functional groups such as carboxyls, hydroxyls and amines, which are able to attach to the metal ions (Choi and Yun 2006).

Heavy metal accumulation in aquatic organisms, which is an active process involving metabolic activity within living organisms, has been studied by several researchers since 1978 (Braek et al. 1980; Duddridge et al. 1980; Hart et al. 1979; Macka et al. 1979; Wong et al. 1978). Biosorption onto biomass, an entirely different process from bioaccumulation, was pioneered by Volesky's group from McGill University in 1981 (Tsezos and Volesky 1981). At present, the biosorption field has been enriched by a vast amount of studies published in different journals. Although at the beginning most researchers focused their efforts upon heavy metal accumulation and concentration within living organisms (Lesmana et al. 2009), upon noticing that dead biomass possesses high metal-sorbing potential (Volesky 1990), their interest shifted to biosorption (Selatnia et al. 2004; Yetis et al. 2000; Zhou 1999; Bossrez et al. 1997; Asthana et al. 1995; Volesky and Prasetyo 1994; Holan et al. 1993; Volesky et al. 1993; Niu et al. 1993; Fourest and Roux 1992). The research efforts directed towards the use of inactive and dead biomass for removal of heavy metals from aqueous solution then resulted in viable method for removing these pollutants. Nonliving biomass of algae, aquatic ferns and seaweeds, waste biomass originated from plants and mycelial wastes from fermentation industries are potential biosorbents for removal of heavy metals from aqueous solution and wastewater. Their efficiency depends on the capacity, affinity and specificity including physico-chemical nature.

Several reviews are available that discuss the use of biosorbents for the treatment of water and wastewater containing heavy metals (Demirbas 2008; Nurchi and Villaescusa 2008; Vijayaraghavan and Yun 2008; Romera et al. 2006; Davis et al. 2003; Kratochvil and Volesky 1998; Zouboulis et al. 1997; Lovley and Coates 1997; Veglio and Beolchini 1997; Volesky and Holan 1995; Wan Ngah and Hanafiah 2008). Biosorbents for the removal of metals mainly come under the following categories: bacteria, fungi, algae, plants, industrial wastes, agricultural wastes and other polysaccharide materials. In general, all types of biomaterials have shown good biosorption capacities towards all types of metal ions. Most studies of biosorption for metal removal have involved the use of either laboratory-grown microorganism or biomass generated by the pharmacology and food processing industries or wastewater treatment units (Agarwal et al. 2006; Chang and Hong 1994; Rao et al. 1993; Macaskie 1990; Rome and Gadd 1987; Townsley et al. 1986; Tsezos and Volesky 1981). Therefore, this promotes environment eco-friendliness. The physiological state of the organism, the age of the cells, the availability of micronutrients during their growth and the environmental conditions during the biosorption process (such as pH, temperature, and the presence of certain co-ions) are important parameters that affect the performance of a living biosorbent. Potent metal biosorbents under the class of bacteria are represented by genera including *Bacillus, Pseudomonas* and *Streptomyces* and fungi including *Aspergillus, Rhizopus* and *Penicillium* etc. Since these microorganisms are used widely in different food/pharmaceutical industries, they are generated as waste, which can be attained free or at low cost from these industries. Another important biosorbent, which has gained momentum in recent years, is seaweed. Marine algae, popularly known as seaweeds, are biological resources, which are available in many parts of the world. Algal divisions include red, green and brown seaweed; of which brown seaweeds are found to be excellent biosorbents). This is due to the presence of alginate, which is present in gel form in their cell walls. Also, their macroscopic structure offers a convenient basis for the production of biosorbent particles that are suitable for sorption process applications. Recently, numerous approaches have been made for the development of low-cost sorbents from industrial and agricultural wastes. Of these, crab shells, activated sludge, rice husks, egg shell and peat moss deserve particular attention (Vijayaraghavan and Yun 2008). The efficiency of metal concentration on the biosorbent is also influenced by chemical features of metal solution.

Equilibrium studies, that give the capacity of the adsorbent and the equilibrium relationships between adsorbent and adsorbate are described by adsorption isotherms which is usually the ratio between the quantity adsorbed and the remaining in solution at fixed temperature at equilibrium. Freundlich and Langmuir isotherms are the earliest and simplest known relationships describing the adsorption equation (Hussein et al. 2005).

Excellent removal capabilities were apparent for several biomasses. More than a few factors, such as pH, temperature, adsorbent dose, etc. significantly affect the biosorption capacities. On the other hand, utilization of them in industrial-scale applications is still some distance from reality. While most available biomasses have the capability to sequester heavy metals from solutions, not all of them fit as alternative adsorbents in real wastewater treatment plants. Several vital characteristics are available and need to be listed to render the materials valuable enough as an industrial adsorbent.

1. High adsorption capacity.
2. Available in large quantities at one location.
3. Low economic value and less useful in alternative products.
4. Attached metals can be easily recovered while biosorbent is reusable.

There is no doubt that many biosorbents and/or alternative adsorbents as mentioned in a scientific literature have a high adsorption capacity to the extent that even some are better than commercially available adsorbents. Looking from this perspective only, it seems that most biosorbents and/or alternative adsorbents have potential for industrial application. Yet, several biomasses that have low binding capacity in nature still widely exist. Their adsorption capacity normally can be improved by pretreatment or modification using physical or chemical methods. Chemically, modification is usually performed by adding some chemicals such as acid, alkali or other oxidizing and organic chemicals, while in the physical method, pretreatment is facilitated by heat, autoclaving, freeze-drying and boilling. Unfortunately chemical activation methods are not favorable, because the advantage of environmentally friendly (waste for waste treatment) and cost effective procedure, is lost. Unused chemicals represent more serious problems and commonly necessitate expensive waste treatment facilities (Lesmana et al. 2009).

Huge markets already exist for cheap biosorbents. Electroplating and metal finishing operations, mining and ore processing operations, smelters, tanneries and printed circuit board manufacturers are a few of the industries in which metal-bearing effluents pose a problem. The potential application for biosorption appears to be enormous. It can easily be envisaged that cheaper biosorbents would open up new, particularly environmental, markets so far non-accessible to ion-exchange resins because of their excessive costs, which make them prohibitive for clean-up operation applications. These considerations clearly demonstrate the economic feasibility and potential of the biosorption process for heavy metal removal/recovery purposes. It should be pointed out that there is a potential added benefit of metal-recovery as an *additional source of revenue* generated by a water treatment that *must* be carried out anyway (from a regulatory and environmental point of view).

2.3 Conclusions

The use of microbial and plant biomass and other biological mechanisms naturally used for heavy metal detoxification and removal, offer promising alternatives to traditional technologies in the treatment of heavy metals. The new biological-based technologies need not necessarily replace conventional treatment approaches but may complement them.

At present, information on different technological approaches is inadequate to accurately define parameters for scale up of processes and design perfection including reliability and economic feasibility. To provide an economically viable treatment, the appropriate choice of technology and proper operational conditions have to be identified.

Probably one of the most studied approaches is biosorption which offers an economically feasible technology for efficient removal and recovery of metal(s) from aqueous solution. The process of biosorption has many attractive features including removal of metals over quite broad range of pH and temperature, its rapid kinetics of adsorption and desorption and low capital and operation cost. Biosorbent can easily be produced using inexpensive growth media or obtained as a by-product from industry (Ahluwalia and Goyal 2007). Biosorption allows significant cost savings in comparison with existing technologies, can be more effective in many cases than its closest rival, ion exchange can be easily converted to the biosorption process. Additional cost reduction results from the possible recovery of heavy metals (Volesky and Naja 2007). Also being aware of the hundreds of biosorbents able to bind various pollutants, sufficient research has been performed on various biomaterials to understand the mechanism responsible for biosorption. Therefore, through continued research, especially on pilot and full-scale biosorption process, the situation is likely to change in the near future, with biosorption technology becoming more beneficial and attractive than currently used technologies (Vijayaraghavan and Yun 2008).

Acknowledgements This chapter was processed with the help of grant Centrum 1M06030 and MSM 6046137305 and Z405505 from the Czech Ministry of Education, Youth and Sports.

References

Agarwal GS, Bhuptawat HK, Chaudhari S (2006) Biosorption of aqueous chromium(VI) by *Tamarindus indica* seeds. Biores Technol 97:949–956

Ahluwalia SS, Goyal D (2007) Microbial and plant derived biomass for removal of heavy metals from wastewater. Biores Technol 98:2243–2257

Asku Z, Kutsal T, Gun S, Haciosmanoglu N, Gholminejad M (1991) Investigation of biosorption of Cu(II), Ni(II), and Cr(VI) ions to activated sludge bacteria. Environ Technol 12:915–921

Asthana RK, Chatterjee S, Singh SP (1995) Investigations on nickel biosorption and its remobilization. Process Biochem 30:729–734

Bossrez S, Remacle J, Coyette J (1997) Adsorption of nickel on *Enterococcus hirae* cell walls. J Chem Technol Biotechnol 70:45–50

Braek GS, Malnes D, Jensen A (1980) Heavy-metal tolerance of marine phytoplankton. Combined effect of zinc and cadmium on growth and uptake in some marine diatoms. J Exp Mar Biol Ecol 42:39–54

Chang JS, Hong J (1994) Biosorption of mercury by the inactivated cells of *Pseudomonas aeruginosa* pu21 (rip84). Biotechnol Bioeng 44:999–1006

Choi SB, Yun YS (2006) Biosorption of cadmium by various types of dried sludge:an equilibrium study and investigation of mechanisms. J Hazard Mater 138:378–383

Davis TA, Volesky B, Mucci A (2003) A review of the biochemistry of heavy metal biosorption by brown algae. Water Res 37:4311–4330

Demirbas A (2008) Heavy metal adsorption onto agro-based waste materials: a review. J Hazard Mater 157:220–229

Duddridge JE, Wainwright M (1980) Heavy-metal accumulation by aquatic fungi and reduction in viability of gammarus-pulex fed Cd^{2+} contaminated mycelium.Water Res 14:1605–1611

Fourest E, Roux JC (1992) Heavy-metal biosorption by fungal mycelial byproducts—mechanisms and influence of pH. Appl Microbiol Biotechnol 37:399–403

Gadd GM (1993) Interactions of fungi with toxic metals. New Phytol 124:25–60

Gadd GM (2009) Heavy metal pollutants: environmental and biotechnological aspects. In: Schaechter M (ed) Encyclopedia of microbiology, 3rd edn. Elsevier, Oxford, pp 321–334

Gadd GM (2010) Metals, minerals and microbes: geomicrobiology and bioremediation. Microbiology 156:609–643

Gadd GM, White C (1990) Biosorption of radionuclides by yeast and fungal biomass. J Chem Technol Biotechnol 49:331–343

Giller KE, Witter E, McGrath SP (2009) Heavy metals and soil microbes. Soil Biol Biochem 41:2031–2037

Hart BA, Bertram PE, Scaife BD (1979) Cadmium transport by *Chlorella pyrenoidosa*. Environ Res 18:327–335

Holan ZR, Volesky B, Prasetyo I (1993) Biosorption of cadmium by biomass of marine-algae. Biotechnol Bioeng 41:819–825

Hussein H, Ibrahim SF, Kandeel K, Moawad H (2005) Biosorption of heavy metals from waste water using *Pseudomonas* sp. Electron J Biotechnol 7. doi:10.2225/vol7-issue1-fulltext-2

Kotrba P, Najmanova J, Macek T, Ruml T, Mackova M (2009) Genetically modified plants in phytoremediation of heavy metal and metalloid soil and sediment pollution. Biotechnol Adv 27:799–810

Kratochvil D, Volesky B (1998) Advances in the biosorption of heavy metals. Trends Biotechnol 16:291–300

Lesmana SO, Febriana N, Soetaredjo V, Sunarso J, Ismadji S (2009) Studies on potential applications of biomass for the separation of heavy metals from water and wastewater. Biochem Eng J 44:19–41

Lovley DR, Coates JD (1997) Bioremediation of metal contamination. Curr Opin Biotechnol 8:285–289

Macaskie LE (1990) An immobilized cell bioprocess for removal of heavy metals from aqueous flows. J Chem Technol Biotechnol 49:330–334

Macek T, Uhlik O, Jecna K, Novakova M, Lovecka P, Rezek J, Dudkova V, Stursa P, Vrchotova B, Pavlikova D, Demnerova K, Mackova M (2009) Advances in phytoremediation and rhizoremediation. In: Singh A, Kuhad RC, Ward OP (eds) Advances in applied bioremediation. Springer, Berlin, pp 257–277

Macka W, Wihlidal H, Stehlik G, Washuttl J, Bancher E (1979) Uptake of $_{203}Hg^{++}$ and $_{115M}Cd^{++}$ by *Chlamydomonas reinhardi* under various conditions. Chemosphere 8:787–796

Muraleedharan TR, Leela I, Venkobachar C (1991) Biosorption: an attractive alternative for metal removal and recovery. Curr Sci 61:379–385

Niu H, Xu XS, Wang JH, Volesky B (1993) Removal of lead from aqueous-solutions by *Penicillium* biomass. Biotechnol Bioeng 42:785–787

Nurchi VM, Villaescusa I (2008) Agricultural biomasses as sorbents of some trace metals. Coord Chem Rev 252:1178–1188

Parvathi K, Nagendran R (2007) Biosorption of chromiumfromeffluent generated in chrome-electroplating unit using *Saccharomyces cerevisiae*. Sep Sci Technol 42:625–638

Pavasant P, Apiratikul R, Sungkhum V, Suthiparinyanont P, Wattanachira S, Marhaba TF (2006) Biosorption of Cu^{2+}, Cd^{2+}, Pb^{2+}, and Zn^{2+} using dried marine green macroalgae *Caulerpa lentillifera*. Biores Technol 97:2321–2329

Rao CRN, Iyengar L, Venkobachar C (1993) Sorption of copper(II) from aqueous phase by waste biomass. J Environ Eng Div ASCE 119:369–377

Roane TM, Pepper IL, Miller RM (2005) Microbial remediation of metals. In: Crawford RL, Crawford DL (eds) Bioremediation principles and applications. Cambridge University Press, Cambridge, pp 312–340

Rome L, Gadd GM (1987) Copper adsorption by *Rhizopus arrhizus, Cladosporium resinae* and *Penicillium italicum*. Appl Microbiol Biotechnol 26:84–90

Romera E, Gonzalez F, Ballester A, Blazquez ML, Munoz JA (2006) Biosorption with algae: a statistical review. Crit Rev Biotechnol 26:223–235

Selatnia A, Boukazoula A, Kechid N, Bakhti MZ, Chergui A, Kerchich Y (2004) Biosorption of lead (II) from aqueous solution by a bacterial dead *Streptomyces rimosus* biomass. Biochem Eng J 19:127–135

Townsley CC, Ross IS, Atkins AS (1986) Biorecovery of metallic residues from various industrial effluents using filamentous fungi. Process Metall 4:279–289

Tsezos M, Volesky B (1981) Biosorption of uranium and thorium. Biotechnol Bioeng 23:583–604

Uhlík O, Jecná K, Leigh M-B, Macková M, Macek T (2009) DNA-based stable isotope probing: a link between community structure and function. Sci Total Environ 407:3611–3619

Veglio F, Beolchini F (1997) Removal of metals by biosorption: a review. Hydrometallurgy 44:301–316

Vijayaraghavan K, Yun YS (2008) Bacterial biosorbents and biosorption. Biotechnol Adv 26:266–291

Vilar VJP, Botelho CMS, Boaventura RAR (2007) Modeling equilibrium and kinetics of metal uptake by algal biomass in continuous stirred and packed bed adsorbers. Adsorption 13:587–601

Violante A, Huang PM, Gadd GM (2008) Biophysicochemical processes of heavy metals and metalloids in soil environments. Wiley, Chichester

Volesky B (1990) Biosorption of heavy metals. CRC Press, Boca Raton

Volesky B (2001) Detoxification of metal-bearing effluents: biosorption for the next century. Hydrometallurgy 59:203–216

Volesky B (2007) Biosorption and me. Water Res 41:4017–4029

Volesky B, Holan ZR (1995) Biosorption of heavy metals. Biotechnol Prog 11:235–250

Volesky B, Naja G (2007) Biosorption technology: starting up an enterprise. Int J Technol Transf Commer 6:196–211

Volesky B, Prasetyo I (1994) Cadmium removal in a biosorption column. Biotechnol Bioeng 43:1010–1015

Volesky B, May H, Holan ZR (1993) Cadmium biosorption by *Saccharomyces cerevisiae*. Biotechnol Bioeng 41:826–829

Wan Ngah WS, Hanafiah MAKM (2008) Removal of heavy metal ions from wastewater by chemically modified plant wastes as adsorbents: a review. Biores Technol 99:3935–3948

Wang JL, Chen C (2006) Biosorption of heavy metals by *Saccharomyces cerevisiae*: a review. Biotechnol Adv 24:427–451

Wang JL, Chen C (2009) Biosorbents for heavy metals removal and their future a review. Biotechnol Adv 27:195–226

Wong PTS, Chau YK, Luxon PL (1978) Toxicity of amixture of metals on freshwater algae. J Fish Res Board Can 35:479–481

Yetis U, Dolek A, Dilek FB, Ozcengiz G (2000) The removal of Pb(II) by *Phanerochaete chrysosporium*. Water Res 34:4090–4100

Zhou JL (1999) Zn biosorption by *Rhizopus arrhizus* and other fungi. Appl Microbiol Biotechnol 51:686–693

Zouboulis AI, Matis KA, Hancock IC (1997) Biosorption of metals from dilute aqueous solutions. Sep Purif Meth 26:255–295

Chapter 3
The Mechanism of Metal Cation and Anion Biosorption

Ghinwa Naja and Bohumil Volesky

Abstract The passive, not metabolically mediated, biosorption uptake of metals by (dead) biomass appears as a powerful tool for somewhat selectively removing heavy metals from solution. Immobilization of dissolved toxic heavy metals and their physical removal by biosorption in a water puricication process is not only technically feasible but it may prove to be economically quite attractive. In order to effectively optimize such a process, the mechanisms involved in metal biosorption need to be well understood and the metal speciation in aqueous solutions has to be taken into consideration as it plays an important role.

As phenomena of complexation, coordination, chelation, ion exchange, adsorption, inorganic microprecipitation may all be involved in the overall metal uptake by biosorption, the configuration and state of the active binding site in the biomass have to be well understood. The state and effectiveness of the binding site is, to a large degree, also affected by the environmental conditions such as pH, temperature and ionic strength of the solution. Because of the multiparameter complexity of the sorption system it is most useful to express the interdependence of the key parameters mathematically whereby the set of equations could be organized into a model of the sytem that could be used for predicting its metal uptake performance under different conditions. The elements and fundamentals of the approach are discussed and outlined in the chapter.

When the microprecipitation phenomenon and physical collection of insolubilized metal is excluded, extensive research results indicate that ion exchange tends to be the dominant metal immobilization mechanism in biosorption. The fact that this phenomenon is in most cases reversible offers an attractive possibility of effective wash-release of the deposited metal, resulting in a highly concentrated regeneration solution suitable for some conventional metal recovery and a refreshed biosorbent material ready for another metal uptake cycle. This feature undoubtedly reinforces the feasibility and competitiveness of the metal biosorption process.

G. Naja (✉)
Science Department, Everglades Foundation, 18001 Old Cutler Road,
Palmetto Bay, FL 33157, USA
e-mail: mnaja@evergladesfoundation.org

P. Kotrba et al. (eds.), *Microbial Biosorption of Metals*,
DOI 10.1007/978-94-007-0443-5_3, © Springer Science+Business Media B.V. 2011

Keywords Binding sites • Biosorption • Microorganisms (bacteria, fungi and algae) • Complexation • Hard and soft ions • Metal speciation in solution • Micro-precipitation

3.1 Introduction

Biosorption is an operation that combines the use of biomaterials for sorbing, sequestering and immobilizing organic or inorganic substances from aqueous solutions. Biosorption from other types of environments, namely from organic liquids, has not been explored—as yet.

Biosorption, as it has been defined, perceived and investigated is based on the passive sequestration by non-metabolizing, non-living biomass. Such biomass is a complex chemical substance whose many types of different chemically active groups may show some tendencies to act in binding other chemical substances or ions, attracting them from solution and binding them to the biomass solid substance. In doing so, the solid mass or particles of biomass *sorbent* becomes enriched in those substances of *sorbate(s)* that were attracted and sequestered. The sorbate-laden biomass solids are easily isolated from the liquid, even if an additional solid-liquid separation process may be required. This is the very technological foundation of a very useful and powerful method of sorption that is capable of removing and extracting specific chemical species from solution.

We distinguish 'passive' ***biosorption*** from what may be termed ***bioaccumulation*** which is active, metabolically mediated transport and deposition of chemical species. Bioaccumulation is then a function of a living cell. It is very difficult to assess bioaccumulation quantitatively because the chemical transport may work both ways—into the cell and out of it, across the cell wall and cell membranes, including some organelles (e.g. vacuoli) serving as deposition or storage sites inside the cell. In addition to it, some cells tend to produce extracellular chemicals. The presence and amount of these potentially sorbent-binding substances may vary greatly with the level and type of cellular metabolic activities. The process is therefore complex and, correspondingly, bioaccumulation is thus difficult to study quantitatively. This text focuses on the passive phenomenon of biosorption only.

As will be seen, this focus takes the line of this work very much outside conventional biological sciences which become less relevant from the point of view taken—quantification of sorption performance.

In the general sorption field we encounter two basic terms—"adsorption" and "absorption". ***Adsorption*** is understood to involve the interphase accumulation or concentration of substances at a surface or interface. Such a process can occur at an interface of any two phases, such as liquid-liquid, gas-liquid, gas-solid, or liquid-solid interfaces. ***Absorption***, conversely, is a process in which the molecules or atoms of one phase interpenetrate nearly uniformly among those of another phase to form a "solution" with it. This latter case will NOT be dealt with in this text.

Adsorption, first observed by C.W. Scheele in 1773 for gases and, subsequently, for solutions by Lowitz in 1785, is now recognized as a significant phenomenon in most natural physical, biological and chemical processes. Sorption on solids, particularly active carbon, has become a widely used operation also for purification of waters and wastewaters. The subject of this chapter is thus reflecting the general phenomenon of adsorption with biosorption studies used as an extensive example of it.

While there is a preponderance of solute (sorbate) molecules (atoms) in the solution, there are none in the sorbent particle to start with. This imbalance between the two environments amounts to a concentration difference driving force for the solute species. First to create a sorbate layer on the surface and then the sorbate may gradually penetrate deeper into the solid—if that is practically possible. Looking at it with a 'magnifying glass', depending on the mechanism of sorbate sequestering, we can start distinguishing further between *chemisorption*, which involves chemical binding, and *physisorption* that depends strictly on the surface-based physical forces of interfacial imbalance and attraction (e.g. van der Waals). The jargon in the field has developed such that the term 'adsorption' may already have some form of physical surface-based deposition implied in it. In order to avoid the specific interpretation of adsorption as a specific mechanism of *physical* sorption, a more general term 'sorption' is preferably used throughout this text. Actually, most of the biomaterials, due to their biological nature, display, to a variable extent, a certain degree of permeability. In terms of binding and deposition of substances by biosorption we are usually considering chemisorption. Because the gel-like nature of many biological materials (cells) is responsible for their relatively high permeability, particularly for transfer of small molecules and atomic or ionic species, we cannot also use the concept of the *surface area.* Instead, one has to rely on quantifying the 'number of binding sites'.

3.2 Metal Biosorption and Bioaccumulation

Unlike with other biomass types, the ability of microorganisms to interact with and to accumulate a variety of metal ions from their aqueous environment has been studied extensively in the last two decades due to the danger of heavy metal toxicity. A number of terms such as bio-concentration, bioaccumulation, bio-adsorption and biosorption has been used to describe this phenomenon. It is necessary to distinguish between active, metabolically mediated metal uptake by living cells as opposed to passive metal sequestering by dead biomass, as explained above. The terms bioaccumulation and biosorption appear to be gaining recognition, designating the two very different modes of metal uptake by biological materials. It is perhaps important to add that actively metabolizing cells may, in some instances, even actively repel metal ions, particularly the more toxic ones, as a self defense. The net result being a relatively low metal content of the biomass. When the cells are inactivated or their metabolic activities suppressed, the chemical binding sites of the biomass may attract metal ions from the solution. Metal concentration by biomass through bioaccumulation and biosorption may thus substantially differ. In general, biosorption can be defined as the passive sequestering of metal ions by metabolically inactive biomass. This type of metal uptake may take place by any one or a combination of different processes such as complexation, coordination, chelation, ion exchange or microprecipitation and entrapment.

All these mechanisms are associated with either living or dead microbial cells except the last two. Micro-precipitation and entrapment refer to immobilization of metal species already solidified located usually outside or even inside the cells, as for example in the extracellular polymeric capsule or cytoplasmic components. The use of dead biomass for metal sequestration (and immobilization) offers some advantages over living cells in that it would be immune to toxicity and non-biotic external conditions. Moreover, since the biomass is metabolically inactive, there could be precise control of the metal-removal process in reactors specifically operated and optimized solely for this purpose. In the procedure aimed at the removal of dissolved metals from solution, the first step is to "immobilize" the metal by binding it from its dissolved form to the solid particle which is easier to separate from the solid-liquid suspension system. Metal ions, removed from the solution by being deposited in the (dead) biomass solids, can easily be removed from the system together with the solid biomass by utilizing any of the feasible solid/liquid separation operations such as settling, flotation, filtration, centrifugation, etc. In case of trickling columns the biomass solids are retained within the column while the liquid flows freely through it. Biosorbed metals remain in the solid phase. Their fate depends on further processing of the metal-loaded solids: the bound metals can be either washed off the solids or the (organic) solids could even be combusted. The metal load would then become concentrated mainly in the small amount of inorganic ash left over. Other alternatives can also be exploited.

3.3 Speciation of Elements in Solution

Metals are recognized as commodity materials. Enormous quantities of different types of metals are required to support our lifestyle. Their extraction and widespread use is the reason for increasing levels of metals found in the environment. From natural deposits as their source, through human technolological activities metals become 'mobilized' and tend to reach unexpected levels in natural cycles. Far from being inert, they are persistent and pose a relatively recently recognized and acknowledged serious threat to natural balances, and ultimately, to human health. Due to using either natural, renewable or even waste biomaterials *biosorption* appears as an economically attractive process particularly for inexpensively removing metals, when they are present even at low levels, from industrial solutions and effluents.

Evironmental pressures result in requirements of metal removal from even relatively dilute solutions before they can be considered safe for discharge or recycled. Minimizing, if not eliminating, the losses of metals during their extraction as well as during their use is the ultimate goal. In order to do so effectively and economically, we need to understand their behavior also in their dissolved form in aqueous solutions. This is in addition to knowing and exploiting their physical and chemical properties and taking advantage of them in metal processing and technological applications.

It is important not to overlook that there are two phases in a sorption process: a solid and a liquid one. The sorbate, first dissolved in the solution, becomes eventually sequestered on the solid phase (sorbent). The properties and behavior of sorbate as well as of sorbent in solution will affect the sorption performance of the system. In this text, we remain focused on the general example of metals as the sorbate species. While the properties of different sorbents in solution often remain to be assessed, ample information exists on the behavior of metals in solution. Upon dissolution of metallic salts, they become dissociated, whereby the metal moieties could appear as cations or anions, sometimes in their complexion or oxoion forms. In this respect the reader should consult good basic text-book literature on solution chemistry of metals. Solution pH plays an important role in this process. Most of the common metals, when dissolved, occur in the solution as positively charged cations. Among the more toxic heavy metals, widespread are cationic forms of e.g. Pb, Hg, Cd, Cu, Zn, Ni, U, Th, etc.

When the sorption process is predominantly based on ion exchange, as the case is in biosorption of metals, there are several possible ways in which the solution pH can influence the sorption process. First of all, as noted, the speciation of metals in the solution depends on the pH. On the other hand, the state of the active binding sites on the biomass may also change at different pH values. For example, in *Sargassum* (seaweed) biomass, the binding groups are acidic and the availability of free sites depends on the solution pH. At low pH, protons would compete for active binding sites with metal ions (Greene et al. 1986; Tobin et al. 1984). The protonation of active sites thus tends to decrease the metal sorption. At a low enough pH, all the binding sites may be protonated, thereby desorbing all originally bound metals from the biomass (Yang 2000; Aldor et al. 1995). More common negatively charged anionic metal species are, often complexed, e.g. As, Se, V, Mn, etc. A case of a toxic Cr should perhaps be mentioned separately because it occurs as a very toxic anionic complex of Cr^{+6} (chromic acid) and only a somewhat less hazardous straightforward cation Cr^{+3}. The reduction of Cr^{+6} to Cr^{+3} may or may not take place in the treatment process.

3.3.1 Speciation Examples: Anions and Cations

3.3.1.1 Chromate in Solution

Chromate (CrO_4^{2-}) is a typical divalent heavy metal anion. It is prone to protolysis in aqueous solution. Chromate exists in different ionic forms and as neutral acid as well in aqueous solutions. The distribution of species is dependent on the total chro-

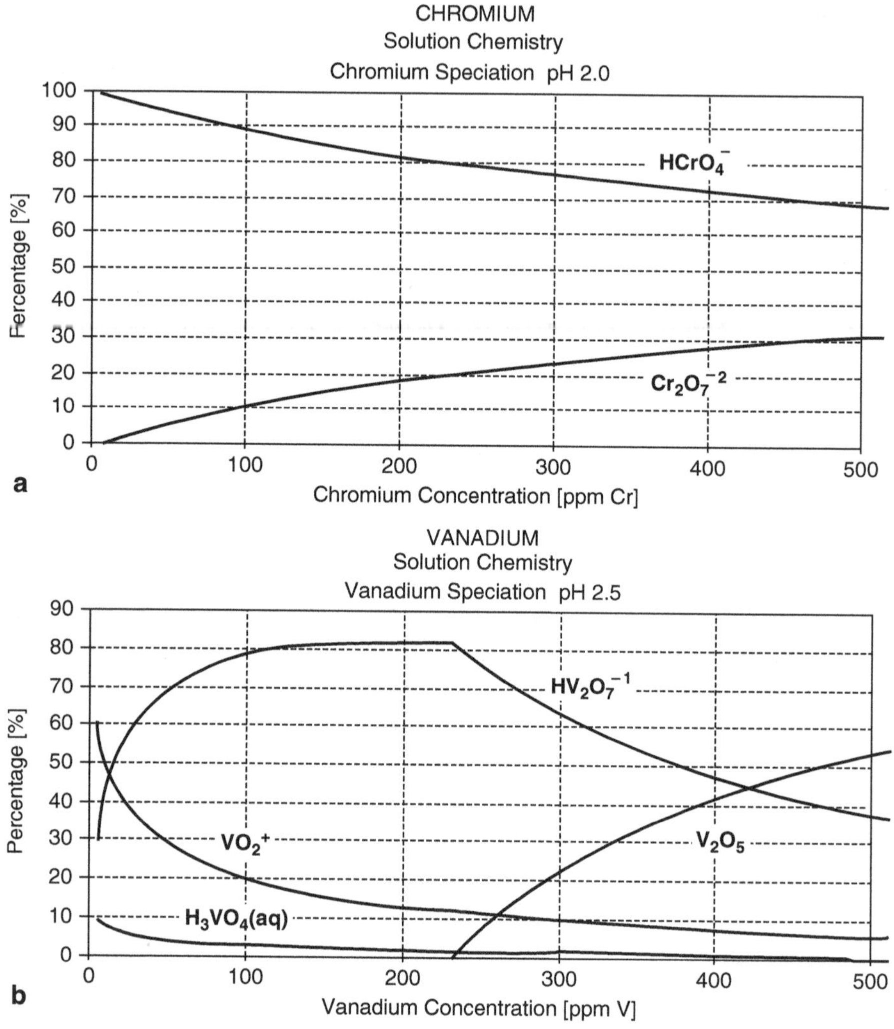

Fig. 3.1 **a** Very toxic Chromium occurs mainly as two predominant species (MINEQL+ output). **b** Vanadium in solution could be in many different ionic forms (MINEQL+ output)

mate concentration and the pH of the solution (Fig. 3.1a). The governing equilibria (at 25°C) are:

$$H_2CrO_4 \leftrightarrow HCrO_4^- + H^+ \quad K_{Cr1} = \frac{\{HCrO_4^-\}\{H^+\}}{\{H_2CrO_4\}} = 10^{0.26}\ (\text{mol/l}) \quad (3.1)$$

$$HCrO_4^- \leftrightarrow CrO_4^{2-} + H^+ \quad K_{Cr2} = \frac{\{CrO_4^{2-}\}\{H^+\}}{\{HCrO_4^-\}} = 10^{-5.9}\ (\text{mol/l}) \quad (3.2)$$

$$2HCrO_4^- \leftrightarrow Cr_2O_7^{2-} + H_2O \quad K_{Cr3} = \frac{\{Cr_2O_7^{2-}\}}{\{HCrO_4^-\}^2} = 10^{2.2}\ (\text{mol/l})^{-1} \tag{3.3}$$

$$HCr_2O_7^- \leftrightarrow Cr_2O_7^{2-} + H^+ \quad K_{Cr4} = \frac{\{Cr_2O_7^{2-}\}\{H^+\}}{\{HCr_2O_7^-\}} = 10^{0.85}\ (\text{mol/l}) \tag{3.4}$$

K_{Cri} is the corresponding protolysis constant. { } represents the activity of the species.

Since the equilibrium constant for the first order dissociation of $H_2Cr_2O_7$ is too large, the equilibrium of this protolysis reaction is not considered. Chromate is reducible. With reducers such as sulphur dioxide, sodium bisulfite, sodium metabisulfite and ferrous sulfate, Cr(VI) could be rapidly reduced to Cr(III) at low pH. This characteristic forms the fundamentals for Cr(VI) removal by precipitation. Only when Cr(VI) is reduced to Cr(III), the precipitation of Cr could be conducted through forming $Cr(OH)_3$.

3.3.1.2 Vanadate in Solution

Vanadium (vanadate VO_4^{3-}), yields a multi-valent metal anion that appears in more complicated forms than the above metals in aqueous solution. At pH13, colorless vanadium (V) is orthovanadate VO_4^{3-}. As pH decreases, other forms of anionic species of V(V) such as $VO_3(OH)^{2-}$, $V_2O_7^{4-}$, $V_4O_{12}^{4-}$, $V_3O_9^{3-}$, $VO_2(OH)_2^-$, $V_{10}O_{27}(OH)^{5-}$, and $V_{10}O_{26}(OH)_2^{4-}$ occur. At the lower pH range of 1–4, there are even cationic VO_2^+ and neutral V_2O_5 and $VO(OH)_3$ produced.

The distribution of species of V(V) presenting in the solution depends on solution pH and vanadium concentration at specific temperatures (Fig. 3.1b). The species distribution relationship and equilibrium constants have been well documented. V(V) could be reduced to V(IV), V(III) and V. However, in the presence of air, V(V) is the most stable oxidized state of vanadium in aqueous solution.

3.3.1.3 Gold-Cyanide in Solution

Gold-cyanide complex ($Au(CN)_2^-$) is very stable, the dissociation constant of Au from the cyanide complex is 38.9. Under conventional conditions (room temperature), it does not dissociate readily. Even at 85°C, it has to be with the catalyst to make Au dissociate from the complex. $Au(CN)_2^-$ could exist as a stable monovalent anion in the aqueous solution. The form of $HAu(CN)_2$ was found only under very acidic conditions such as 0.5–1N H_2SO_4. In the gold leaching process, gold is concentrated by steam activated carbon containing strong or weak-bases. Gold-cyanide could be reducible in the presence of a strong reducer such as Zn, which is used to eventually precipitate Au from a cyanide complex to obtain elemental Au.

3.3.1.4 Uranium Speciation and Complex Binding

As an example, uranium sorption by *Sargassum* biomass will be examined here in more detail. The solution pH significantly influences the ionic speciation of uranium.

At a lower solution pH, the metal ion is the predominant form in the solution. But metal hydrolysis occurs at higher solution pH, whereby these hydroxide complexes represent a significant percentage of the overall speciation.

The hydrolysis of the uranyl ions in aqueous solution is significant at higher solution pH values, such as pH 4.0. Numerous hydroxides of U(VI) are known, and $(UO_2)_2(OH)_2^{2+}$. H_2O is the stable species in the presence of water at 25°C (Baes and Mesmer 1976). An illustration of the distribution of the uranium hydrolysis products at 0.1 M and 10^{-5} M uranium concentration in the aqueous solution over a pH range is shown in Fig. 3.2a, b (Baes and Mesmer 1976). At 0.1 M concentration, the percentage of UO_2^{2+} decreases while proportions of $(UO_2)_2(OH)_2^{2+}$ and

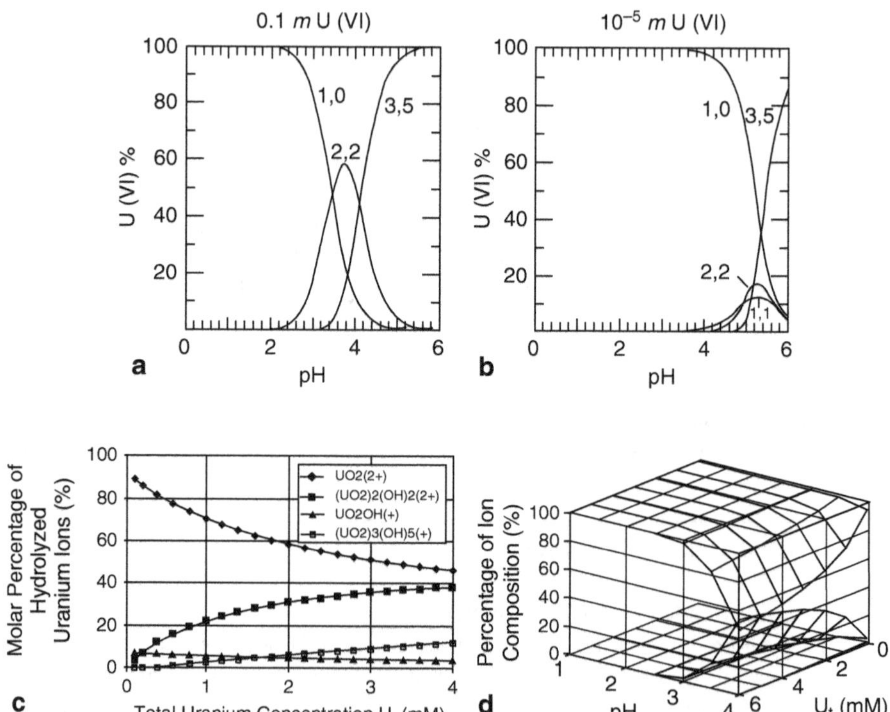

Fig. 3.2 **a**, **b** Distribution of uranium hydrolysis products. (Reproduced from Baes and Mesmer 1976). **c** Ionic composition of hydrolyzed uranium ions at pH 4.0. (Obtained from running program MINEQL+ (Schecher 1998)), $UO_2^{2+}+NO_3^-+H_2O+H^+$ system, pH 4.0, multiple run for total concentration). **d** Speciation of hydrolyzed ionic uranium: top mesh: UO_2^{2+} middle mesh: $(UO_2^{2+})_2(OH)_2^{2+}$ bottom mesh: $(UO_2^{2+})(OH)^+$

$(UO_2)_3(OH)_5^{2+}$ increase with an increase in the solution pH. At pH 4.0, the concentration of $(UO_2)_2(OH)_2^{2+}$ can constitute approximately 60% of the total uranium concentration. It needs to be noted that this complexion contains *twice* as much uranium in it. However, both diagrams in Fig. 3.2a, b do not depict well the distribution of the ionic composition for the concentration range likely to be encountered in environmental studies (0–6.0 mM).

An extensively used chemical equilibria calculation program MINEQL+ (Schecher 1998) can be applied in order to obtain the uranium ionic composition distribution in aqueous solution in the uranium concentration range of 0–4.0 mM under pH 4.0. The results are illustrated in Fig. 3.2c. The hydrolyzed uranium complex ions, especially $(UO_2)_2(OH)_2^{2+}$, contributes 10–40% of the total uranium concentration within the range of 0.5–4.0 mM at pH 4.0. The uranium complexes may have even higher affinity for the biomass binding sites in some instances which results in an enhancement of biosorption performance at higher solution pH (Yang and Volesky 1999; Stumm and Morgan 1996a). The binding of the hydrolyzed uranium complex ions, and thus their 'disappearance' from the solution, may drive the hydrolysis reaction toward the formation of hydrolyzed complex ions: The much higher than theoretically expected uranium biosorption uptakes that were observed at pH 3.5 and 4.0, compared to those of other cations (Cd, Cu, Zn), may be attributed to the influence of this hydrolysis. This is because the binding of hydrolyzed ions results in actually much more uranium bound by the biosorbent (Yang and Volesky 1999).

Uranium Speciation and Distribution in Solution

The distribution of hydrolyzed uranium ions in aqueous solution is dependent on both the solution pH and the total uranium concentration. It is calculable from the hydrolysis equilibrium constants as outlined below in an abbreviated form.

The hydrolysis equilibria of uranium metal ions obey the following stoichiometric relationships (Baes and Mesmer 1976):

$$2UO_2^{2+} + 4H_2O \leftrightarrow (UO_2)_2(OH)_2^{2+} + 2H_3O^+ \quad pK = 5.62 \tag{3.5}$$

$$UO_2^{2+} + 2H_2O \leftrightarrow UO_2OH^+ + H_3O^+ \quad pK = 5.80 \tag{3.6}$$

$$3UO_2^{2+} + 10H_2O \leftrightarrow (UO_2)_3(OH)_5^+ + 5H_3O^+ \quad pK = 15.63 \tag{3.7}$$

where *pK*s are the negative logarithms of the equilibrium constants.

As indicated by the *pK* value in Eq. 3.7, the contribution of the uranium complex ion $(UO_2)_3(OH)_5^+$ to the overall uranium biosorption can eventually be neglected in subsequent model development. The hydrolysis equilibrium constants for Eqs. 3.5 and 3.6 are expressed with activities of the ions in the following manner:

$$K_{ey}^{a} = \frac{a_Y a_H^2}{a_X^2} = \left(\frac{\gamma_Y \gamma_H^2}{\gamma_X^2}\right)\left(\frac{[Y][H]^2}{[X]^2}\right) = 10^{-5.62} \tag{3.8}$$

$$K_{ez}^{a} = \frac{a_Z a_H}{a_X} = \left(\frac{\gamma_Z \gamma_H}{\gamma_X}\right)\left(\frac{[Z][H]}{[X]}\right) = 10^{-5.80} \tag{3.9}$$

Combining Eqs. 3.8 and 3.9 with HIEM model Eq. 3.10 (developed in Volesky 2003)

$$[X] = \frac{[H]^2 \left\{ \sqrt{\left(\frac{K_{ez}}{[H^+]} + 2\right)^2 + \frac{8K_{ey}}{[H^+]^2}\left([NO_3^-] + [A^-] + \frac{10^{-14}}{[H^+]} - [H^+] - [Li^+]\right)} - \left(\frac{K_{ez}}{[H^+]} + 2\right) \right\}}{4K_{ey}} \tag{3.10}$$

where $[X]$ is the concentration of free uranyl ions (UO_2^{+2}) in solution, the concentrations of the complex ion species UO_2^{2+}, $(UO_2^{2+})_2(OH)_2^{2+}$ and $(UO_2^{2+})(OH)^+$ may eventually be calculated from the solution pH and the total uranium concentration. The results of the calculations are presented in Fig. 3.2d, where the x-axis represents solution pH, the y-axis uranium normality and the 3-D surfaces the percentage of various uranium ionic species in aqueous solution as a function of solution pH and uranium normality.

For solution pH values below pH 2.7, UO_2^{2+} is the predominant cation. Above pH 2.7, the percentage of UO_2^{2+} starts to decrease with solution pH while that of the $(UO_2)_2(OH)_2^{2+}$ concentration increases with pH and the total uranium concentration. At pH 4.0 and high uranium normality, more than 50% of uranium species exist as the complex $(UO_2)_2(OH)_2^{2+}$.

Effect of Simultaneous Sorption on Stoichiometric Equilibria

The monovalent UO_2OH^+ represents less than 1% of U_t even at a higher pH and total uranium concentrations. According to Collins and Stotzky (1992), hydrolyzed species are better sorbed than the free hydrated ions. Thus the binding of hydrolyzed ionic species to biomass should drive the hydrolysis reaction towards the formation of more hydrolyzed species at a fixed pH, which is maintained by adding alkali to neutralize the released protons. This in turn results in more binding of hydrolyzed ions.

The "hydrolysis equilibrium" which is to say the equilibrium that exists between the hydrolyzed uranium ions and the biomass displays a different stoichiometry than is typically displayed during the ion exchange of non-hydrolyzable species. This is due the fact that the hydrolyzed species contain more uranium per equivalent charge. This is in contrast to a 1:2 stoichiometric relationship for the uranyl complex ion (UO_2^{2+}). In the case of uranium hydrolyzed species the ratio of uranium/proton would be 1:1 for $[(UO_2)_2(OH)_2^{2+}]_{0.5}$–[SORBENT] and $[(UO_2)(OH)^+]$–[SORBENT] complexes. The maximum molar uranium uptake therefore becomes higher than would be expected on the basis of previous ion exchange models and assumptions, whereby the value for the total binding capacity of the given studied biosorbent (*Sargassum*) is 2.25 meq/g.

According to Eqs. 3.5 and 3.6, the formation of one mole of each of the species $[(UO_2)_2(OH)_2^{2+}]_{0.5}$ and $[UO_2OH^+]$ results in the corresponding production of one mole of proton per species upon biosorption. In addition, the binding of either of the two species will result in the exchange and release of one mole of proton to the aqueous phase. The final result is that the binding of one hydrolyzed uranium specie results in an observable increase in proton concentration by two moles of protons: one proton produced by the hydrolysis of uranyl and a second due to ion exchange. The resultant decrease in pH appears to be the same as that for the direct ion exchange of UO_2^{2+} for a proton, despite the fact the mechanisms are different. In order to maintain a constant solution pH of 4.0, two moles of monovalent base (LiOH) were required to neutralize the released protons for every mole of uranium sequestered. This was supported by the relevant experimental results obtained (Yang and Volesky 1999).

3.3.2 Computerized Systems for Assessing Speciation (MINEQL+)

MINEQL+ is a chemical equilibrium model capable of calculating metal ion speciation, solid phase saturation states, precipitation-dissolution, and adsorption. MINEQL+ is a data driven program, so there is no programming to do. In the simplest scenario, one creates systems by selecting chemical components from a menu, scanning the thermodynamic database and running the calculation. MINEQL+ also provides tools to allow one to take control of reaction data, create a personal thermodynamic database, perform synthetic titrations and automatically process multiple samples (such as field data). An extensive thermodynamic database is included in the model. As much more information is available in the User's Manual, this section will only attempt a short introduction to the program and its fundamentals.

The program can basically model any type of a chemical aqueous system. The program uses equilibrium constants to calculate the speciation of the different species (dissolved gases, adsorbed complexes, solid precipitates or dissolved complexes) that can possibly coexist in a given system at equilibrium. It can also perform some adsorption modeling. Mass balance constraints are used to insure that the sum of all species for a given component is equal to the total concentration input by the user. For each problem, there is a choice between either setting the pH at a fixed value or have the program calculate it, or even perform a pH titration (or any type of titration). An open or closed system can be simulated. The results can then be displayed graphically on-screen or printed out.

MINEQL+ gets its power from two sources: *First*, its numerical engine is a modified version of the original MINEQL developed at MIT in the mid 1970s. This numerical approach has become the standard for many other chemical equilibrium models. *Second*, MINEQL+ uses a thermodynamic database that contains the entire US-EPA MINTEQA2 database plus data for chemical components that the EPA did not include, so all calculations will produce results compatible with EPA specifications. MINEQL+ is easy because it provides a standard Windows user interface coupled with the powerful tools listed below. MINEQL+ is an intuitive and state-of-the-art modeling system. Examples of MINEQL+ outputs for speciation of some metals in aqueous solutions are shown in Fig. 3.3a, b.

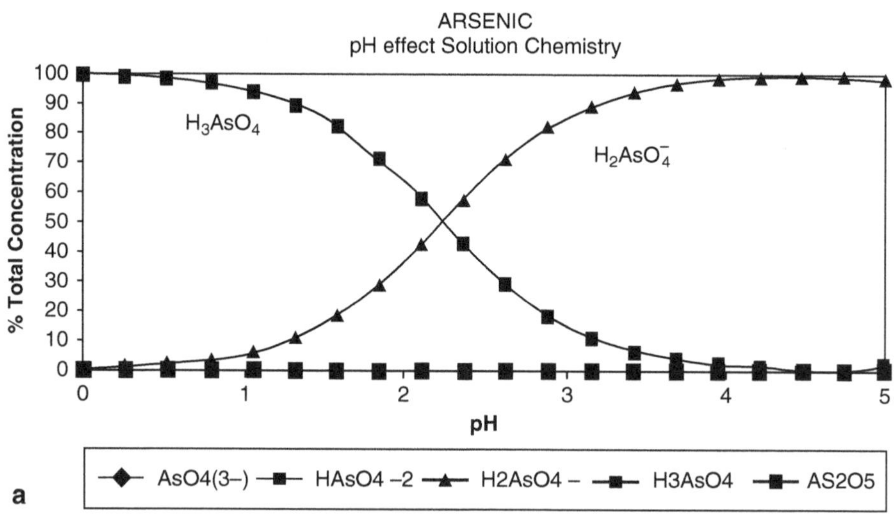

a

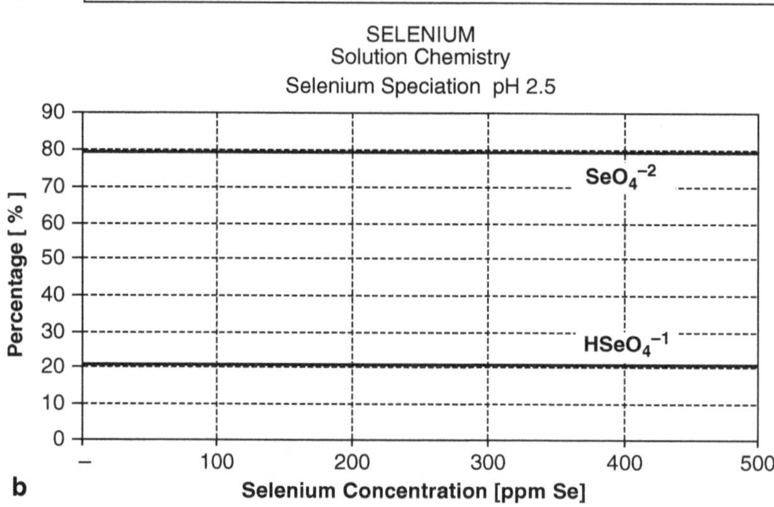

b

Fig. 3.3 Examples of the MINEQL+ output graph for speciation for arsenic (**a**) and selenium (**b**)

MINEQL+ can handle a fairly large number of components (25 can be selected at a time) and then determine all the properties of the new species resulting from any chemically possible combination of components. This powerful Windows-based program can handle multiple run calculations and thus many parameters can be varied at the same time. It is compatible with other graphical programs (e.g. MS-Excel, Lotus-1-2-3, etc.) and supports the *Cut-and-Paste* features allowing the use of the Windows clipboard for transferring data to third party software.

3.4 Sorption Mechanims of Anionic and Cationic Toxic Compounds in Solution

It is necessary to realize that the binding of metal ions to the solid biomass can be either physical or chemical. Correspondingly, while the term *sorption* remains non-specific, the term *adsorption* implies physical attraction and 'surface' deposition as opposed to *chemisorption*, which is based on the chemical attractive mechanisms active throughout the material permeable to ionic species. A brief review of the mechanisms involved in the sequestration of metals in the biomass follows.

3.4.1 Metal Complexation and Chelation

3.4.1.1 Complexation

Complexation is defined as the formation of a species by the association of two or more species (Fig. 3.4a). When one of the species is a metal ion, the resulting entity is known as a metal complex. Mononuclear complexes are formed between a metal ion and a number of anions, or ligands. As a general rule, the metal atom occupies a central position in a complex as exemplified by cobalt, platinum and copper complexes respectively shown.

However, there are complexes, known as polynuclear complexes, which contain more than one metal atom center. The metal-centered structure may carry a positive, negative or zero charge (neutral), depending on the charge and the number of anions involved. It has been proposed that particularly nitrogen and oxygen ligands in microbial cell walls contribute to complexation of transition-metal ions (Fig. 3.4b).

A more elaborate example of complexation is shown in Fig. 3.4c. Cadmium will form the following complexes with chloride ions: $CdCl^+$, $CdCl_2^0$, $CdCl_3^-$, $CdCl_4^{2-}$.

The formation of these complexes can be described by appropriate equations and the equilibrium constants associated with the reactions can be expressed by using standard definitions from reaction kinetics.

$$K_1 = \frac{(CdCl^+)}{(Cd^{2+})(Cl^-)} \tag{3.11}$$

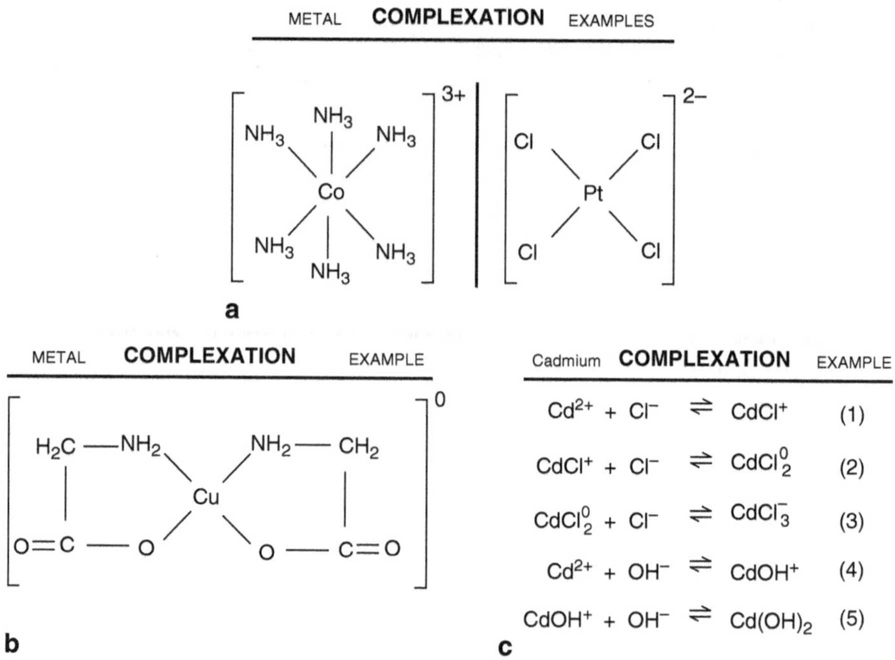

Fig. 3.4 **a** Binding by complexation. **b** Example of copper complexation. **c** Complex formation—stoichiometric relationships

$$K_2 = \frac{(\mathrm{CdCl_2^0})}{(\mathrm{CdCl^+})(\mathrm{Cl^-})} \tag{3.12}$$

$$K_3 = \frac{(\mathrm{CdCl_3})}{(\mathrm{CdCl_2^0})(\mathrm{Cl^-})} \tag{3.13}$$

$$K_4 = \frac{(\mathrm{CdOH^+})}{(\mathrm{Cd^{2+}})(\mathrm{OH^-})} \tag{3.14}$$

$$K_5 = \frac{(\mathrm{Cd(OH)_2})}{(\mathrm{CdOH^+})(\mathrm{OH^-})} \tag{3.15}$$

In relation to Eqs. 3.11–3.15, the degree of complexation will depend on the values of the equilibrium constants K_1–K_5, that are represented by Eqs. 3.11–3.15, respectively. The brackets denote the activities, as opposed to concentrations, of the molecules in aqueous solution. From the expression for these equilibrium constants, it can be seen that as the activity of the free chloride increases, the concentration of the cadmium-chloride complex increases.

According to the published equilibrium constants for metal-chloride complex formation, cadmium will form complexes more readily than lead and iron with chloride ions. In addition to this, nickel and copper form relatively weak complexes with chloride ions.

3.4.1.2 Coordination

When the central metal atom of a complex is bound to its immediate neighbors by covalent bonds formed as the result of the metal atom accepting an electron pair from each non-metal atom, the latter is called the donor and the former the acceptor atom. Alternatively, the non-metal atom is called the coordinating atom and the bond between it and the metal atom a coordinate bond. Compounds in which such bonds are present are widely known as coordination compounds (Fig. 3.5a)—perhaps more often than as metal complexes.

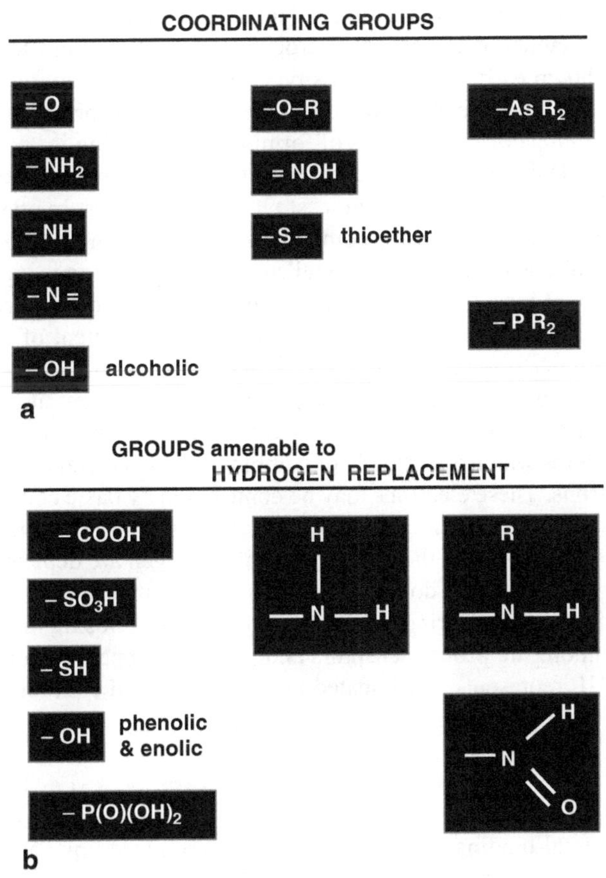

Fig. 3.5 **a** Some coordinating groups. **b** Common chemical groups amenable to hydrogen replacement

Although the term metal complex and coordination compound are frequently used synonymously, they are not, strictly speaking, identical in their coverage since the latter term includes compounds in which no metal atom is present. For example, in the compound formed by the combination of trimethylamine with phosphorus fluoride $[(CH_3)_3N+PF_3]$, there is a coordinate bond formed by the overlapping of the lone pair orbital of nitrogen with the vacant *sp* hybrid orbital of phosphorus. In other words, a metal complex is a particular kind of a coordination compound. Once it is formed, there is no difference between a coordinate bond and an ordinary covalent bond.

3.4.1.3 Chelation of Metals

The term ligand has been used in two different senses. It is sometimes applied to the particular atom in a molecule by means of which the molecule is attached to a central metal atom, e.g. the nitrogen atom in ammonia, or it may be applied to the molecule as a whole. Where there is any risk of ambiguity, it may be avoided by using the term ligand atom or donor atom to denote the atom attached to a metal.

Some ligands are attached to a metal atom by more than one donor atom in such a manner as to form a heterocyclic ring of the kind found in the copper complex shown earlier as (III). This type of a ring has been given a special name—chelate ring—and the molecule or ion from which it is formed is known as a chelating agent or chelator. The process of forming a chelate ring is known as chelation.

Thus, metal chelates are metal complexes where there is an organic compound bound to the metal by at least two available sites. In other words, metal chelate is a special kind of metal complex because one can find non-chelate metal complexes too. The most common metal complexes occurring in aqueous solutions are aquated metal ions or aquocomplexes. For the most part, it is from complexes of this kind that metal chelates are formed by the replacement of water molecules. Some examples of metal chelates are shown in Fig. 3.5b.

If a molecule is to function as a chelating agent, it must fulfill at least two conditions. First, it must possess at least two appropriate functional groups, the donor atoms which are capable of combining with a metal atom by donating a pair of electrons. These electrons may be contributed by basic coordinating groups such as NH or groups functioning as acids by losing a proton. Some groups that combine with metal atoms by the replacement of hydrogen are depicted in Fig. 3.5b.

Second, the donor atoms must be so situated in the molecule as to permit the formation of a ring with a metal atom as the closing member. In solution, chelating anions are proton acceptors and protons compete with metal ions for the anions. If HL represents a protonated ligand, the overall equilibrium with divalent metal ion may be represented as:

$$Me^{2+} + 2HL \Leftrightarrow [MeL_2] + 2H^+ \tag{3.16}$$

Metal-binding extracellular polymers produced by some bacteria have been noted for their metal chelating capabilities.

3.4.2 Biosorbents

Certain types of microbial biomass have been identified for their high metal-sorbing capacity. The uptake of metal ions is due solely to the chemical composition of biomass which for biosorption applications is dead and therefore metabolically inactive. With biosorption applications in mind it makes sense to screen microbial biomass types that are readily available in large quantities—the material can be available very inexpensively. There are basically two types of biomass sources that can practically be considered with low costs and availability in mind:

- industrial waste biomass generated as a by-product of large-scale fermentation processes. With virtually no uses for it, it often poses a disposal problem.
- seaweed biomass generated in large quantities in the ocean. It can be easily collected or harvested as raw material for biosorbents.

Microbial biomass in particular can be grown extremely fast and in many instances there are large quantities of it posing even a serious disposal problem. The least expensive biomass sources also include 'macroscopic' seaweeds—a renewable nuisance or resource. While there are copious quantities of waste activated sludge from wastewater treatment plants all over the world, the metal-sorbing capacities of these sludges, representing very mixed and heterogeneous microbial populations, are usually rather low. There may be some possibilities for improving their metal-sorbing capacity but the heterogeneity of the biomass makes this difficult.

Some types of industrial fermentation waste biomass are excellent metal sorbers. They offer good prospects for practical utilization of their metal biosorption properties. As a potential competition for synthetic ion exchange resins which do the same 'job', the costs of biosorbents must be maintained very low in order for these materials to have an edge. That could be guaranteed by low-cost raw material and minimum of processing. It is necessary to realize that some "waste" biomass is actually a commodity, not a waste: this applies particularly for ubiquitous brewer's yeasts sold on the open market for a price, usually as an animal fodder.

For preparation of suitable biosorbent materials from industrial biomass for application in large-scale sorbing equipment, the consistency of the biomass will have to be altered. Normally, its original consistency is of wet 'mud' or dry cake or powder. It would have to be processed into durable small granules to withstand the conditions of the sorption process.

Seaweed biomass, on the other hand, has a rigid structure of its own and in some instances it has been revealed to offer excellent metal-sorbing properties. Certain ocean locations offer plentiful and very fast growing seaweeds. At some locations overabundant seaweeds threaten the tourist industry by spoiling pristine environments and fouling beaches. Turning seaweeds into a resource has already proven quite beneficial for some local economies. From simple collection of the seaweed biomass the trend is toward progressing to more advanced, knowledge-based and organized aquaculture methods as the demand for seaweeds increases. As a fall-back, high metal-sorbing biomass could even be specifically economically propagated in fermentors using low-cost or waste carbohydrate-containing growth media (e.g. molasses or cheese whey).

There is no evidence that microbial resistance to metals would be connected with high metal biosorption. However, some metal biosorption studies have been conducted with these types of biomass. Even if metal biosorption was found high in some of these cultures, their biomass is not readily available for application purposes. It would have to be specifically cultivated at a cost which would definitely make the material uneconomical.

The potential price advantage of biosorbent materials is of crucial importance for environmental applications and it must be preserved because synthetic ion exchange resins are certainly capable of effective metal sorption. However, in most cases it is their high price that makes their routine application in wastewater treatment uneconomical.

Screening of microbial biomass types for metal biosorption constitutes an important, albeit tedious, way of identifying the most promising types of biomass. As there has been no suitable guidance developed so far to aid in the search for high metal-biosorption, considerable efforts go into testing many different materials in order to assess their metal-sorbing potential. This is done mainly based on simple batch *equilibrium* sorption tests. For the sake of expediency at this stage of work many errors have been committed and even reported in the literature by those who do not quite understand equilibrium sorption concepts.

Biosorbent materials are derived from raw microbial, seaweed or even some plant biomass through different kinds of simple procedures. Biosorbents intended for application need to be derived usually as granules of classified size ranges between 0.1 and 3 mm with a desired rigidity, so as to resist pressure in the column, and water permeability. They may be chemically pretreated for better performance and/or suitability for process applications. Biosorbents are capable of directly sorbing metal ionic species from aqueous solutions.

The ability to directly sorb heavy metals from (process) solutions or wastewater is important because it eliminates the need for costly and cumbersome chemical pretreatment of these metal-loaded effluent streams. These procedures most often result in the production of toxic sludges that eventually cause problems:

- The classification of metal-containing sludges as "hazardous substances" makes the costs of sludge handling and disposal high.
- The recovery of metals from chemical sludges is more difficult and usually uneconomical.

Toxic metal removal can be accomplished quite cost-effectively by biosorption technology that also minimizes the volume of hazardous waste sludges to landfill. Concentration of metals through the biosorption process enables their eventual easy recovery and recycling for resale and reuse. Biosorbent materials contain metal-binding sites not only on the surface but throughout the material (granules, fibers) itself. The high metal-collection performance, low cost and the possibility of multiple reuse of biosorbents makes them quite effective new wastewater treatment alternative. While there are copious quantities of waste activated sludge from wastewater treatment plants all over the world, the metal-sorbing capacities of these sludges, representing very mixed and heterogeneous microbial populations, are usually rather low.

3.4.2.1 Biosorbent Metal Selectivity

Being derived from different natural raw materials, the new family of biosorbent products represents a wide variety of possibilities due to their individual metal-sequestering properties. They invariably feature specifically high affinities for heavy metals (often considered toxic) and there is very little or no significant sorption interference from non-toxic alkaline earth metals (Ca, Na, K, Mg). The broad-spectrum biosorbent materials are not selective for the heavy metals they sorb. They tend to simultaneously remove several different hazardous metals from the solution regardless of their differing concentrations. Since accumulating evidence seems to be pointing at the fact that biosorption involves ion exchange to a high degree, some biosorbents may be more selective toward cations (e.g. Cd, Cu, Ni, Pb, Cr^{3+}, etc.) or anions (e.g. As, Se, V, Cr^{6+}, etc.). It is notable that Cr can appear in the solution as both types of ions.

Some biosorbent materials have been observed to be more specific in their choice of metal they bind. In general, heavier metals (Pb, U, Au) are better sorbed, with exception for Al. Gold, however, was bound quite selectively only in a specific cationic form Ar^{3+}. A relevant discussion on the ion-selective nature of sorption by brown algal biomass, for instance, is in the recent review article by Davis et al. (2003b).

In most cases choices have to be made with regard to the number of metals of interest. For reasonably well executed studies, the volume of biosorption experiments increases almost exponentially with the number of metallic species present in the solution. Single heavy-metal systems are reasonable to handle and they are usually explored first. The uranium uptake when examined for a high-sorber *Sargassum* was observed (Yang and Volesky 1999) to exceed the stoichiometric ion-exchange prediction, highlighting thus the importance of considering carefully the solution chemistry of sequestered metals. A widely available computer program MINEQL+ is extremely useful for establishing the ionic speciation of metals in solution.

3.4.2.2 Biosorption by Bacteria

Bacterial biomass (e.g. *Bacillus, Streptomyces, Citrobacter*) can be obtained as waste products from fermentation industries which makes it a cheap raw material. However, the raw biomass may contain residual chemicals that affect metal binding and the product may be of variable quality due to variations in the fermentation conditions. It may also be necessary to immobilize the biomass before application in reactors, which adds to the cost.

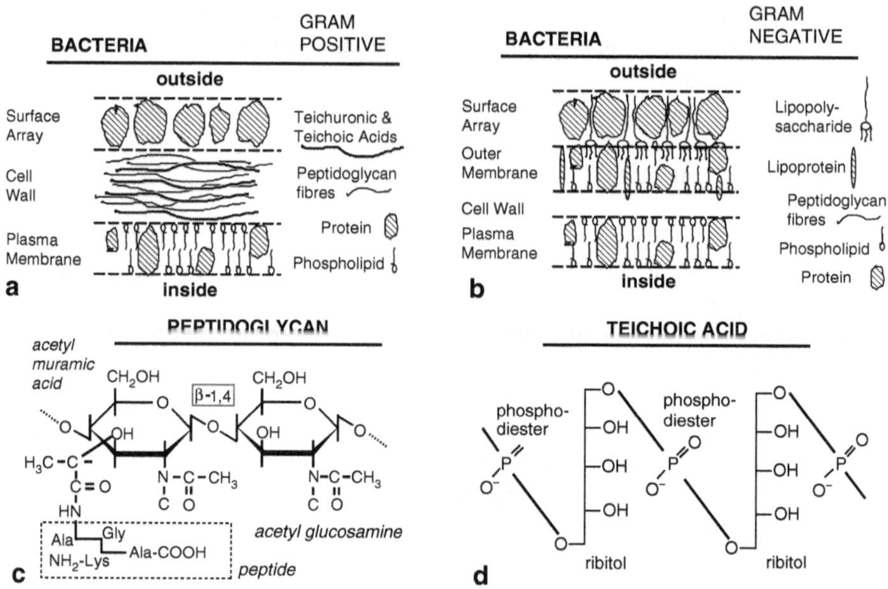

Fig. 3.6 **a** The cell wall structure of gram-positive bacteria. **b** The cell wall structure of gram-negative bacteria. **c** Peptidoglycan is in bacterial cell walls. **d** Teichoic acid—a component of bacterial cell wall

Micro-precipitation is a common phenomenon in metal binding by bacteria (McLean and Beveridge 1990) but complexation by extracellular substances or by N and O ligands in the cell wall, as well as electrostatic attraction to charged groups in the cell wall may also occur (Brierley 1990a). Micro-precipitation is often preceded by binding to specific sites which provides nucleation points (McLean and Beveridge 1990). The higher metal binding capacity of gram-positive, as compared to gram-negative, bacteria can be traced back to their cell wall makeup (Fig. 3.6a, b): gram-positive bacteria possess a thicker peptidoglycan layer (Brierley 1990a).

The gram-positive cell wall features an about 20–30 nm thick layer (~25 PG molecules) of peptidoglycan (PG) into which teichoic acids (TA) and teichuronic acids (TUA) are embedded (McLean and Beveridge 1990; Beveridge 1986). The total cell wall can be 50–150 nm thick (Remacle 1990). PG is reported to represent 40–90% of the cell wall (Remacle 1990), other sources mention that TA and TUA can constitute up to 80% of the wall (Beveridge 1986). Peptidoglycan is a linear polymer of alternating glucosamine and muramic acid with peptide side chains (Fig. 3.6c). These side chains bear one carboxyl group at the terminal amino acid and additional functional groups on certain intermediate amino acids like asparagine, lysine, cysteine, or aspartic acid (McLean and Beveridge 1990). Teichoic acid contains phosphodiesters (Fig. 3.6d), teichuronic acids feature carboxyl groups both of which contribute to the negative charge of the biomass and enable ion exchange

(Brierley 1990a). For *Bacillus subtilis* the major importance of carboxyl groups of the peptidoglycan as well as a minor contribution of phosphate groups in metal uptake was demonstrated by blocking experiments, amine groups did not appear to be relevant (Brierley 1990a).

Gram-negative bacteria have a much thinner (only 1–3 molecules thick (Beveridge 1986)) PG layer which makes up about 10% of the weight of the total cell wall that can be 30–80 nm thick (Remacle 1990). The PG layer of gram-negative bacteria does not contain TA or TUA. Therefore, they offer less negatively charged carboxyl groups which is a reason for their lower biosorptive capacity (Brierley 1990a; McLean and Beveridge 1990).

On the other hand, a characteristic of these bacteria is an outer membrane which contains lipopolysaccharides (LPS) and phospholipids. Their phosphonate groups, creating a negative surface charge conducive to cation binding, were confirmed to be the primary metal binding site in *E. coli* (McLean and Beveridge 1990; Beveridge 1986).

Proteinaceous surface arrays or "S-layers" are present in many bacteria (McLean and Beveridge 1990; Beveridge 1986). Extracellular polymers of the capsule or slime layer contain carboxyl and occasionally phosphonate or sulfonate groups (McLean and Beveridge 1990). Certain bacteria produce SO_4^{2-} or S^{2-} or enzymes that liberate HPO_4^{2-}. These ligands can form micro-precipitates with metal cations (Brierley 1990a; McLean and Beveridge 1990). Although application-oriented work on metal biosorption by waste bacterial biomass from *Bacillus subtilis* has not reached expectations in the early commercial thrust (Brierley 1990b), the biosorption promise and the quantities of bacterial biomass by-product do exist.

3.4.2.3 Biosorption by Fungi

Fungi can be inexpensive and readily available as industrial waste products. *Aspergillus niger* is used in the production of citric acid and of the enzyme glucamylase, *Saccharomyces cerevisiae* is employed in food and beverage industry, *Rhizopus arrhizus* produces the enzyme lipase just to name a few examples of fungi and yeasts that have been employed in biosorption studies (Naja et al. 2005a; Gadd 1990). Some filamentous fungi such as *Aspergillus niger* grow in pellets that could be useful in the recovery or retention of the metal laden biosorbent (Gadd 1990). Other types of fungi create problems of solid/liquid separation and are not easily filterable (Volesky 1990). Another potential disadvantages of the use of fungi are impurities, due to adhering fermentation broth residues that may eventually affect metal uptake.

Similarly to algae and bacteria, the cell wall is the main site of metal deposition in fungi (Volesky 1990). Polysaccharides constitute up to 90% of the fungal cell wall (Remacle 1990). Figure 3.7a shows the fungal cell wall architecture. The inner micro-fibrillar layer of the wall usually consists of chitin (Fig. 3.7b), but cellulose, or in rarer cases, non-cellulosic β-glucan (Hemiascomycetes, e.g. *Saccharomyces*) can take its place, depending on the taxonomic group (Remacle 1990; Mueller and

Fig. 3.7 **a** Composition of one type of a fungal cell wall (*Euascomycete*). **b** Some fungi of specific interest to biosorption

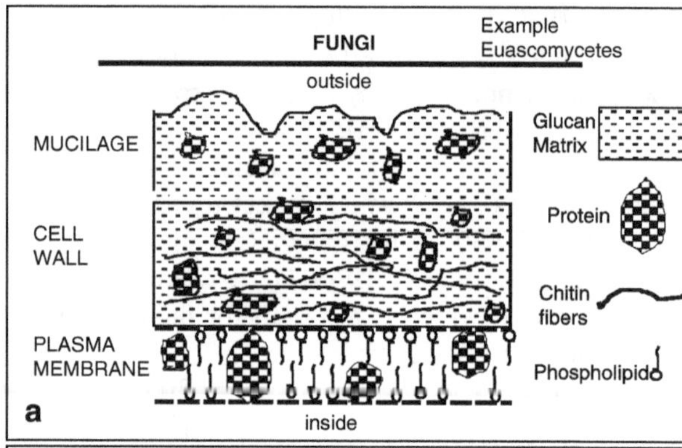

Common Fungi
examined for metal **Biosorption**

Rhizopus species	HIGH	enzyme production steroid transformation
Absidia orchidis	HIGH	steroid transformation
Penicillia species	MEDIUM	antibiotics
Saccharomyces species (yeasts)	variable MEDIUM	baker's yeasts beer, etc.
Aspergillus species	LOW	citric acid food, etc.

b

Loeffler 1976). The outer more amorphous layer is made up of mostly a glucans but can also contain mannan, galactans, chitosan (Fig. 3.8b) (Zygomycetes, e.g. *Mucor, Rhizopus*) or glycogen (Remacle 1990). As comparison of Fig. 3.8a, b shows, chitin is acetylated chitosan. The main components of cell walls of different groups of fungi are listed in Table 3.1 (Schiewer and Volesky 2000), the composition of macromolecules is explained in Table 3.2. Phosphated polysaccharides may also occur (Gadd 1990), the phosphate content in *Mucor* can exceed 20% of the cell wall dry weight (Remacle 1990).

Phosphate and carboxyl groups (of glucuronic acid) are thought to be responsible for negative charge in the fungal wall, amine groups of the chitosan create a positive charge (Naja et al. 2005b; Remacle 1990). Apart from electrostatic attraction to these charged groups, complexation with N or O donors (e.g. of chitin) may occur (Naja et al. 2005b). Since the protein content of the fungal cell wall is only about 10% (Remacle 1990; Volesky 1990), the importance of amino acid functional groups in metal uptake is small. As in the case of bacteria, released metabolites can lead to micro-precipitation (oxalates due to oxalic acid, sulfides due to H_2S) or chelation (citric acid, siderophores).

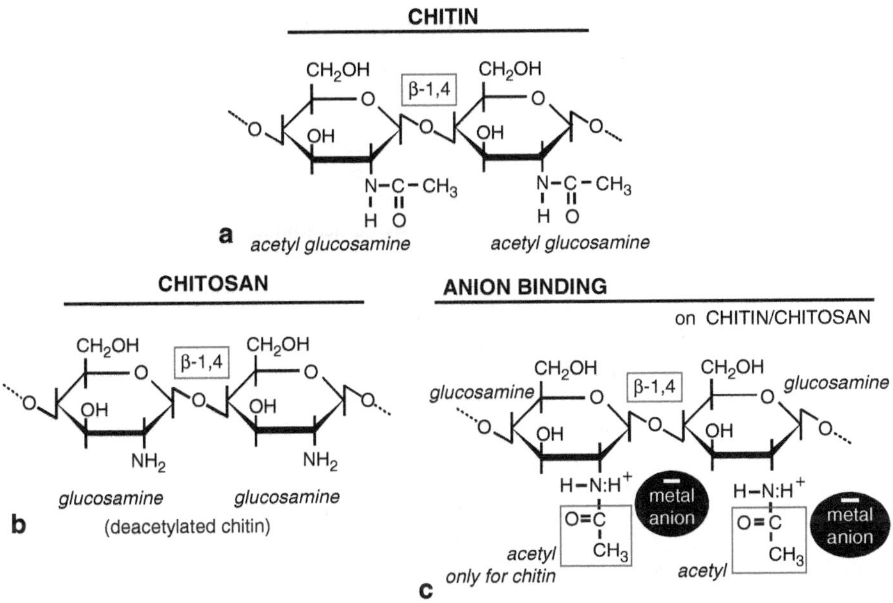

Fig. 3.8 **a** The units of a chitin polymer molecule. **b** Chitosan is deacetylated chitin. **c** Binding of metal anions on chitin or chitosan

Table 3.1 Cell wall biomolecules in fungal biomass

Polymer biomass	Monomers polysaccharides			Group (2nd) amine	
	Cell	Glu	Man	Cto	Cti
Myxomycetes	f				
Acrasiomycetes (I)	f	+			
Plasmodiophoromycetes					f
Hyphochytridiomyc. (III)	f				f
Chytridiomycetes (V)		+			f
Oomycetes (II)	f	+			
Trichomycetes (VIII)	Galactan				Galactosamine
Zygomycetes (IV)				+	f
Euascomycetes (V)		+			f
Hemiascomycetes (VI)		f	+		
Deuteromycetes (V)		+			f
Homobasidiomycetes (V)		+			f
Heterobasidiomycetes (VII)			+		f

Compiled by Schiewer and Volesky (2000)—fungal wall types according to Bartnicki-Garcia are in roman numerals

Cell Cellulose, *Glu* non-cellulosic Glucan, *Man* Mannan, *Cto* chitosan, *Cti* chitin, + present, *f* in inner fibrillar skeleton

Table 3.2 Composition of cell wall biopolymers in fungi and bacteria

Polymer	Link	Monomers	Group
Nitrogenated biopolymers			
Chitosan	β-1,4	D glucosamine	Amine
Chitin	β-1,4	*N*-acetyl-D-glucosamine	2nd amine, acetyl
Peptidoglycan[alt]	β-1,4	*N*-acetyl-D-glucosamine	2nd amine, acetyl
	β-1,4	Muramic acid with peptide side chain	2nd amine, acetyl carboxyl etc.
Phosphated biomolecules			
Teichoic acid	Ribitol or glycerol phosphate	Phosphodiester	
Lipopolysaccharides	Phosphorylated lipid, oligosaccharide core and polysaccharide side chain	Phosphodiester	
Phospholipids	Phosphatidyl ethanol amine and fatty acids	Phosphonate	

Compiled by Schiewer and Volesky (2000)
alt alternating monomers

Good evidence now exists showing that the biomass of the order of Mucorales represents a good biosorbent material particularly for a wide range of heavy metals. The metal-binding sites are predominantly associated with the cell wall structure of these molds. Both *Rhizopus arrhizus* and *Rhizopus nigricans* contain chitin and chitosan in their cell walls. While these compounds may be active, they may not necessarily be the key components in biosorption (Fig. 3.8c). The earlier reports on *Penicillium* biosorption (Jilek et al. 1979) of uranium and lead (Niu et al. 1993) expressed the metal uptakes on a weight basis and came out overly optimistic, since the two elements are especially heavy. Also, common types of *Aspergillus* never performed very well.

The fungal biomass of *Absidia orchidis* exhibited an excellent uptake of lead (Holan and Volesky 1995). Although this has never been ascertained, there are several possible reasons why this fungus, industrially used in transhydroxylation reactions of methylpyridines or steroids (Gai et al. 1981), could bind heavy metals well since there are some prerequisites for this behavior:

- ionic groups abound in its cell wall such as those on guluronic acid in linear (1-4)-linked β-D-glucuronans, and phosphates (Campos-Takaki et al. 1983)
- hydroxyl groups (*cis*-oriented) in α-D-mannans (Yamada et al. 1982) capable of forming chelate complexes with metals
- SH groups may be more abundant in steroid transforming fungi with elevated levels of cytochrome P_{450} performing the transhydroxylation reactions. The formation of strong covalent bonds between lead and SH groups is well known

- *p*-Toluenesulfonic acid is known to be used in the process of microbial hydroxylation of steroids—when incorporated, it could enhance metal uptake
- presence of chitin and glucan, the main structural components of this and other fungi, can offer ample binding sites and, in addition to this, metals could also be entrapped in the inter- and intrafibrillar capillaries in both of these polymers.

All these aspects are worth noting when looking for good biosorption materials among the fungi. Species of *Penicillium* lack such a diversity of structures and this perhaps may be reflected in its lower metal biosorption performance. Quantities of fungal mycelial by-product available from industrial fermentation processes continue to attract interest in the search for good metal biosorbents.

3.4.2.4 Sorption by Chitinuous Biomass

Metal Uptake by Chitinous Biomass

Research on anion biosorption gained momentum relatively recently. Since biosorption is to a large degree based on ion exchange, there are specific biomass types that offer chemical groups particularly suited for sequestering anionic species. Due to the amine moieties they contain, biomass with high amounts of chitin and chitosan is specifically promising for biosorption of anions. While the extraction of chitin or chitosan (Fig. 3.8a, b) is costly, natural biomaterials with chitin have been recognized as effective enough biosorbents for metal removal (Guibal et al. 1998). Chitin (Fig. 3.8a) is a natural polysaccharide consisting of (1,4) 2-acetamide-2-deoxy-D-glucose units, some of which are deacetylated (chitosan) (Roberts 1992a). The ability of chitin/chitosan to form complexes with metal ions, particularly transition and post-transition metal ions, is well documented (Roberts 1992a; Muzzarelli 1977). The binding of molybdate (MoO_4^{2-}) by chitosan or chitin has been examined (Guibal et al. 1999) (Fig. 3.8c). In order to avoid the dissolution of biosorbent beads under acidic conditions, the chitosan was partially cross-linked with glutaraldehyde and processed into small beads suitable for effective sorption applications. Dambies et al. (1999) studied arsenic sorption on molybdate-impregnated chitosan gel beads. The sorption capacity of raw chitosan for As(V) was increased by impregnation with molybdate.

Chitin can be obtained from fungi, insects, lobster, shrimp and krill, but the most important commercial sources are the exoskeletons of crabs obtained as waste from seafood industrial processing (Roberts 1992a). It is estimated that millions of tons of crab shell waste material is disposed of by the seafood industry annually (Roberts 1992a).

Chitin and Chitosan

Chitin is one of the most abundant organic materials in crab shells. It is a natural polysaccharide consisting of (1, 4) 2-acetamide-2-deoxy-D-glucose units, some of which are deacetylated (chitosan) (Roberts 1992a). The structure of chitin is depicted in Fig. 3.8a.

While the names "chitin" and "chitosan" are widely used in the literature, neither term represents a unique chemical structure. The terms chitin and chitosan describe a continuum polymorphic form of copolymers of *N*-acetyl-D-glucosamine and D-glucosamine residues, the two being distinguished by insolubility or solubility in dilute aqueous acid solutions (Roberts 1992a). The *N*-acetylation degree, i.e. the percentage of acetylated amine, could be determined by titration, by determining the overall N/C ratio in chitin, or by NMR spectroscopy (Roberts 1992a).

There are three categories of chitin (Roberts 1992a): α-, β- and γ-chitin. In α-chitin, the chains are anti parallel, in β chitin, they are parallel, and in γ chitin, two chains come "up" to each chain "down". The most abundant form is α-chitin found where extreme hardness is required, as in arthropod cuticle, and is frequently associated with sclerotised protein (chitin proteoglycan), or inorganic materials or both. The strength of chitin is determined by the *N*-acetylation degree, i.e. the percentage of acetylated amine group (called amide). Natural chitin usually exhibits high *N*-acetylation degree (Roberts 1992a). The acetyl group of chitin can be dissociated under strongly basic conditions aided by heating. As *N*-acetylation degree is increased, the solubility of chitin is increased in acidic solutions. Similar to conventional amine family, the N atom of the chitin amide group or chitosan amine group is able to donate and share the lone electron pair with the empty orbit of other cations, featuring thus with weak-base characteristics. The conjugated acid dissociation constant pK_a of chitosan is 6.5 (Domard 1987), and that of chitin is <3.5 (Roberts 1992a). Therefore, only when the solution pH is lower than the corresponding pK_a, the amine/amide sites could be effectively protonated with a positive charge, and an anion could thus be bound.

3.4.3 Biosorption Mechanisms

3.4.3.1 Ion Exchange

Ion exchange (Fig. 3.9a) is the interchange of ions which are formed by molecular or atomic species either losing or gaining electrons. The ion exchange properties of certain natural polysaccharides have been studied in detail and it is well established that bivalent metal ions exchange with counterions from active groups of polysaccharides such as alginic acid (ALG):

$$2\text{NaALG} + \text{Me}^{2+} \Leftrightarrow \text{Me(ALG)}_2 + 2\text{Na} \qquad (3.17)$$

Relatively recently, it has been confirmed that ion exchange is predominantly involved in metal biosorption by algal biomass (Naja and Volesky 2006; Davis et al. 2003b).

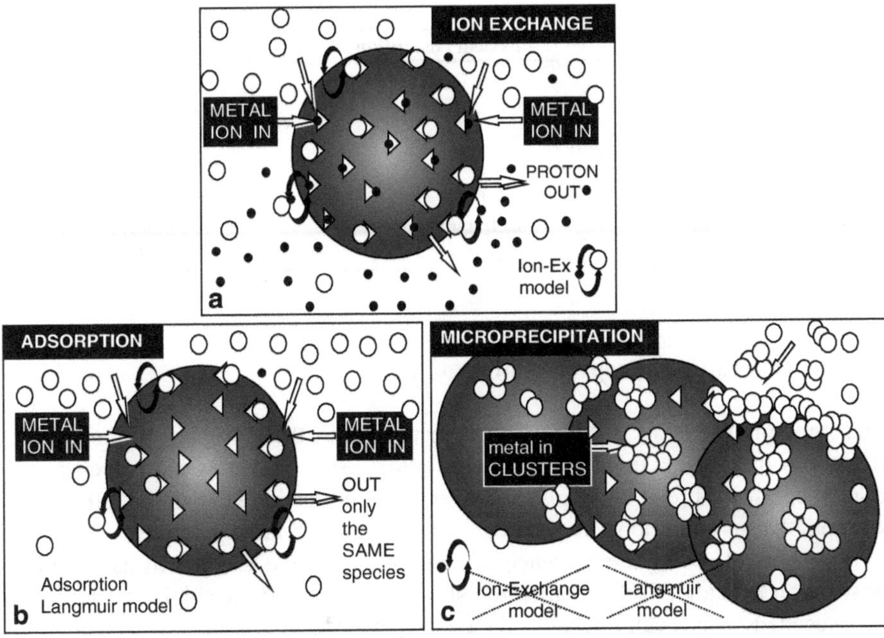

Fig. 3.9 **a** The principle of ion exchange. **b** The principle of adsorption. **c** With microprecipitation the sorbate is deposited in clusters and filtration may gain predominance

3.4.3.2 Adsorption

Adsorption (Fig. 3.9b) is a process by which molecules adhere to solid surfaces. Note the fact that the definition implies no mechanical aspects of the nature of binding. The attraction may often be based on electrostatic charges. Negative adsorption is the adsorption of positive species by negative adsorption sites and vice versa for positive adsorption. While the term adsorption implies a surface phenomenon, the actual sequestration may take place based on either physical phenomena (physical adsorption) or through a variety of chemical binding means (chemisorption).

Physical adsorption is non-specific. The forces attracting the molecules to the solid surface are relatively weak. The energy of activation for physical adsorption is usually no more than 1 kcal/g mol. Since the forces involved are weak the amount of physical adsorption decreases rapidly as the temperature is raised and is generally very small above the critical temperature of the adsorbed component.

Chemisorption is specific and involves forces much stronger than in physical adsorption. According to Langmuir's pioneering work (1918), the adsorbed molecules are held to the surface by valence forces of the same type as those occurring between atoms in molecules. Examples of chemisorption are metal complexation and chelation.

The differences between chemisorption and physical adsorption are summarized in Table 3.3.

Table 3.3 Fundamentals of sorption methods

Parameter	Physical adsorption	Chemisorption
Temperature range	Lower	Higher
Heat of adsorption	Lower	Higher
Order of H.	Condensation	Reaction
Rate	Fast	Non-activated
Activation Energy	Low E	Low E
Coverage	Multilayer possible	Monolayer
Reversibility	High	Often irreversible

3.4.3.3 Inorganic Microprecipitation

Microprecipitation (Fig. 3.9c) of metals results when the solubility of the sorbate reaches its limit (Naja et al. 2005a; Raize et al. 2004). This may happen even due to local conditions (e.g. on or inside the sorbent), not necessarily in the bulk of the solution. These conditions may be created by local deviations in physical conditions such as pH or by the presence of materials from the biosorbent itself. When biosorption is studied, special attention should be paid that the solubility limits are not exceeded even locally because the consequence would be that the metal is not removed from solution by sorption but by precipitation. On the other hand, microprecipitation in the actual biosorption process could contribute to the overall metal removal efficiency a great deal whereby the metal microprecipitate becomes collected by the solid phase and thus immobilized and separated from the solution itself. In the flow-through sorption column arrangement the column and its filling function as an in-depth filtration device. While this is good for increasing the overall performance of the (metal removal) process, naturally, a different mechanism of sorbate collection and sequestration is at work. The mechanism would not matter all that much when the process is being designed empirically but for studying the process it is of relevance since the sorbate sequestration mechanism differs rather significantly.

3.4.3.4 The Mechanism of Biosorption

Understanding the mechanism of biosorption is not merely a question of academic interest, there are also practical benefits gained. The main objective of studying biosorption is obviously to optimize its application. Rather than establishing optimal conditions by a lengthy and expensive trial and error process, one should aim for conceptual understanding that allows predictions to be made. For example, knowing that the mechanism of biosorption is largely based on ion exchange implies that changes in the ionic strength of the solution will affect metal uptake. For predicting quantitatively how much a given factor (such as ionic strength) influences metal uptake we have to employ mathematical models. These will be more reliable if they are not arbitrary mathematical correlations but rather based on the actual mechanism. Furthermore, the choice of desorption technique also depends on the mechanism involved. Metal binding to acidic groups can, for example, be reversed by lowering the pH and thereby protonating these groups.

Unfortunately, knowledge of the biosorption mechanism is not easily obtained since we are not dealing with simple, clearly defined chemical compounds. Biosorbents comprise different types of cells with a highly complex structure whose various building blocks consist of a multitude of different molecules which in turn can display several binding sites. Moreover, even one binding site can participate in different binding mechanisms: carboxyl groups can, for example, engage both in complexation and electrostatic attraction of metal cations. Consequently, several mechanisms often act in combination. The mechanism may also vary with external conditions such as pH.

As for overall metal binding mechanisms, we can distinguish between ion exchange, sorption of electrically neutral material (soluble metal-ligand complexes) to specific binding sites, and micro-precipitation. These main mechanisms are based on sorbate/sorbent or solute/solvent interactions which, in turn, rely on some combination of covalent, electrostatic and van der Waals forces.

The particular amount of metal bound depends not only on the given biosorbent but also on the type of the metal ion, its concentration as well as other physico-chemical factors such as the solution temperature, pH, ionic strength and ion interference by other metals present.

Temperature Effect

It was noted that the temperature can influence the sorption process (Fig. 3.10a) (Naja et al. 2006). Based on Kuyucak and Volesky (1989a), the binding of Co by the brown alga *Ascophyllum nodosum* increased by 50–70% when the temperature was raised from 4 to 23°C. Temperature increase to 40°C caused only a slight binding increase, whereas temperatures of 60°C or more caused a change in the texture of the sorbent and a loss in the sorption capacity due to the material deterioration.

The effect of temperature on biosorption depends on the adsorption heat (enthalpy change). The intrinsic adsorption equilibrium constant K_{int} could be described thermodynamically as follows (Stumm and Morgan 1996b):

$$K_{int} = \exp\left(\frac{-\Delta H^0 + T\Delta S^0}{RT}\right) = \exp\left(\frac{-\Delta H^0}{RT}\right)\exp\left(\frac{-\Delta S^0}{R}\right) \quad (3.18)$$

where ΔH^0 is the enthalpy change (adsorption heat), ΔS^0 is the entropy change, R is gas constant and T is the temperature. For physical adsorption, adsorption heat $\Delta H^0 < 0$, adsorption reaction is exothermic and preferred at lower temperatures. For chemisorption, $\Delta H^0 > 0$, adsorption reaction is endothermic and favored at higher temperatures. This corresponds to the observation of Haug and Smidsrod (1970) for alkaline earth metal binding to alginate where the reaction was exothermic. For the

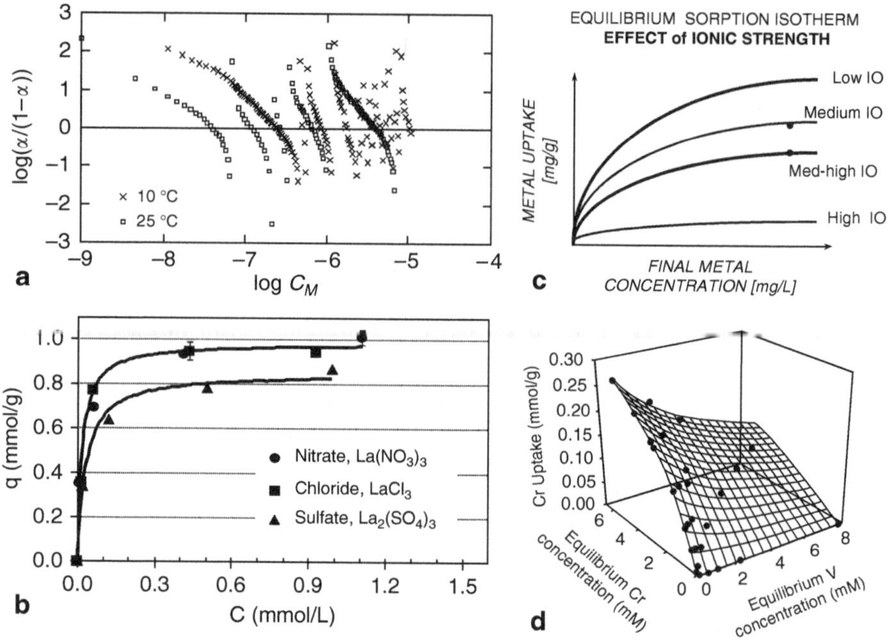

Fig. 3.10 **a** Effect of Temperature on lead sorption by *Rhizopus arrhizus*. (Naja et al. 2006). **b** Effect of counterions on lanthanum biosorption by *Sargassum*. (Diniz and Volesky 2005). **c** Ionic strength suppresses the primary uptake. **d** Example of Cr-V interference

binding of Cu, however, the reaction exhibited a positive enthalpy change (i.e. was endothermic), since the equilibrium constant rose with temperature.

Biomass usually contains more than one type of sites for metal binding. The effect of temperature on each kind of site can thus contribute to the overall metal uptake. This was confirmed in Cu adsorption by potassium-saturated microbial biomass (Weppen and Hornburg 1995). For most metals, the heat of reaction was constant, independent of the degree of site occupation. For Cu, however, the heat of reaction decreased with increasing degree of site occupation from 27 to 14 kJ/mol, indicating the involvement of different binding sites or the formation of different types of Cu complexes with the biomass. For other heavy metals, the heat of adsorption reaction was between ~7 and 11 kJ/mol, for light metals between ~2.1 and 6 kJ/mol (Weppen and Hornburg 1995).

In conclusion, for practical applications of biosorption a reasonably narrow temperature range can be expected (Schiewer 1996). In this range, the effect of temperature is small as compared to other influencing factors.

Influence of pH

Of great importance in both cation and anion biosorption is the pH value of the solution. However, the optimum pH for anion biosorption is opposite to that of cation

Table 3.4 Comparison of the values of the total organic acidity and the lead uptake (me g^{-1}) at pH 4, 5, 6 by *C. glutamicum* and *R. arrhizus*. Calculation of the occupied sites (% of A_{TO}). (Naja et al. 2005b)

pH	*Corynebacterium glutamicum*			*Rhizopus arrhizus*		
	A_{TO} (me g^{-1})	Pb uptake (me g^{-1})	Occupied sites (%)	A_{TO} (me g^{-1})	Pb uptake (me g^{-1})	Occupied sites (%)
4	0.32	0.11	34	0.38	0.14	37
5	0.54	0.20	37	0.58	0.35	60
6	0.73	0.28	38	0.75	0.48	64

biosorption. While cation biosorption is favored at increased pH>4.5 (Kratochvil 1997), anion adsorption is preferred in a lower pH range of pH 1.5–4 (Roberts 1992b). This was determined based on the characteristics of the biomass as well as considering the speciation of metals in the solution (Table 3.4).

There are three ways how the pH can influence metal biosorption:

- the state of chemically active sites could be changed by the solution pH. When the metal binding groups are weakly acidic or basic, the availability of free sites is dependent on the pH. In the case of dye with $-SO_3^-$ adsorption by chitin, only when chitin amide groups were protonated with a positive charge, the dye could be effectively bound through its anionic sulfate group onto positively charged chitin amide groups (Giles and Hassan 1958; Giles et al. 1958).
  Chromate adsorption by chitosan was enhanced by lowering pH to 5.7 and was relatively independent of pH from pH 2.5 to pH 5.7. The logarithm of the conjugation acid dissociation constants (pK_a) could be one of the key parameters in determination of the optimum pH for charging the sites. It was determined that pK_a of chitin is lower than 3.5 and that of chitosan is 6.5 (Domard 1987). For chitinous materials, if the anion binding is through electrostatic attraction, when the solution pH is increased, the bound anions could be eluted from the sorbents.
- extreme pH values, as they are employed for regeneration (desorption) of the sorbent, may damage the structure of the biosorbent material. Distortion of cells was observed under the microscope, accompanied by significant biomass weight loss and decrease in the sorption capacity (Kuyucak and Volesky 1989b).
- the speciation of the metal in solution is pH dependent, whereas metals in aqueous solutions occur as hydrolyzed ions when pH is low, especially metal anions of high charge and a small size (Beveridge 1990). The speciation of metal anions is shown in Fig. 3.1a, b using chromate and vanadate as examples. Biosorption behavior of those anionic metal species systems is expected to be affected by the anionic speciation.

Ionic Strength Effect

The influence of ionic strength on biosorption had not been established until Schiewer and Volesky (1997) systematically studied the effect of ionic strength on the biosorption of cations such as Zn, Cd, Cu and Na (Fig. 3.10b). The increased ionic strength suppressed biosorption as a result of the increased electrostatic charge. It was also established that the effect of ionic strength on adsorption is relevant to the way that the metal is bound. Electrostatic attraction based adsorption, such as that of Na, is severely affected by increased ionic strength (Hayes and Leckie 1987).

Presence of Other Anions

Other sorbable ions in the solution may compete with the metals of interest for sorption sites. The binding of the primary metal ion is then decreased.

For cation biosorption, it was concluded that the light metals bind less strongly than the heavy metal ions. The overall affinity sequence follows that established in the metal adsorption by the ion exchange resin (Schiewer 1996). Diniz and Volesky (2005) studied the effect of counterions on lantanum biosoption by Sargassum and the authors concluded that sulfate had the stronger interference during the process (Fig. 3.10c).

For anion biosorption, recent studies systematically addressed the interference of other anions on the adsorption of anions of interest (Fig. 3.10d). However, the study of anion exchange established that the selectivity of anion exchanger could be enhanced by the counterion of higher valence, with the smaller (solvated) equivalent volume and greater polarizability, and interacting more strongly with the fixed ionic groups on the matrix and participating least in complex formation with the co-ion. The established affinity is as follows (Helfferich 1995): $SO_4^{2-} > I^- > NO_3^- > CrO_4^{2-} > Br^- > SCN^- > Cl^- > F^-$. Among the three strong acids, SO_4^{2-}, NO_3^- and Cl^-, Cl^- has the lowest affinity for the resin. Therefore, it would be appropriate to use strong electrolytic Cl^- salts as background electrolyte for ionic strength control.

3.4.3.5 Overall Mechanisms—Ion Exchange, Adsorption, Micro-precipitation

While some authors consider only an exchange of electrostatically bound ions as ion exchange, we will here adopt a broader definition of this term. The term ion exchange will be used when the charge of ions taken up equals the charge of ions released (so that the particle's charge neutrality is maintained), regardless of whether these ions are bound electrostatically or by complexation.

Adsorption and ion exchange can be the results of three kinds of binding forces. One is a chemical force, another is physical force, and the third is the combination of both. Chemical forces extend over a very short distance (0.1–0.2 nm) (Myers 1991). This type of bond is rather strong, ranging from 20–900 kJ/mol (Smith 1981). Covalent bonds are formed by merging electron clouds such that a non-ionic molecule is formed. These bonds are directional (characteristic bond angles and lengths) and localized (Myers 1991).

Physical forces can be subdivided into electrostatic and London-van der Waals forces (Myers 1991). The energy of physi-sorption is reported as 2–20 kJ/mol and

20–40 kJ/mol by Smith (1981) and Pagenkopf (1978), respectively. In the resulting bonds, the electrons stay in their original systems. Electrostatic (or coulombic) forces between ions or between ions and dipoles extend over a long range and are the strongest among the physical bonds (Myers 1991) with energies >>40 kJ/mol (Stumm and Morgan 1970). The interaction is repulsive for ion charges of the same sign and attractive for unlike charges. The magnitude of the force is proportional to the charge of each ion and inversely proportional to the square of the distance between the ions.

London-van der Waals forces can be divided into three categories: dipole-dipole interaction (creating orientational energy), dipole-induced dipole interaction and the London dispersion force (Myers 1991). The first two are closely related to coulombic forces while the last one is of a quantum-mechanical nature and acts over a long range of up to ~10 nm (Myers 1991). The energy of the dispersion force (8–40 kJ/mol) is larger than the one of orientational or induced dipole (or: polarization) energy (<8 kJ/mol) (Stumm and Morgan 1970). An example of a typically strong (almost ionic) dipole interaction is hydrogen bonding. It occurs between molecules in which H is bound to a very electronegative atom such as N or O (Russell 1980).

The driving force of ion exchange is mostly the attraction of the biosorbent for the sorbate (metal). Metals can be bound electrostatically or by complexation. Interactions between the solute (metal) and the solvent (usually water) play a role in so far as less hydrophilic (and consequently lesser hydrated) molecules have less affinity for the liquid phase and are therefore sorbed more easily. The importance of ion exchange in biosorption has frequently been reported. The amounts of ions from the natural environment (Na, K, Ca, Mg, H) and from biosorbent pretreatment (such as protonation) which are released during biosorption balance the heavy metal uptake by algae (Naja and Volesky 2006), bacteria (Plette et al. 1993), fungi (Naja et al. 2005b) and peat (Crist et al. 1996). Desorption can, in many cases, also be interpreted in terms of ion exchange. Such 'competitive' desorption can be achieved by acids (e.g. HCl, H_2SO_4) and/or salt solutions (e.g. $CaCl_2$) (Aldor et al. 1995). Cations (H, Ca) then compete with the bound metal ions for the binding sites, replacing them if the concentration of the desorption agent is high enough.

We use the terms adsorption and micro-precipitation to describe the accumulation of electrically neutral material which does not involve the release of a stoichiometric amount of previously bound ions. The difference between adsorption and micro-precipitation is that in the former case, affinity between sorbent and sorbate (metal complex) and in the latter case limited solubility (i.e. an interaction between the solute and solvent) represents the main driving force. In micro-precipitation, the metal cation and an anion (e.g. SO_4^{2-}, S^{2-}, oxalic acid, HPO_4^{2-}), itself often a metabolic product of certain biomass types, form insoluble aggregates (salts, complexes) such as sulfides, carbonates, oxides, oxalates and phosphonates (Remacle 1990). Changed local pH or redox potential can also influence the occurrence of precipitation. Micro-precipitation does not necessarily involve a bond between biomass and metal. The process may, however, be nucleated by metal initially bound to active sites in the biomass (Naja et al. 2006). This means that a two-stage process takes place—binding to specific sites is followed by micro-precipitation. The latter process is not limited by the number of binding sites but can occur in multiple layers (Macaskie et al. 1992). Sorption of neutral complexes was thought to be responsible for Cu binding to peat at high concentrations (Chen et al. 1990).

3.4.3.6 Contribution of Electrostatic Attraction and Complexation

Ligands in the biomass (such as carboxyl groups) can form complexes (or: coordination compounds) with metal ions. Chelation, i.e. binding of one metal ion to two coordinating atoms in the same biomolecule, may also occur. Complex formation involves both covalent and electrostatic components whose relative contribution can be estimated by investigating how specific the binding is. When purely electrostatic attraction occurs, the binding strength should correlate with the charge density (z^2/r_{hyd}). Ions of the same charge z and hydrated radius r_{hyd} should therefore be bound with equal strength. Major deviations of the binding strength from the z^2/r_{hyd} correlation indicate a tendency towards a covalent bond character.

The nature of ions released provides information about the bond type. Ions that are bound electrostatically cannot displace covalently bound ions. It was observed that proton release only occurred during heavy metal uptake, not during light metal uptake (Crist et al. 1981; Haug and Smidsrod 1970). Since protons are mainly bound covalently, the binding of heavy metals must have been to a higher degree covalent than that of light metal ions. Similarly, the more Na (that only binds electrostatically) reduces the uptake of other ions, the higher is the contribution of electrostatic attraction in the binding of those ions (Schiewer and Volesky 1997).

The bond character in biosorption can partially be explained by the concept of hard and soft acids and bases introduced by Pearson 1967. So called "hard" (or: type "a") ions (such as the alkaline (earth) metals and Mn) participate in ionic bonds. Easily polarizable "soft" (or: type "b") ions such as (Ag^+, Au^+, Hg^+, Hg^{2+}, Cu^+) tend to form covalent bonds. Many transition metals (e.g. Fe^{2+}, Co^{2+}, Ni^{2+}, Cu^{2+}, Zn^{2+}, Pb^{2+}, Cd^{2+}) as well as protons have intermediate characteristics whereby Zn^{2+} tends toward "hard", Pb^{2+} towards "soft" (Nieboer and Richardson 1980; Pearson and Songstad 1967; Ahrland et al. 1958). Among ligand atoms, O or F are considered hard, S, P and As soft, while N is classified as intermediate.

Different correlations have been proposed to describe the increase of ionic bond character with increasing electronegativity difference between the two bonding atoms. For typical elements in biological ligands (O, N, S) ionic bond character increases therefore with the electronegativity of the metal. It follows from the concept of hard and soft acids and bases that Pb or Cu can display more covalent bond character and thus stronger binding than the "hard" ions Na or Ca.

3.4.3.7 Binding Sites

Numerous chemical groups have been proposed to contribute to biosorption metal binding by e.g. algae (Crist et al. 1981), bacteria (Brierley 1990a) or biopolymers (Hunt 1986), including hydroxyl, carbonyl, carboxyl, sulfhydryl, thioether, sulfonate, amine, imine, amide, imidazole, phosphonate, phosphodiester groups. Whether any given group is important for biosorption of a certain metal by a certain biomass depends on factors as:

- quantity of sites in the biosorbent material
- accessibility of the sites
- chemical state of the site, i.e. availability
- affinity between site and metal, i.e. binding strength

For covalent metal binding even an already occupied site is theoretically available. To what extent the site can be used by the metal in question depends on its binding strength and concentration as compared to the metal already occupying the site. For electrostatic metal binding a site is only available if it is ionized.

The major binding sites in biosorption are acidic. Many groups (hydroxyl, carboxyl, sulfhydryl, sulfonate, phosphonate) are neutral when protonated and negatively charged when deprotonated. When the pH of the solution exceeds their pK_a these groups become mostly available for the attraction of cations. Amine, imine, amide and imidazole groups on the other hand are neutral when deprotonated and positively charged when protonated. Therefore, they attract anions if the pH is lowered such that the groups are protonated. Structural formulae and pK_a values of binding groups are summarized in Table 3.5. The occurrence of these groups in different types of biomass is discussed in another section.

For the freshwater alga *Chlorella* the charge was positive, favoring anion binding, at pH<3 (probably due to amine groups) and negatively charged (mostly carboxyl groups probably) at higher pH so that electrostatic attraction of cations occurs (Greene et al. 1987). The charge of the biosorbent does not only depend on the pH value. Covalent binding of metals can "consume" negatively charged groups. Groups become charge neutral that would otherwise have been negatively charged in metal free solution of the same pH.

It was confirmed that the metal cation biosorption by *Sargassum* involved ion exchange (Naja and Volesky 2006). In the case of heavy metals, the binding force is chemical–sorption through the formation of covalent bonds (Figueira et al. 1997). However, in the case of most alkaline metal cations, the binding is electrostatic. The mechanism of anion biosorption, only studied relatively recently, was briefly reviewed by Volesky (2003). Kratochvil (1997) had proposed a mechanism of chromate adsorption by *Sargassum*, whereby parts of anionic chromate was bound through acid adsorption:

$$\text{Biomass} + H^+ + HCrO_4^- \leftrightarrow \text{Biomass-}H_2CrO_4 \tag{3.19}$$

some of the chromate was reduced by *Sargassum* to Cr(III) that was then bound onto the acidic groups of *Sargassum*.

In $AuCl_4^-$ adsorption by *Sargassum*, Au(III) was reduced to Au(I) and elemental gold. Biosorption mechanism involved the redox, ion exchange as well as micro-precipitation (Kuyucak and Volesky 1989c). Giles et al. (1958) described dye with—SO_3^- adsorption by chitin:

$$\text{chitin-NH} + H^+ + \text{Dye-}SO_3^- \leftrightarrow \text{chitin-}NH_2^+ - SO_3^-\text{-dye} \tag{3.20}$$

Table 3.5 Major binding groups for biosorption

Binding group	Structural formula	pK_a	HSAB classif.	Ligand atom	Occurrence in selected biomolecules
Hydroxyl	–OH	9.5–13	Hard	O	PS, UA, SPS, AA
Carbonyl (Ketone)	>C=O	–	Hard	O	Peptide bond
Carboxyl	–C=O | OH	1.7–4.7	Hard	O	UA, AA
Sulfhydryl (Thiol)	–SH	8.3–10.8	Soft	S	AA
Sulfonate	O ‖ –S=O ‖ O	1.3	Hard	O	SPS
Thioether	>S	–	Soft	S	AA
Amine	$-NH_2$	8–11	Int	N	Cto, AA
Secondary Amine	>NH	13	Int	N	Cti, PG, peptide bond
Amide	–C=O | NH_2	–	Int	N	AA
Imine	=NH	11.6–12.6	Int	N	AA
Imidazole	–C–N–H ‖ >CH H–C–N	6.0	Soft	N	AA
Phosphonate	OH | –P=O | OH	0.9–2.1 6.1–6.8	Hard	O	PL
Phosphodiester	>P=O | OH	1.5	Hard	O	TA, LPS

PS Polysaccharides, *UA* Uronic acids, *SPS* Sulfated PS, *Cto* Chitosan, *PG* PeptidoGlycan, *AA* Amino Acids, *TA* Teichoic Acid, *PL* PhosphoLipids, *LPS* LipoPS

They attributed the dye adsorption by chitin amide to the electrostatic attraction. However, the anion adsorption mechanism is determined not only by the functional groups on the sorbents but also by the characteristics of anionic metal solutes. It was determined that metal speciation in the solution and the functional groups on the biosorbents are both relevant to the metal binding mechanism (Volesky et al. 2001).

3.4.4 Instrumental Analysis

Transmission electron microscopy (TEM), coupled with a microanalysis apparatus such as the electron dispersive spectroscope (EDS), can provide a valuable input in determining distribution of the metal-biomass binding throughout the cell structure. When the objective is to obtain structural as well as analytical information about the metal-biomass interactions, the Fourier-transform infrared (FTIR) and the x-ray photoelectron (XPS) spectroscopies are very helpful, especially considering the simple sample preparation procedures. Infrared spectroscopy has proven to be a powerful tool for studying biological molecules and its application in obtaining structural and bonding information on complex and large molecules has been fruitful (Nakamoto 1997). The more expensive and sophisticated NMR technique is also capable of revealing a great deal about molecular structures of biopolymers and mechanisms involved in selective (metal) binding (Davis et al. 2003a).

The possibility of studying surface chemistry of solids in a non-destructive manner by XPS analysis made it one of the most powerful analytical techniques available to chemists. The use of this technique to investigate the valence of a metal in metal-containing proteins (Leibfritz 1972) is just one example of its applications. Other information such as the nature of attachment of biomolecules to metallic surfaces can also be obtained by the means of XPS analysis (Brizzolara et al. 1997).

Quantitative evaluation of the experimental data indicated that ion-exchange is the most common and prevailing mechanism of metal uptake by most biosorbent materials examined so far. However, when the chemical groups responsible for metal accumulation in the biomass are studied, the image-based instrumental techniques (Transmission Electron Microscopy—TEM, Electron Dispersive Spectroscopy—EDS, X-ray Photoelectron Spectroscopy—XPS) hardly reveal much relevant information (Figueira et al. 1999). Other instrumental analysis such as EPR (Electron Paramagnetic Resonance) could provide more information on the particular conformation of the sites around the metal ion and a computational molecular modeling approach could help in visualizing the structure of the molecule and especially the environment around the sequestered ion. The FTIR (Fourier Transform Infra-Red Spectroscopy) analysis seems to have yielded valuable information on the chemical groups involved in metal biosorption (Naja et al. 2008; Niu and Volesky 2003).

References

Ahrland S, Chatt J, Davis NR (1958) The relative affinities of ligand atoms for acceptor molecules and ions. Q Rev Chem 12:265–276

Aldor I, Fourest E, Volesky B (1995) Desorption of cadmium from algal biosorbent. Can J Chem Eng 73:516–522

Baes CFJ, Mesmer RE (1976) The hydrolysis of cations. Wiley, New York

Beveridge TJ (1986) The immobilization of soluble metals by bacterial walls. In: Ehrlich HL, Holmes DS (eds) Biotechnology and bioengineering symposium no. 16: biotechnology for the mining, metal-refining, and fossil fuel processing industries. Wiley, New York, pp 127–140

Beveridge TJ (1990) Interactions of metal ions with components of bacterial cell walls and their biomineralization. In: Poole RK, Gadd GM (eds) Metal-microbe interactions. IRL Press, Oxford, pp 65–83

Brierley CL (1990a) Metal immobilization using bacteria. In: Ehrlich HL, Brierley CL (eds) Microbial mineral recovery. McGraw-Hill, New York, pp 303–324
Brierley JA (1990b) Production and application of a *Bacillus*-based product for use in metals biosorption. In: Volesky (ed) Biosorption of heavy metals. CRC Press, Boca Raton, pp 305–312
Brizzolara RA, Boyd JL, Tate AE (1997) Evidence for covalent attachment of purple membrane to a gold surface via genetic modification of bacteriorhodopsin. J Vac Sci Technol A 15:773–778
Campos-Takaki GM, Beakes GW, Dietrich SMC (1983) Electron microscopie X-ray microprobe and cytochemical study of isolated cell walls of Mucoralean fungi. Trans Br Mycol Soc 80:536–541
Chen XH, Gosset T, Thevenot DR (1990) Batch copper ion binding and exchange properties of peat. Water Res 24:1463–1471
Collins YE, Stotzky G (1992) Heavy metals alter the electrokinetic properties of bacteria, yeast and clay minerals. Appl Environ Microbiol 58:1592–1600
Crist RH, Oberholser K, Shank N, Nguyen M (1981) Nature of bonding between metallic ions and algal cell walls. Environ Sci Technol 15:1212–1217
Crist RH, Martin JR, Joseph C (1996) Uptake of metal ions on peat moss: and ion-exchange process. Environ Sci Technol 30:2456–2461
Dambies L, Roze A, Roussy J, Guibal E (1999) As(V) removal from dilute solutions using MICB (molybdate-impregnated chitosan beads). In: Amils R, Ballester A (eds) Biohydrometallurgy and the environment toward the mining of the 21st century (Part B). International biohydrometallurgy symposium—proceedings. Elsevier, Amsterdam, pp 277–287
Davis TA, Llanes F, Volesky B, Diaz-Pulido G, McCook L, Mucci A (2003a) A 1H-NMR spectroscopic characterization of sodium alginates extracted from *Sargassum* spp. and its relevance to heavy metal biosorption. Appl Biochem Biotechnol 110:75–90
Davis TA, Mucci A, Volesky B (2003b) A review of the biochemistry of heavy metal biosorption by brown algae. Water Res 37:4311–4330
Diniz V, Volesky B (2005) Effect of counterions on lanthanum biosorption by *Sargassum polycystum*. Water Res 39:2229–2236
Domard A (1987) pH and c.d. measurements on a fully deacetylated chitosan: application to Cu^{II}—polymer interactions. Int J Biol Macromol 9:98–104
Figueira MM, Volesky B, Ciminelli VST (1997) Assessment of interference in biosorption of heavy metals. Biotechnol Bioeng 54:344–350
Figueira MM, Volesky B, Mathieu HJ (1999) Instrumental analysis study of iron species biosorption by *Sargassum* biomass. Environ Sci Technol 33:1840–1846
Gadd GM (1990) Fungi and yeasts for metal binding. In: Ehrlich HL, Brierley CL (eds) Microbial mineral recovery. McGraw-Hill, New York, pp 249–276
Gai Z, Gao Z, Peng B, Yu X (1981) Microbial hydroxylation of 6a-methyl-17a-hydroxyprogesterone and 6a-methyl-17a-hydroxy-11-deoxycorticosterone-21-acetate Yaoxue Xuebao 16:342–348; Chem Abstr (1982) 97:90333
Giles CH, Hassan ASA (1958) A study of the adsorption of dyes and other organic solutes by cellulose and chitin. J Soc Dyers Colour 74:846–857
Giles CH, Hassan ASA, Subramanian RVR (1958) Adsorption at organic surfaces. J Soc Dyers Colour 74:681–688
Greene B, Henzl MT, Hosea JM, Darnall DW (1986) Elimination of bicarbonate interference in the binding of U(VI) in mill waters to freeze-dried *Chlorella vulgaris*. Biotechnol Bioeng 28:764–767
Greene B, McPherson R, Darnall DW (1987) Algal sorbents for selective metal ion recovery. In: Patterson JW, PasinoR (eds) Metals speciation separation and recovery. Lewis Chelsea, Rome, pp 315–338
Guibal E, Milot C, Tobin JM (1998) Metal-anion sorption by chitosan beads: Equilibrium and kinetic studies. Ind Eng Chem Res 37:1454–1463
Guibal E, Milot C, Roussy J (1999) Molybdate sorption by cross-linked chitosan beads: dynamic studies. Water Environ Res 71:10–17
Haug A, Smidsrod O (1970) Selectivity of some anionic polymers for divalent metal ions. Acta Chem Scand 24:843–854

Hayes KF, Leckie JO (1987) Modelling ionic strength effects of cation adsorption at hydrous ohide/solution interfaces. J Colloid Interface Sci 115:564–572

Helfferich F (1995) Ion exchange. Dover Publications Inc, New York, p 168

Holan ZR, Volesky B (1995) Accumulation of cadmium, lead and nickel by fungal and wood biosorbents. Appl Biochem Biotechnol 53:133–142

Hunt S (1986) Diversity of biopolymer structure and its potential for ionbinding applications. In: Eccles H, Hunt S (eds) Immobilisation of Ions by bio-sorption. Ellis Horwood, Chichester, pp 15–45

Jilek R, Docekalova H, Slovak Z (1979) Selective sorbents based on *Penicillium chrysogenum* mycelium. Sorption of cationts (*in Czech*). Chem Prum 29:139–143

Kratochvil D (1997) Biosorption of heavy metals by *Sargassum* seaweed biomass. Ph.D. thesis, chemical engineering, McGill University, Montreal, Canada

Kuyucak N, Volesky B (1989a) Accumulation of cobalt by marine alga. Biotechnol Bioeng 33:809–814

Kuyucak N, Volesky B (1989b) The mechanism of cobalt biosorption. Biotechnol Bioeng 33:823–831

Kuyucak N, Volesky B (1989c) The mechanism of gold biosorption. Biorecovery 1:219–1235

Langmuir I (1918) The adsorption of gases on plane surfaces of glass, mica and platinum. J Am Chem Soc 40:1361–1402

Leibfritz D (1972) X-Ray photoelectron spectroscopy of iron-containing proteins—the valency of iron in ferredoxins. Angew Chem Internat Edit 11:232–234

Macaskie LE, Empson RM, Cheetham AK, Grey CP, Skarnulis AJ (1992) Uranium bioaccumulation by a *Citrobacter* sp. as a result of an ezymatically mediated growth of polycrystalline HUO_2PO_4. Science 257:782–784

McLean RJC, Beveridge TJ (1990) Metal binding capacity of bacterial surfaces and their ability to form mineralized aggregates. In: Ehrlicg HL, Brierley CL (eds) Microbial mineral recovery. McGraw-Hill, New York, pp 185–222

Mueller E, Loeffler W (1976) Mycology. Thieme, Stuttgart

Muzzarelli RAA (1977) Chitin. Pergamon Press, London

Myers D (1991) Surfaces, interfaces, colloids principles and applications. VCH, Weinheim

Naja G, Volesky B (2006) Multi-metal biosorption in a fixed-bed flow-through column. Colloid Surf A-Physicochem Eng Asp 281:194–201

Naja G, Mustin C, Berthelin J, Volesky B (2005a) Lead biosorption study with *Rhizopus arrhizus* using a metal-based titration technique. J Colloid Interface Sci 292:537–543

Naja G, Mustin C, Volesky B, Berthelin J (2005b) A high resolution titrator: a new approach to studying binding sites of microbial biosorbents. Water Res 39:579–586

Naja G, Mustin C, Volesky B, Berthelin J (2006) Association constants of Pb^{2+} with binding sites of fungal biomass using metal-based titrations. Environ Technol 27:109–117

Naja G, Mustin C, Volesky B, Berthelin J (2008) Biosorption study in a mining wastewater reservoir. Int J Environ Poll 34:14–27

Nakamoto K (1997) Infrared and Raman spectra of inorganic and coordination compounds. Part B: Application in coordination organometallicand bioinorganic chemistry. Wiley, New York

Nieboer E, Richardson DHS (1980) The replacement of the nondescript term “heavy metals” by a biologically and chemically significant classification of metal ions. Environ Poll B 1(1):3–26

Niu H, Volesky B (2003) Characteristics of anionic metal species biosorption with waste crab shells. Hydrometallurgy 71:209–215

Niu H, Xu XS, Wang JH, Volesky B (1993) Removal of lead from aqueous solutions by *Penicillium* biomass. Biotechnol Bioeng 42:785–787

Pagenkopf GK (1978) Introduction to natural water chemistry. Marcel Dekker, New York

Pearson RG (1967) Hard and soft acids and bases. Chem Br 3:103–107

Pearson RG, Songstad J (1967) Application of the principle of hard and soft acids and bases to organic chemistry. J Am Chem Soc 89:1827–1836

Plette ACC, Haanstra L, Van Riemsdijk WV (1993) Influence of pH on sorption of cadmium and zinc by soil bacteria. In: Bauda P (ed) Metals-microorganisms relationships and applications—FEMS symposium. Societe Francaise de Microbiologie, Paris, France, p 8

Raize O, Argaman Y, Yannai S (2004) Mechanisms of biosorption of different heavy metals by brown marine macroalgae. Biotechnol Bioeng 87:451–458
Remacle J (1990) The cell wall and metal binding. In: Volesky B (ed) Biosorption of heavy metals. CRC Press, Boca Raton, pp 82–92
Roberts GAF (1992a) Chitin chemistry. Macmillan, London, p 213
Roberts GAF (1992b) Chitin cvhemistry. Macmillan, London, p 495
Russell JB (1980) General chemistry. McGraw-Hill, New York, pp 314–316
Schecher WD (1998) MINEQL+: a chemical equilibrium modeling system, user's manual, version 4.0. Environmental Research Software, Hallowell, ME, USA
Schiewer S (1996) Ion exchange in metal biosorption. Ph.D. thesis, chemical engineering, McGill University, Montreal, Canada
Schiewer S, Volesky B (1997) Ionic strength and electrostatic effects in biosorption of divalent metal ions and protons. Environ Sci Technol 31:1863–1871
Schiewer S, Volesky B (2000) Biosorption process for heavy metal removal. In: Lovley DR (ed) Environmental microbe-metal interactions. ASM Press, Washington, pp 329–362
Smith JM (1981) Chemical engineering kinetics. McGraw-Hill, New York, pp 310–322
Stumm W, Morgan JJ (1970) Aquatic chemistr. Wiley, New York, p 583
Stumm W, Morgan JJ (1996a) Aquatic chemistry. Wiley, New York, p 103
Stumm W, Morgan JJ (1996b) Aquatic chemistry. Wiley, New York, pp 49–51
Tobin JM, Cooper DG, Neufeld RJ (1984) Uptake of metal ions by *Rhizopus arrhizus* biomass. Appl Environ Microbiol 47:821–824
Volesky B (1990) Biosorption by fungal biomass. In: Volesky B (ed) Biosorption of heavy metals. CRC Press, Boca Raton, pp 139–172
Volesky B (2003) Sorption and biosorption. BV Sorbex Inc., Montreal, pp 72–80
Volesky B, Yang J, Niu H (2001) Biosorption of metal cations and anions. In: SenGupta AK, Marcus Y (eds) Ion exchange and solvent extraction, vol 14. Marcel Dekker, New York, pp 119–168
Weppen P, Hornburg A (1995) Calorimetric studies on interactions of divalent-cations and microorganisms or microbial envelopes. Thermochim Acta 269/270:393–404
Yamada H, Oshima Y, Miyazaki T (1982) Characterization of fucomannopeptide and mannoprotein from Absidia cylin- drospora. Carbohydr Res 110:113–126
Yang J (2000) Biosorption of uranium and cadmium on *Sargassum* seaweed biomass. Ph.D. thesis, chemical engineering, McGill University, Montreal, Canada
Yang J, Volesky B (1999) Biosorption of uranium by *Sargassum* biomass. Water Res 33:3357–3363

Chapter 4
Equilibrium, Kinetic and Dynamic Modelling of Biosorption Processes

Francesca Pagnanelli

Abstract This chapter presents an overview of the different modelling approaches used to represent the equilibrium data of heavy metal biosorption, the kinetics in batch reactors and the dynamics in continuous-flow configurations. For each category the attention was focused on the model architecture as a function of the degree of complexity adopted for the representation of the mechanisms involved.

Equilibrium distribution of metals between solid biophase and aqueous phase strictly depends on the operating conditions of the system (mainly pH, ionic strength and solution composition), which influence the state of dissociation of the active sites, the intensity of electrostatic effects, the speciation of metals in solution and their competition.

Biosorption isotherms can be represented by empirical and mechanistic models. The first are simple mathematical relations (such as Langmuir and Freundlich isotherms and their extensions) able to represent experimental trends but without any interpretative or predictive intent. The latter are theoretically derived assuming a set of reactions between biosorbent active sites and ionic species in solution: these models can not only represent but also interpret and predict the effect of the most influencing factors on equilibrium metal distribution. Mechanistic models can also include electrostatic effects due to the electric double layer at the interface, heterogeneity of biosorbent sites and non ideal competition among metals.

Time profiles of metals in both batch and continuous configurations can be determined by different limiting rate steps depending on the nature of the metal-biosorbent system and the specific operating conditions adopted during the tests.

Metal biosorption generally occurred by the following steps: bulk transport; film diffusion through the hydrodynamic boundary layer around the biosorbent surface; intraparticle diffusion through the biophase; chemical reaction of binding with the active sites.

Kinetic data in batch reactors are generally represented by empirical models neglecting mass transfer effects (pseudo-first and pseudo-second-order models), while

F. Pagnanelli (✉)
Department of Chemistry, Sapienza University of Rome, P.le A. Moro 5, 00185 Rome, Italy
e-mail: francesca.pagnanelli@uniroma1.it

P. Kotrba et al. (eds.), *Microbial Biosorption of Metals*,
DOI 10.1007/978-94-007-0443-5_4, © Springer Science+Business Media B.V. 2011

phenomenological models including the description of mass transport steps limiting the global rate of sorption are less diffuse.

Continuous applications of heavy metal biosorption are generally performed in fixed-bed reactors: mechanisms operating in these systems are axial dispersion in the direction of the liquid flow, film diffusion resistance, intraparticle diffusion resistance, and sorption kinetic at the adsorbent surface. Rigorous models taking in consideration all these mechanisms present mathematic and numerical difficulties (especially due to the non linearity associated to equilibrium models) and require independent experiments and/or reliable engineering correlations to estimate the numerous equilibrium, transport and kinetic parameters to avoid the loose of the physical significance of the mechanistic parameters. For these reasons approximate modelling approaches have been widely used (such as Thomas, and Adams-Bohart models), which allow to model breakthrough behaviour without the need of numerical solution and with immediate practical benefits in process development and design.

Keywords Empirical models • Mechanistic models • Equilibrium models • Kinetic models • Dynamic models

4.1 Introduction

The simplest way of imaging heavy metal-matrix interactions is in term of electrostatic interactions as for strong cationic resins: negatively charged groups originated from strong acids (such as sulfates) giving ion exchange reactions. This simplistic idea is complicated by further chemical and electrical aspects.

First of all active sites involved in heavy metal biosorption can be extremely heterogeneous according to the wide variety of biosorbents used for this application. Bacterial biomasses are characterised by carboxylic, phosphordiesters, phosphoric, amines and hydroxyl sites that are typical of the peptidoglycan, teichoic and teichuronic acids of the membrane cell wall (Wang and Chen 2009; Volesky 2007; Cox et al. 1999). Seaweed, and especially brown seaweed containing high amount of alginate (polymer mannuronic and guluronic acids), are rich of carboxylic, amines, phosphates, hydroxyl and sulphate groups (Wang and Chen 2009; Volesky 2007; Chen et al. 1997). Wastes from agriculture, wood and paper industry are mainly made up of cellulose and lignin containing different kinds of polyphenolic and polyhydroxyl groups, which are active in metal removal (Demirbas 2008; Bailey et al. 1999; Williams et al. 1998).

Physical adsorption, ionic exchange, surface complexation and surface microprecipitation can take place simultaneously and to different extends depending on the nature of the active sites and the operating conditions in the system.

In addition electrostatic double layer originated at the interface between charged solid and solution could significantly affect metal distribution: the charged metallic species in solution move in the potential field originated from the charged sites on the surface matrices and then their bulk concentration should be corrected.

Furthermore heavy metals can be present in solution as different species (metal speciation) due to the variability of the environmental conditions (total metal concentration, pH, redox potential, ionic strength, presence of organic and inorganic ligands) affecting metal availability for interactions with solid matrices (i.e. negative charged metallic species can be also present in solution and react with positively charged groups on the matrix).

Heavy metal-matrix interactions are then strictly related to site heterogeneity and metal speciation in solution which in turn are dependent on the operating conditions and especially pH and ionic strength.

Finally according to the structure of the biosorbent and the fuido-dynamic conditions of the systems the kinetics of metal biosorption can be determined simultaneously by different transport mechanisms (such as external film diffusion and intraparticle pore diffusion).

In this complex scenario the choice of modelling approach should be strictly related to the specific aim of developed models. Models can be developed only to have a mathematical representation of data to compare performances of biosorbents in different operating conditions: in this case simple empirical equations should be preferred requiring the minimum number of parameters to adequately represent the experimental data. In other cases models are used not only to represent data but also to interpret them and predict performances in various operating conditions and reactor configurations. In this case mechanisms determining the effects of significant factors on biosorption performances should be included in the model structure in order to interpret those effects and predict them. These interpretative models are specifically devoted to the definition and quantification of the specific mechanisms involved in metal-solid interactions and metal transport and can be used to guide and optimize the choice of operating conditions in the design of the decontamination technology.

In this chapter an overview of the different modelling approaches used for representing equilibrium, kinetic and dynamic data of heavy metal biosorption was reported. Attention was focused on the model architecture as a function of the degree of complexity adopted for the representation of the mechanisms involved.

Model overview was organised by considering three main categories according to the most widely used classification reported in the literature concerning heavy metal biosorption

- equilibrium models,
- kinetic models in batch reactors,
- dynamic models in continuous or semi-continuous processes.

4.2 Equilibrium Models

The main problem in defining heavy metal interactions with natural matrices is that the evaluation of the metal speciation in solid phase (as sorbed species according to different mechanisms) cannot be determined quantitatively or by direct experimental measurements.

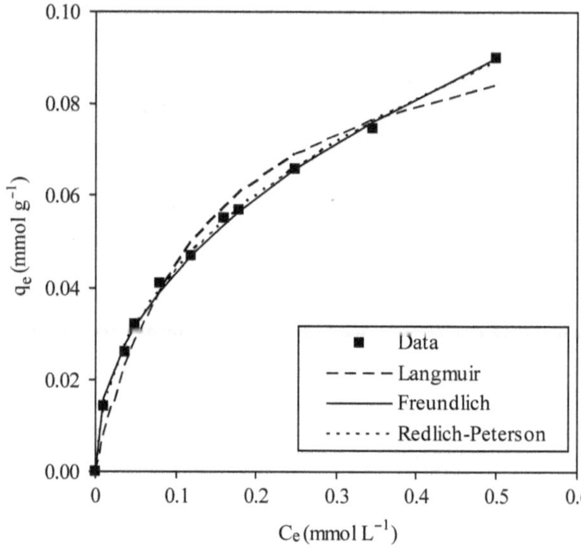

Fig. 4.1 Experimental data and model predictions for copper biosorption onto olive pomace wastes at pH=5

The conventional method used to build isothermal curves only quantify the total amount of sorbed metal without distinguishing the nature of metal-solid binding. In fact metal concentration in solid phase is obtained by a mass balance in the system as the difference among initial and final metal concentration in the liquid phase (Eq. 4.1)

$$q_e = \frac{C_0 V_0 - C_e V_e}{m} \tag{4.1}$$

where q_e is the equilibrium metal concentration in solid phase, C_e is the equilibrium metal concentration in liquid phase, C_0 is the initial metal concentration, V_0 and V_e are the initial and final suspension volume, and m is the sorbent weight.

Varying the initial metal concentration allows to obtain final metal concentrations in solid phase for different values of the equilibrium concentration in liquid phase and then what is called the isotherm (Fig. 4.1).

4.2.1 Empirical Models

Empirical models used for representing equilibrium biosorption data of single- and multi-metallic systems are simple mathematical relations between equilibrium metal concentration in solid phase and equilibrium concentration in liquid phase of the different metals present in solution.

These models are generally characterised by a limited number of adjustable parameters, and required a reduced experimental work to be determined. Nevertheless the application of empirical models have no predictive or interpretative value and then no information about operating mechanisms can be deduced from their parameters. In addition these models generally do not include the effect of any variable environmental factor and then conditional parameters are obtained which are characteristic of the operating conditions used during biosorption experiments.

Criteria for choosing the best empirical model are dictated only by the goodness of data representation, which can be determined by simple statistical discrimination as described in the follow.

4.2.1.1 Empirical Models for Single Metal Systems

Biosorption data of single-metal systems are usually represented by various simple models which are purely empirical (i.e. the Freundlich isotherm), or derived from theoretical models originally developed for different systems on the base of assumptions that are quite simplistic for biological systems (i.e. the Langmuir isotherm for gaseous adsorption on planar surfaces). These isotherms have been widely applied since they are simple, give a good description of experimental behaviour in a large range of fixed operating conditions, and are characterised by a limited number of adjustable parameters.

Langmuir Isotherm

The most widely applied isotherm for equilibrium data modelling (see Febrianto et al. 2009 for some examples of application) is the Langmuir model (Langmuir 1918) as given by Eq. 4.2

$$q_e = \frac{q_{max} b C_e}{1 + b C_e} \tag{4.2}$$

The easy interpretation of Langmuir parameters made the fortune of this model: in fact q_{max} is often used to compare biosorbent performances in terms of maximum capacity, while b, which characterise the initial slope of the isotherm, is taken as a measure of the biosorbent affinity for a metal. Nevertheless considering the assumptions used for theoretical derivation of Langmuir isotherm (monolayer coverage; all sites are alike; each site can bind only one adsorbed species and the adsorption of a single species is not influences by neighbouring occupied sites) the interpretation of these parameters should not be done in such simplistic way.

In fact previous assumptions are hardly satisfied in aqueous suspensions of biomasses in which heterogeneous nature of sites, electrostatic effects and different stoichiometry can occur and should be taken into account using another kind of models as described in the follow (Sect. 4.2.2).

Langmuir isotherm is a non linear model, whose parameters can be found by using non linear regression. Alternatively model can be linearised and linear least-square method used. Langmuir model can be linearised in different ways:

$$\text{Langmuir linearization} \quad \frac{C_e}{q_e} = \frac{1}{q_{max} b} + \frac{1}{q_{max}} C_e \tag{4.3}$$

$$\text{Lineweaver-Burk linearization} \quad \frac{1}{q_e} = \frac{1}{q_{max}} + \frac{1}{q_{max} b} \frac{1}{C_e} \tag{4.4}$$

$$\text{Eadie-Hoffsie linearization} \quad q_e = q_{max} + \frac{1}{b} \frac{q_e}{C_e} \tag{4.5}$$

$$\text{Scatchard linearization} \quad \frac{q_e}{C_e} = q_{max} b - b q_e \tag{4.6}$$

Langmuir parameters obtained from the four linearization can differ among each other due to an improper weight associated to the linearised variables in linear regression (Himmelblau 1978). This can determine a distortion in the fit resulting in prediction errors. Even though the recent advancement in computer and software allows the use of non linear method providing better results in the determination of isothermal parameters (Ho 2006a), lsinear least-square method is often preferred due to its simple application requiring only narrow understanding of data fitting process.

In Fig. 4.2 the histograms of the relative frequency of Langmuir parameters reported in the literature for different kinds of biomasses and heavy metals were reported. Literature data were taken reporting Langmuir parameters referred to equilibrium pH values in the range of 4–6 units. It should be noted that reporting the maximum adsorption capacity as mmol g^{-1} the strong discrepancies observed for different metals disappeared and independently of the biosorbent nature, 70% of q_{max} values are below 1 mmol g^{-1}. As for biomass "affinity", except the case of lead (typically denoting high affinity for biomasses), b values are mainly concentrated below the value of 5 l $mmol^{-1}$.

Freundlich Isotherm

The Freundlich isotherm is the other widely used model for equilibrium modelling of single metal system (see Febrianto et al. 2009 for some examples of application)

$$q_e = K_F C_e^{1/n} \tag{4.7}$$

where K_F and n are the mono-component Freundlich constants, generally related to the binding capacity and biosorption intensity, respectively. Even though an interpretation was given about the nature of systems well-represented by this equation (heterogeneous surfaces with a logarithmic diminution of the affinity during surface coverage), Freundlich isotherm is a purely empirical model and then the assignment of physical meaning to parameters should be avoided.

Freundlich parameters can be regressed by non linear methods or after linearization (taking the logarithmic form) by least-square method.

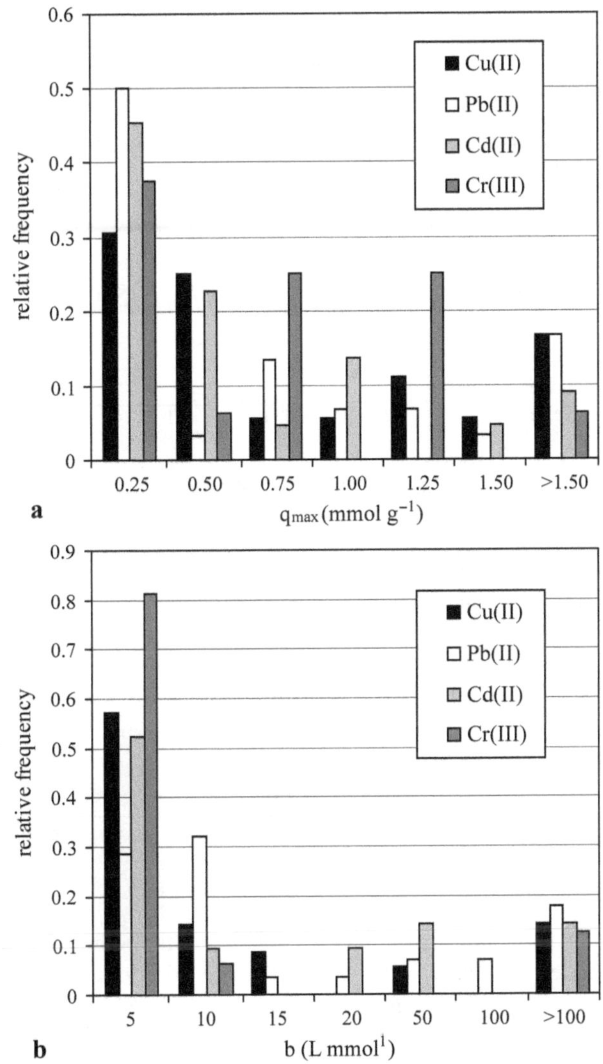

Fig. 4.2 Relative frequency of Langmuir parameters (q_{max} (**a**) and b(**b**)) for copper, lead, cadmium and chromium biosorption for pH ranging from 4 to 6

Other Two- and Three-Parameter Models

In the literature a wide number of models initially developed for gas adsorption was applied to represent biosorption of single metal systems onto biomasses. In Table 4.1 a lists of two- and three- parameters models used for isothermal data of single metal systems were reported (Dang et al. 2009; Febrianto et al. 2009; Papageorgiou et al. 2009; Quintelas et al. 2009; Vaghetti et al. 2009; Chen et al. 2008; Namasivayam and Sureshkumar 2008; Sari et al. 2008; Ko et al. 2004; Pagnanelli et al. 2001; Al-Asheh et al. 2000; Aksu et al. 1999b; Webster et al. 1997).

Table 4.1 Two- and three-parameter isotherms for empirical modelling of single-metal systems

Isotherm	Equation	Notes
Temkin	$q_e = \frac{RT}{b_T} \ln (a_T C_e)$	
Flory-Huggins	$\log\left(\frac{C_0 - C_e}{C_0}\right) = \log K_{FH} + n_{FH} \log\left(\frac{C_e}{C_0}\right)$	
Halsey	$q_e = \left(\frac{K_H}{C_e}\right)^{1/n_H}$	Proposed for representing a multilayer condensation at relatively large distance from the surface
Dubinin-Radushkevich	$q_e = q_D \exp\left[-b_D\left(RT \ln\left\{1 + \frac{1}{C_e}\right\}\right)^2\right]$	Originally proposed for organic compounds sorption in gas phase in porous solids
Brunauer-Emmer-Teller (BET)	$q_e = \frac{q_{BET} b_{BET} C_e}{(C_e - C^*)[1 + (b_{BET} - 1)(C_e/C^*)]}$	Describes multilayer adsorption at the adsorbent surface and assumes that a Langmuir isotherm applies to each layer
Redlich-Peterson	$q_e = \frac{a_{RP} C_e}{1 + b_{RP} C_e^{n_{RP}}}$	Describes adsorption equilibria over a wide concentration range
Sips	$q_e = \frac{q_{mS}(b_S C_e)^{n_S}}{1 + (b_S C_e)^{n_S}}$	Proposed to solve the problem of continuing increase of the adsorbed amount with rising concentration observed for Freundlich isotherm
Toth	$q_e = \frac{K_T C_e}{(1 + (K_T C_e)^{n_T})^{1/n_T}}$	Describes sorption onto heterogeneous surfaces

C_0 initial metal concentration, R gas constant, T absolute temperature, C^* theoretical saturation concentration

As for three-parameter models it should be noted that, in the ambit of empirical modelling, there is no need of using additional adjustable parameters if simpler models can represent data reasonably well. Even in this case statistical discrimination of models can clearly evidence if enough compensation was gained by using additional parameters.

4.2.1.2 Empirical Models for Multi-Metal Systems

Waste waters usually contain not one but many ions making difficult the description of metal adsorption. When several ions are present in solution, interference and competition phenomena for adsorption sites can occur and lead to more complex mathematical formulations. Several isotherms have been proposed to describe equilibrium competitive adsorption for such systems.

In the follow empirical models for multi-metal systems are classified according to the sets of data (single or multi-metal systems) used for parameter regression.

Predictive Models

The simplest way to represent multi-metal system data is a prediction using mathematical equations, which contain parameters obtained only from single metal isotherms.

The predictive extension of the Langmuir isotherms (also named as extended Langmuir model) (Papageorgiou et al. 2009; Al-Asheh et al. 2000; Ho and McKay 2000; Chang and Chen 1998) can be written as:

$$q_{e,i} = \frac{q_{max,i} b_i C_{e,i}}{1 + \sum_{a=1}^{N} b_a C_{e,a}} \tag{4.8}$$

where $q_{e,i}$ is the uptake of the i-th component in the N-component system, $C_{e,i}$ is the equilibrium concentration of each component, $q_{max,i}$ and b_i were obtained from single isotherm (Eq. 4.2) of each metal ions.

Jain and Snoeyink (1973) studied the competitive sorption of metals on activated carbon and developed a model that predicts sorption equilibria in binary solutions. They started from the assumption that the Langmuir theory for multi-sorbate systems is not based on competition. In fact a simple competitive Langmuir model simulates a system in which the presence of other metals in solution can affect only the apparent affinity of the metal for the active site. This means that the shape of the adsorption curve changes when there are other metals in solution (growing slowly), but the asymptotical maximal capacity for large metal concentration in solution remains the same. On the other side the experimental evidence is that not only the isothermal shape, but also the maximal capacity changes passing from a single- to a multi-component system. Jain and Snoeyink (1973) proposed then to add another term into the classical competitive Langmuir isotherm for binary systems:

$$q_{e,1} = \frac{(q_{max,1} - q_{max,2}) b_1 C_{e,1}}{1 + b_1 C_{e,1}} + \frac{q_{max,2} b_1 C_{e,1}}{1 + b_1 C_{e,1} + b_2 C_{e,2}} \tag{4.9}$$

$$q_{e,2} = \frac{q_{max,2} b_2 C_{e,2}}{1 + b_1 C_{e,1} + b_2 C_{e,2}} \tag{4.10}$$

The first term, proportional to ($q_{max,1} - q_{max,2}$), takes into account the fraction of type 1 ions that adsorbs without competition, while the second term, proportional to $q_{max,2}$, represents the type 1 ions that adsorb in competition with type 2 ions. All the parameters reported in the previous formula were determined by single component isotherms.

In Table 4.2 some predictive extensions of other empirical models taken from the literature were reported (Papageorgiou et al. 2009; Wang and Chen 2009; Gavrilescu 2004; Ko et al. 2004; Al-Ashes et al. 2000).

Table 4.2 Predictive and semi-predictive models for empirical modelling of multi-metal system biosorption: for predictive models all parameters are determined by single metal system data, while for semi-predictive models correction parameters regressed using multi-metal system data are reported in the column notes

Type	Isotherm	Equation	Notes
Predictive	Freundlich extension	$q_{e,1} = \frac{n(K_{F,1}/n_1)^{1/n_1} C_{e,1}}{\left[(K_{F,1}/n_1)^{1/n_1} C_{e,1} + (K_{F,2}/n_2)^{1/n_2} C_{e,2}\right]^{1-n}} + \Delta_{F2}$	For binary solution Δ_{F2} and n are functions of single-metal system parameters and residual metal concentrations
Predictive	Sips extension	$q_{e,i} = \frac{q_{mS,i} b_{S,i} C_{e,i} \left(\sum_{a=1}^{N} b_{S,a} C_{e,a}\right)^{1/n_i - 1}}{1 + \left(\sum_{a=1}^{N} b_{S,a} C_{e,a}\right)^{1/n_i}}$	
Predictive	Redlich-Peterson extension	$q_{e,i} = \frac{a_{RP,i} C_{e,i}}{1 + \sum_{a=1}^{N} b_{RP,a} C_{e,a}^{\ n_{RP,a}}}$	
Predictive	Combined Langmuir-Freundlich	$q_{e,i} = \frac{q_{LF,i} b_{LF,i} C_{e,i}^{\ 1/n_{LF,i}}}{1 + \sum_{a=1}^{N} b_{LF,a} C_{e,a}^{\ 1/n_{LF,a}}}$	
Predictive	IAST	$\frac{1}{q_{tot}} = \sum \frac{z_i}{q_{e,i}}$	q_{tot} is the total amount of sorbed species, z_i is the mole fraction of component i in solid phase, and $q_{e,i}$ is the solid concentration in equilibrium for single-metal systems
Semi-Predictive	Langmuir extension	$q_{e,i} = \frac{q_{\max,i} b_i C_{e,i}}{1 + b_i C_{e,i}} [1 - F_i] \quad F_i = \frac{\sum_{a=1, a \neq i}^{N} Ki_a C_{0,a}}{\sum_{a=1}^{N} Ki_a C_{0,a}^{o}}$	Ki_j are correction parameters and $C_{0,j}$ are the initial concentration
Semi-Predictive	Freundlich extension	$q_{e,1} = \frac{K_{F,1} C_{e,1}^{1/n_1 + b_{11}}}{C_{e,1}^{b_{11}} + a_{12} C_{e,2}^{b_{12}}}$ $q_{e,2} = \frac{K_{F,2} C_{e,2}^{1/n_2 + b_{22}}}{C_{e,2}^{b_{22}} + a_{21} C_{e,1}^{b_{21}}}$	restricted to binary mixtures b_{11}, a_{12}, b_{12}, b_{22}, a_{21} and b_{21}) are correction coefficients
Semi-Predictive	Freundlich extension	$q_{e,i} = K_{F,i} C_{e,i} \left(\sum_{a=1}^{N} a_{ia} C_{e,a}\right)^{1/n_i - 1}$	a_{ia} are correction coefficients

The IAST (ideal adsorbed solution theory) (Radke and Prausnitz 1972) is another approach to develop predictive models for multi-component sorption isotherms using single component sorption data. Model development is based on the thermodynamic equivalence of the spreading pressure (π) for each single component, where the spreading pressure is defined as the difference between the interfacial tension of the pure solvent-solid interface and that of the solution-solid interface at the same temperature. The IAST can be combined with different adsorption equations in order to obtain the mole fraction of sorbate in solid phase. IAST models have been developed for Langmuir, Freundlich, Langmuir-Freundlich and Sips isotherms (Papageorgiou et al. 2009; Ko et al. 2004; Cheung et al. 2003; Al-Asheh et al. 2000).

Application of predictive models denoted that in most cases, when suitable multi-component adsorption isotherms models are selected, single isotherm data can be used to predict multi-metal system data in selected range of competition conditions (Papageorgiou et al. 2009; Ko et al. 2004; Al-Asheh et al. 2000; Sag and Kutsal 1996). When predictive models fail to represent multi-metal systems the simplest solution is adding correction parameters regressed on multi-metal system data.

Semi-Predictive Models

In the literature there are several examples of empirical models, which comprehend parameters derived from single system data, but also adjustable correction parameters that take into account the interaction between the two metals in solution and that are fitted on binary data.

A widely used modified extended Langmuir model (Papageorgiou et al. 2009; Pagnanelli et al. 2001, 2002; Aksu et al. 1997, 1999a; Sag et al. 1998; Chong and Volesky 1996) can be written as

$$q_{e,1} = \frac{q_{max,1} b_1 \frac{C_{e,1}}{\eta_1}}{1 + b_1 \frac{C_{e,1}}{\eta_1} + b_2 \frac{C_{e,2}}{\eta_2}} \tag{4.11}$$

$$q_{e,2} = \frac{q_{max,2} b_2 \frac{C_{e,2}}{\eta_2}}{1 + b_1 \frac{C_{e,1}}{\eta_1} + b_2 \frac{C_{e,2}}{\eta_2}} \tag{4.12}$$

where b_1, b_2, $q_{max,1}$ and $q_{max,2}$ are the single metal system Langmuir adsorption constants of the first and second metallic ion, while η_1 and η_2 are the multi-component Langmuir adsorption constants of the first and the second metallic ion, respectively, which are estimated from binary adsorption data.

The addition of further correction coefficients into the extended Langmuir model (Eq. 4.8) makes this equation more flexible and representative of the complexity of multi-metal systems.

In Table 4.2 some semi-predictive extensions of empirical models taken from the literature were reported (Papageorgiou et al. 2009; Aksu and Gulen 2002; Pagnanelli et al. 2001; Sag et al. 1998, 2001; Aksu et al. 1999a, b; Chang and Chen 1998).

Competitive Models

Empirical models for multi-metal systems can be also characterised by parameters fitted directly on multi-metal system data. Among these the competitive Langmuir model (Perez-Marin et al. 2008; Fiol et al. 2006; Sag et al. 2001; Figueira et al. 1997; Chong and Volesky 1996) is one of the most used

$$q_{e,i} = \frac{q_{max} b_i C_{e,i}}{1 + \sum_{a=1}^{N} b_a C_{e,a}} \tag{4.13}$$

where all the parameters (q_{max} and all the b_i) were determined using both single and multi-metal system data. It is noteworthy that the maximal capacity, q_{max}, is unique for all metals, while in the predictive Langmuir model each metal has its specific $q_{max,i}$ (Eq. 4.8) representing the maximum specific uptake determined for the single system. This is not a secondary aspect because a unique q_{max} means that all the metals obey to the fundamental hypothesis of the Langmuir model that the surface is uniform and all the solutes compete for the same sites without interaction and competition among the ions. By using the competitive Langmuir model (Eq. 4.13), the q_{max} parameters independently determined for different multi-component systems should be always the same according to the theoretical assumption of the Langmuir isotherm that all the sites are equal and their concentration is constant. When the maximal capacities change according to the different multi-metal systems, a possible explanation is that the active sites are not homogeneous, but metal-specific. This observation is not consistent with the basic assumption of the Langmuir model.

Competitive Langmuir model can be derived assuming purely competitive competition among different ionic species in solution (M_1 and M_2) for the active site (S)

$$M_1 + S \overset{K_1}{\leftrightarrow} SM_1 \tag{4.14}$$

$$M_2 + S \overset{K_2}{\leftrightarrow} SM_2 \tag{4.15}$$

where K_1 and K_2 are the equilibrium complexation constants.

Combining the site mass balance with the equilibrium constants above, the equilibrium concentration of each metal in solid phase is obtained

$$[S]_{Tot} = [S] + [SM_1] + [SM_2] = [SM_1]\left(1 + K_1\left[C_{e,1}\right] + K_2\left[C_{e,2}\right]\right) \tag{4.16}$$

$$[SM_1] = \frac{[S]_{Tot} K_1 [C_{e,1}]}{1 + K_1\left[C_{e,1}\right] + K_2\left[C_{e,2}\right]} \tag{4.17}$$

In the literature other examples of Langmuir-type isotherms for multi-metal systems were reported assuming different type of competition among ionic species in solution such as the formation of ternary complexes SM1M2 (Apiratikul and Pavasant 2006; Sanchez et al. 1999; Chong and Volesky 1995).

Even though a reaction scheme is supposed these Langmuir-type models have been classified here as empirical because no chemical meaning can be assigned to the regressed parameters lacking a biomass characterisation for the identification of the active sites really involved in heavy metal biosorption.

4.2.1.3 Statistical Discrimination of Empirical Models

When empirical models are used, discrimination among different available equations can be done only by means of statistical methods, known as model discrimination (Himmelblau 1978). This analysis has the purpose to select the most appropriate model among many feasible ones. The best model should satisfy some criteria such as the lowest number of coefficients and the simplest form consistent with reasonable error, rational physical consistency and the minimum sum of squares of deviations between predicted and empirical value.

Except few examples (Vaghetti et al. 2009; Basha et al. 2008, 2009; Chen et al. 2008; Pagnanelli et al. 2001; Ho and McKay 2000), in the large majority of the works empirical models are simply discriminated according to the values of the regression coefficient (R^2)

$$R^2 = \frac{SS_R}{SS_T} \tag{4.18}$$

where SS_R and SS_T are given by the following expressions:

$$SS_R = \sum_{j=1}^{P} \left(q_{e,j(cal)} - \overline{q}_e\right)^2 \tag{4.19}$$

$$SS_T = \sum_{j=1}^{P} \left(q_{e,j(\exp)} - \overline{q}_e\right)^2 \tag{4.20}$$

where $q_{e,j(cal)}$ and $q_{e,j(exp)}$ are the metal specific uptake calculated by the model and experimentally measured respectively, and $\overline{q}_e$ is the average value of all the P experimental data.

Regression coefficient values near unity are generally assumed for good models. Nevertheless regression coefficient doesn't tell the entire story giving no information about parameters errors and residue distribution.

In the follow a brief introduction to simple statistical tools used to statistically discriminate models is presented.

First of all model parameters should be reported along with their standard deviation or confidence intervals (Himmelblau 1978), which clearly indicate the precision which characterise their determination and then their identifiability. Parameters presenting standard deviation even exceeding the parameter value are the first sign of poorly identifiable parameters (exist different combinations of parameters giving the same representation of the experimental data).

Besides the error parameters, another direct comparison among different models can be obtained by graphing the experimental data versus the predicted values (scatter diagrams). These diagrams output the spreading of the experimental data with respect to the selected model: the more the model is suitable, the more the data are tightly and symmetrically distributed on the diagonal (see Fig. 4.3a for an example).

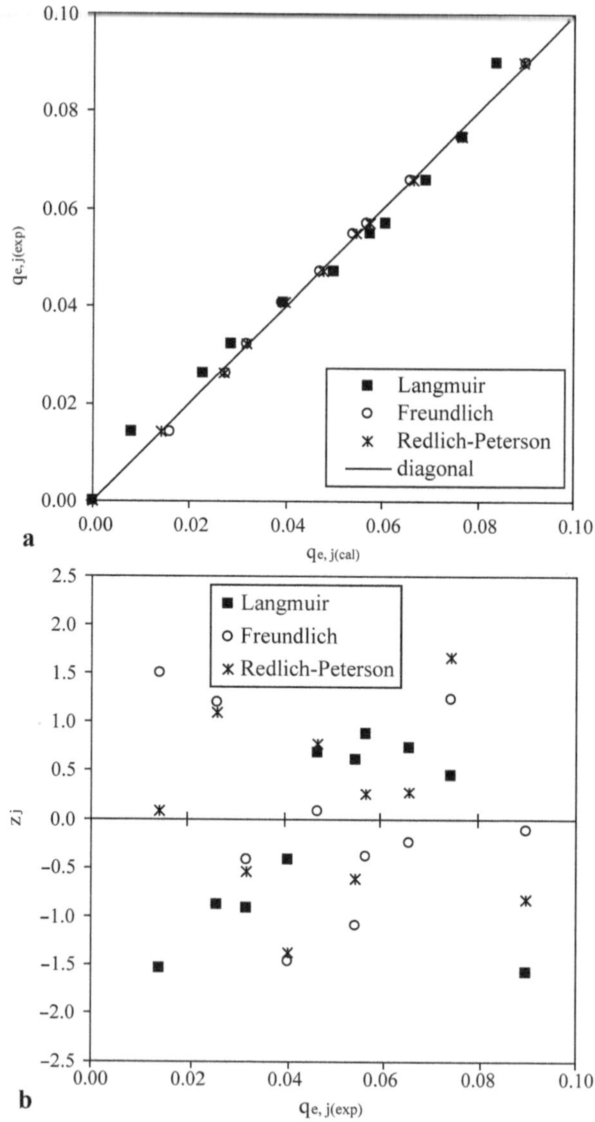

Fig. 4.3 Scatter diagrams (**a**) and residuals (**b**) for Langmuir, Freundlich and Redlich-Peterson models applied to single-metal system copper biosorption onto olive pomace wastes at pH=5

Quantitative parameters to judge scatter diagrams can be obtained by linear regression of data, where the good fit of the model is represented by values of the intercept $d_0=0$ and slope $d_1=1$.

The analysis of the residuals (Eq. 4.21) examining the deviations between experimental and predicted values can detect the presence of model discrepancies in data representation (see Fig. 4.3 for an example)

$$z_j = \frac{q_{e,j(cal)} - q_{e,j(exp)}}{\sqrt{SS_R^2}} \tag{4.21}$$

If a model well fits the experimental data, the residuals should be randomly distributed, while systematic deviations indicate that the model may be not a good representation for the set of data. The residual patterns could also be used to improve and optimize a model.

A quantitative measure of the asymmetry of the z residual distribution is the skewness α_3 defined as follow:

$$\alpha_3 = \frac{\mu_3}{\sigma^3} \tag{4.22}$$

where μ_3 is the third order moment and σ is the standard deviation of experimental data.

The model residual variance MS_R^2 (Eq. 4.23) is a quantitative measure of the deviation between experimental and predicted values:

$$MS_R^2 = \frac{\sum_{j=1}^{P} \left(q_{e,j(\exp)} - q_{e,j(cal)}\right)^2}{P - p} \tag{4.23}$$

where P is the total number of experimental points and p is the number of estimated parameters.

A two-sided F-Test can be used to compare different model residual variances and asses if calculated values are statistically different or not.

All the previous statistics can be calculated using the classical equations reported in literature (Himmelblau 1978). Otherwise dedicated computer programs can be used providing all these statistical values when data regression was performed. The statistics reported above were used to discriminate three commonly used empirical models (Langmuir, Freundlich, Redlich-Peterson) for representing single component data of copper biosorption onto olive pomace wastes at pH=5 (see Figs. 4.1, 4.3 and Table 4.3).

Statistical discrimination table for three empirical models used to represent Error parameters (expressed as standard deviations) denoted that Freundlich model is that characterised by the lowest values of errors, while R^2 values are very good for all three models underlining that this parameter cannot be considered alone as a useful tool to evaluate if a model represents well the experimental data. Linear equations for scatter diagrams denoted that Freundlich and Redlich-Peterson models give data

Table 4.3 Statistical discrimination table for three empirical models (Langmuir, Freundlich, Redlich-Peterson) used to represent single component data of copper biosorption onto olive pomace wastes at pH=5

Model	Regressed parameters		R^2	Scatter diagram diagonals	Skewness	Model residual variances
Langmuir	$q_m \pm \sigma_{qm}$	0.107±0.007	0.9952	y=0.9533+0.0028	−2.38	1.6075E-05
	$b \pm \sigma_b$	7±1		(R^2=0.9829)		
Freundlich	$K_F \pm \sigma_{KF}$	0.123±0.002	0.9995	y=1.0062−0.0003	1.04	1.462E-06
	$n \pm \sigma_n$	2.20±0.04		(R^2=0.9982)		
Redlich-Peterson	$a_{RP} \pm \sigma_{aRP}$	4±2		y=0.9982+0.00001		
	$b_{RP} \pm \sigma_{bRP}$	(3±1)*10	0.9997	(R^2=0.9989)	0.27	9.65E-07
	$n_{RP} \pm \sigma_{nRP}$	0.60±0.02				

more tightly and symmetrically distributed on the diagonal. Residues distribution for Langmuir model is characterised by a high values of skewnees indicating the likelihood of a few large negative residuals having an unduly large effect on the fit, while this parameter is lower (as absolute values) and probably not significant for Freundlich and Redlich-Peterson models. F-test indicated that Langmuir residual variance is significantly larger than Freundlich and Redlich-Peterson ones, while no significant difference can be denoted between Freundlich and Redlich-Peterson models (95% significance). On the base of this discrimination analysis Freundlich model is the best among the two-parameters models (lower parameter errors, larger regression coefficient, better linear trend in the scatter diagram, lower skewnees, lower model residual variance). The use of an additional adjustable parameter with the Redlich-Peterson model does not improve data representation (larger parameter errors, no improvement in model residual variance).

4.2.2 Mechanistic Models

The identification of the physico-chemical interactions between active sites on the adsorbent and metallic species, and the quantification of metal removed according to different mechanisms (physical adsorption, ion exchange, complexation, chelation, surface microprecipitation) is an intriguing target that would allow the precise description of the effect of the environmental factors (e.g., pH, ionic strength, presence of multi-metallic systems and/or ligands in solutions) on equilibrium metal removal. The interpretation of the effect of environmental factors on equilibrium biosorption can be guided by an outstanding approach in equilibrium modelling of sorption phenomena: the development of mechanistic equilibrium models able to describe the effect of the operating conditions in terms of supposed reactions among active sites and metallic species in solution.

Instead of empirical isotherms, mechanistic models generally present complicated forms with several adjustable parameters (equilibrium constants of the supposed reactions, active site concentrations, parameters related to electrostatic field at the interface, to surface heterogeneity, and to cooperative phenomena), whose regression require a large number of experimental tests to be performed. In fact independent sets of experimental data should be used at least for obtaining guess values of adjustable parameters. Otherwise the risk is obtaining parameters with scarce identifiability (different sets of parameters can represent equally well the data) and then without chemico-physical meaning. This would vanish the predictive power of the model and then make it only more complicated than empirical models, but without use.

Mechanistic models can be characterised by different degree of complexity or accuracy in system description.

In the literature a number of models have been reported with different schematizations of the interfacial structure in terms of supposed reactions (Stumm 1987), description of the electric double layer (Westall and Hohl 1980) introduction of cooperative phenomena (Koopal et al. 1994) and continuous distribution for the equilibrium constants among active sites and ionic species in solution (Benedetti et al. 1995). According to the previous observation about parameter identifiability it is noteworthy that the increase of model complexity should be accompanied by an augment degree of knowledge about the systems obtained by performing specific experimental investigations.

In addition model complexity degree should be sized according to the intensity of the effects of the different phenomena considered: i.e. only phenomena that significantly affect the equilibrium in a specific range of operating conditions should be included in models. Preliminary screening of the factors significantly affecting equilibrium sorption can be determined by sequential factorial designs and analysis of variance (Amini et al. 2009; Carmona et al. 2005) also using conventional response surface analysis procedure (Ghorbani et al. 2008; Preetha et al. 2007; Zulkali et al. 2006). By this way factors (and then phenomena) whose effect should be taken in consideration in modelling, can be identified and model complexity chosen properly.

It should be noted that in the case of mechanistic models, model goodness cannot be ascertained by statistical discrimination as in the case of empirical models. Model goodness should be determined according to the predictive power of the model in operating condition different from those used for parameter regression. Successive experimental investigations can be then used to refine model structure in order to improve its predictive capacity.

4.2.2.1 Mechanistic Models Including a Chemical Reaction Scheme

The simplest way of including a mechanism in a model representing metal equilibrium distribution between solid and liquid phases is writing a set of reactions between biomass active sites and ionic species in solution, and then combining the mass balance of the sites with the equilibrium constants of the supposed reactions. By this way an analytical expression for the concentration of metal in solid phases as a function of bulk variables (metal and antagonist ion concentrations) can be obtained.

Fitting of parameters (equilibrium constants of the supposed reactions, and site concentrations) can be then performed by non linear regression of experimental data (q_e) obtained for different values of the bulk variables.

The mathematical approach behind these models is then quite simple. The difficult arises in the identification of the active site and the choice of the set of metal binding reactions in which they are involved. This is a crucial point in mechanistic models which can be solved combining experimental data of biomass characterisation and extrapolating to surface groups the known chemical behaviour of analogues substances in aqueous solutions. For example complexation constants of liquid analogues can be used as guess values during parameter regression, even though phenomena typically occurring during sorption (electrostatic and cooperative effects) can strongly affect complexation constants between active groups on solid adsorbent and ionic species in solution (Pagnanelli et al. 2004).

Complexation theory such as HSAB can be also used to guide the formulation of reaction scheme including at first approximation only binding between ligands (biomass sites) and metal of the same group (Wang and Chen 2009; Gadd 2008; Volesky 2007). Finally the reaction scheme should be written with particular attention to the ion solution chemistry. In fact according to the solution composition (pH, electric potential, ligands) metal can be present in differently charged complexed forms and not only as simple aquo-complexes.

Modelling the Acid-Base Properties of the Active Sites

The first step in mechanistic modelling is the determination of nature and concentration of sites potentially able to bind heavy metals in solution. Biomass characterisation can be performed by simple experimental procedures to determine the exchange capability of strong acid groups by ionic content (Pagnanelli et al. 2000; Schiewer and Volesky 1995) and the acid-base properties of the weakly acidic groups by potentiometric titrations of solid suspensions (Pagnanelli et al. 2000, 2003a, 2004; Davis et al. 2000) involved in surface complexation reactions. Further information can be also derived by other independent characterisations such IR spectroscopic analysis of solid samples (Pagnanelli et al. 2000, 2003a; Figueira et al. 1999; Fourest and Volesky 1996), conductometric titrations (Seki et al. 1998), chemical derivatisation and selective extractions of specific groups (Pagnanelli et al. 2003a).

Even different sophisticated instrumental analyses can be used such as scanning electron microscopy, transmission electron microscopy, X-Ray diffraction, nuclear magnetic resonance, etc. Nevertheless it should be noted that information obtained by many sophisticated and expensive analytical techniques cannot always be adequately used to understand and interpret biosorption phenomenon according to a mechanistic modelling approach. In this view, being pH the most influencing factor in heavy metal biosorption, the determination of acid-base properties of biomass is a priority aspect in characterisation task. In fact weakly acidic nature of groups, typically present on biomasses used for biosorption applications, is a characteristic strictly related to metal binding ability. This can be explained by considering that sites dissociation can generate a negative charge on the biomass facilitating interactions (both electrostatic and chemical) with positively charged metallic species in solution. Potentiometric titrations of biomass suspensions can be used to determine nature and concentration of titrable sites.

Due to the heterogeneous nature of biomasses their titration curves can be of difficult interpretation due to the simultaneous dissociation of different kinds of sites. Application of mechanistic models to titration data can facilitate the identification and quantification of titrable sites.

Titration data were generally reported as charge concentration in solid phase (Q_H) versus the bulk solution pH (Fig. 4.4):

$$Q_H = \frac{\left(C_a - C_b - [H^+] + [OH^-]\right) V}{m} \tag{4.24}$$

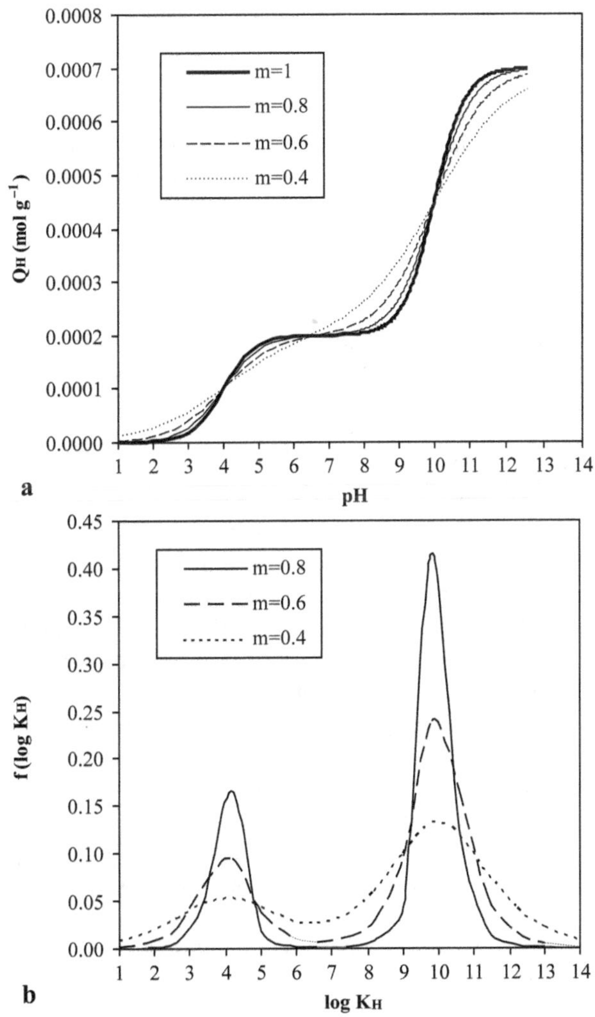

Fig. 4.4 Simulations of potentiometric titrations (**a**) and proton affinities distribution (**b**) for different degree of biomass heterogeneity

where C_a and C_b are the acid and base concentrations for each titrant addition, V is the total suspension volume after each titrant addition ($V=V_o+V_t$ with V_o and V_t initial volume and titrant volume, respectively), and m is the amount of titrated biomass.

The simplest mechanistic approach considers the heterogeneous matrix as a sum of M monoprotic sites, whose number is gradually increased until a good representation of the experimental data is obtained, fitting site concentrations and acid constants on the base of the electro-neutrality condition or charge balance:

$$Q_H = \sum_{k=1}^{M} [S_k^-] \tag{4.25}$$

where $[S_k^-]$ is the solid concentration for the deprotonated form of the *k*-th site according to the protonation reaction (Eq. 4.26) with K_{Hk} as apparent equilibrium protonation constant.

$$S_k^- + H^+ \overset{K_{Hk}}{\leftrightarrow} S_k H \tag{4.26}$$

The mass balance equation for the generic *k*-th site can be combined with the equilibrium constant to obtain a relation among $[S_k^-]$ and $[H^+]$ containing the parameters $[S_k]_{Tot}$ and K_{Hk}

$$[S_i]_{Tot} = [S_k^-] + [S_k H] = [S_k^-]\left(1 + K_{Hk}[H^+]\right) \tag{4.27}$$

The total negative charge concentration in the solid (Q_H) obtained from the charge balance for each titration point (Eq. 4.24) can be then expressed as

$$Q_H = \sum_{k=1}^{M} \frac{[S_k]_{Tot}}{1 + K_{Hk}[H^+]} \tag{4.28}$$

A non linear regression of the experimental data expressed as Q_H versus pH can be then performed to obtain the adjustable parameters ($[S_k]_{Tot}$ and K_{Hk}) for each one of the supposed active site. The number of active sites on the adsorbent can be progressively increased to obtain the best model as the simplest mathematical relation, which adequately describes the observed phenomena and makes quantitative description that can be experimentally tested.

The approach reported above can be easily extended to other reaction schemes assuming anphoteric sites, diprotic sites, triprotic sites and so on, according to the specific acid-base properties evidenced by titration data and other characterisation tests of chemical composition.

Similar elaborations of titration data according to a hypothesised set of protonation reactions were reported in the literature for different kinds of biomasses (algae, agro-industrial wastes, bacteria) showing the versatility of the approach in representing acid-base properties of natural matrices (Han et al. 2006a; Chen and Yang 2006; Chojnacka et al. 2005; Cox et al. 1999; Figueira et al. 2000; Herrero et al.

2005; Yun and Volesky 2003; Kim et al. 1998; Pagnanelli et al. 2000, 2003a, 2004; Seki and Suzuki 1998, 2002; Yun et al. 2001; Seki et al. 1998; Schiewer et al. 1995).

Total sites concentrations determined by titration modelling can be used as guess values for the regression of parameters contained in the mechanistic model for metal removal. In addition equilibrium constants of protonation reactions, once compared with literature data (see Volesky 2007 for a classification of functional group according to pK values) can be used to identified active site nature and then hypothesise reactions consistent with the known chemical behaviour of soluble analogues. This can guide the choice of reactions between active sites and metal in solution by hypothesising reactions between groups and metals with high complexation constants and using the proper stoichiometry.

Including a Mechanism for the Effect of pH on Metal Biosorption

Once biomass sites were characterised for their acid base properties, the inclusion of pH effect in a mechanistic model is straightforward. The set of reactions include not only metal-site interactions but also site-proton interactions. Different type of competition between metals and protons can be assumed also on the base of the chemical nature of the active sites. The general competitive scheme in which each kind of site can bind alternatively proton and metal lead to a sum of competitive Langmuir equations (Eq. 4.13). The difference in this approach with respect to empirical Langmuir models lays in the previous characterisation of biomass by potentiometric titrations, which give information about the number of site types and the relative protonation constants. By this way parameters related to proton binding are determined by independent set of data (i.e. potentiometric titrations) and not by metal isothermal data as in the case of competitive Langmuir isotherm and other Langmuir-type models reported above (Section Competitive Models).

Langmuir-type competition is able to represent isothermal data at different level of equilibrium pH that are characterised by a gradual diminution of the metal affinity for decreasing pH values, but with the same maximum sorption capacity for sufficiently large metal concentrations. This could be the observed case, but often the effect of pH is a diminution of maximum sorption capacity rather than affinity. This behaviour can be described by a different reaction scheme inspired by non competitive inhibition of enzymatic reactions (Esposito et al. 2002). In particular it is assumed that a site can react both with proton and metal according to the following reactions

$$\mathrm{S} + \mathrm{H} \overset{K_H}{\leftrightarrow} \mathrm{SH} \tag{4.29}$$

$$\mathrm{S} + \mathrm{M} \overset{K_M}{\leftrightarrow} \mathrm{SM} \tag{4.30}$$

$$\mathrm{SH} + \mathrm{M} \overset{K_{HM}}{\leftrightarrow} \mathrm{SHM} \tag{4.31}$$

Combining the material balance with the equilibrium constants and the non competitive hypothesis ($K_M=K_{HM}$), a relation can be obtained among the metal specific uptake ($[SM]=q_e$) and metal and hydrogen concentrations ($[M]=C_e$):

$$q_e = \frac{[S]_{Tot} C_e}{\left(1 + \frac{[H^+]}{K_H}\right)(K_M + C_e)} \tag{4.32}$$

This reaction scheme can be reasonable applied to functional groups which can bind ionic species even in the protonated form (such as carboxylic groups).

Other models including the effect of pH in the mechanistic framework account for site protonation reactions and reactions between sites and different ionic species in solution (Su et al. 2007; Vilar et al. 2007; Han et al. 2006a; Herrero et al. 2005; Juang and Shao 2002; Yun et al. 2001; Seki and Suzuki 1998, 2002; Fowle and Fein 2000; Pagnanelli et al. 2000; Yang and Volesky 1999a; Kim et al. 1998).

As an example Pagnanelli et al. (2000) described the effect of pH on heavy metal biosorption onto a bacterial biomass. Potentiometric tests and equilibrium isotherms of single metal systems at different equilibrium pH were then represented by hypothesising the following set of reactions:

$$S_1H \overset{KH1}{\leftrightarrow} S_1^- + H^+ \tag{4.33}$$

$$S_2H \overset{KH2}{\leftrightarrow} S_2^- + H^+ \tag{4.34}$$

$$S_1^- + M^{2+} + OH^- \overset{KM1}{\leftrightarrow} S_1M(OH) \tag{4.35}$$

$$S_2H + M^{2+} + OH^- \overset{KM2}{\leftrightarrow} S_{21}HM(OH)^+ \tag{4.36}$$

Using the equilibrium constants and the site mass balances the following expression for metal concentration in solid phase is obtained

$$\begin{aligned} q_e &= [S_1M\,(OH)] + \left[S_2HM(OH)^+\right] \\ &= \frac{[S_1]_{Tot} K_{M1} C_e}{1 + K_{M1}C_e + [H^+]/K_{H1}} + \frac{[S_2]_{Tot} K_{M2} K_W C_e}{[H^+] + K_{H2} + K_{M2} K_W C_e} \end{aligned} \tag{4.37}$$

where K_w is the water hydrolysis constant.

The six parameters contained in this expression were regressed by using titration data ($[S_1]_{Tot}$, $[S_2]_{Tot}$, K_{H1} and K_{H2}) and metal sorption data at different pH (K_{M1} and K_{M2}). This model scheme was able to represent metal removal up to pH 5. The authors suppose that above this values other mechanisms occur in metal removal not directly related to the acid-base properties of the biomass. In fact metal concentration in solid phase for pH larger than 5 exceeded total concentration of titrable sites determined by potentiometric titration modelling. Known behaviour of bacterial strain used for biosorption tests let the authors suppose that surface microprecipitation occurred near neutral pH values. Mechanistic approach were also developed to include microprecipitation mechanism in heavy metal biosorption (Warren and Ferris 1998).

Including a Mechanism for Competition Effects in Multi-Metallic Systems

The effect of antagonist species can be easily included in a mechanistic scheme of reactions in order to take in consideration the competition among different heavy metals in solution (Matos et al. 2009; Figueira et al. 2000; Fowle and Fein 1999; Schiewer et al. 1995) but also the effect of alkaline and alkaline earth metals (Figueira et al. 2000; Fowle and Fein 1999; Crist et al. 1994) which can be present in solution at concentration largely exceeding heavy metal ones.

Reaction scheme can include ion exchange reactions (Eq. 4.38) (Matos et al. 2009; Yun and Volesky 2003; Figueira et al. 2000; Schiewer et al. 1995; Crist et al. 1994) and surface complexation reactions (Eqs. 4.39 and 4.40) (Lodeiro et al. 2008; Mukhopadhyay 2008; Chen and Yang 2006; Herrero et al. 2005; Figueira et al. 2000; Fowle and Fein 1999; Kim et al. 1998; Crist et al. 1994) between active sites and ionic species in solution.

$$SM_2 + M_1 \overset{K_{21}}{\leftrightarrow} SM_1 + M_2 \tag{4.38}$$

$$S + M_1 \overset{K_1}{\leftrightarrow} SM_1 \tag{4.39}$$

$$S + M_2 \overset{K_2}{\leftrightarrow} SM_2 \tag{4.40}$$

Ion exchange was recognised as the main mechanism operating in metal removal for different kinds of biomasses (Kratochvil and Volesky 1998). The release in solution of counter-ions (both protons and alkaline and alkaline earth metals) during heavy metal binding is a clear experimental evidence of the occurrence of such removal mechanism also denoting the stoichiometry of the exchange reaction.

Nevertheless it should be noted that the difference in supposed mechanism can be only formal. In fact, if the binding reaction for the counter ion is included in the reaction scheme, the ion exchange constant is obtained by simply dividing the complexation constants for the two exchanging ions (Schiewer and Volesky 1996).

The ion exchange scheme in the general case of 1:alpha stoichiometry can be written as

$$S(M_2)_\alpha + M_1 \overset{K_{21}}{\leftrightarrow} SM_1 + \alpha M_2 \tag{4.41}$$

Combining site balance and exchange constant the following ion-exchange isotherm is obtained, which can be arranged according to the general Langmuir isothermal form

$$[SM_1] = \frac{[S]_{Tot} K_{21} \left[C_{e,1}\right]}{\left[C_{e,2}\right]^{\alpha} + K_{12}\left[C_{e,1}\right]} = \frac{[S]_{Tot} K_{12} \frac{\left[C_{e,1}\right]}{\left[C_{e,2}\right]^{\alpha}}}{1 + K_{21}\frac{\left[C_{e,1}\right]}{\left[C_{e,2}\right]^{\alpha}}} \tag{4.42}$$

Due to this strict formal analogy and to the extra-labour necessary for ion exchange models complexation mechanism is still widely used even if ion exchange is recognised as predominant (Crist et al. 1994).

Schiewer and Volesky (1996) extended the previous approach to the case of multimetallic systems at various equilibrium pH obtaining the following expression

$$q_{e,i} = \sum_{k=1}^{M} \left[S_k (M_i)_{1/zi}\right] = \sum_{k=1}^{M} [S_k]_{Tot} \frac{\left(K_{ki}\left[C_{e,i}\right]\right)^{1/zi}}{1 + \sum_{a=1}^{N} \left(K_{ka}\left[C_{e,a}\right]\right)^{1/za}} \tag{4.43}$$

where $C_{e,i}$ is the concentration of the generic ion with charge z_i, reacting with M sites S_k in a system of N competiting metals.

Even though these mechanistic models are quite simplistic because neglect electrostatic effects, surface heterogeneity and non ideal interactions among competiting ions, they can be considered the first step of a modelling sequential procedure based on growing complexity models. Consequently, these models have to be as simple as possible with the major purpose of driving further experimental research and modelling.

4.2.2.2 Mechanistic Models Accounting for Electrostatic Corrections

Surface charges associated to active sites on biosorbent materials determined a structured location of counter-ions in the liquid layer at the interface (Fig. 4.5). In particular acid dissociation of the active sites of the sorbent determined the accumulation of negative charges on the surface or within the solid matrix depending on the structural nature of the sorbent. As a consequence of the presence of this charge accumulation, ionic species in proximity of sorption sites can present concentration different from those measured in bulk solution.

This phenomenon known as electric double layer is characterised by an interface potential profile which starting from a surface value (ψ_0) gradually decreases to zero in the bulk solution (Fig. 4.5).

Mechanistic models accounting for electrostatic corrections due to the presence of this potential profile introduce a correction term for the bulk concentrations including the potential at the plane of the adsorption ψ_s:

$$\left[\overline{H}^{+}\right] = \left[H^{+}\right] e^{-\frac{e\psi_s}{kT}} \tag{4.44}$$

where e is the electron charge, k the Boltzmann constant and T the absolute temperature.

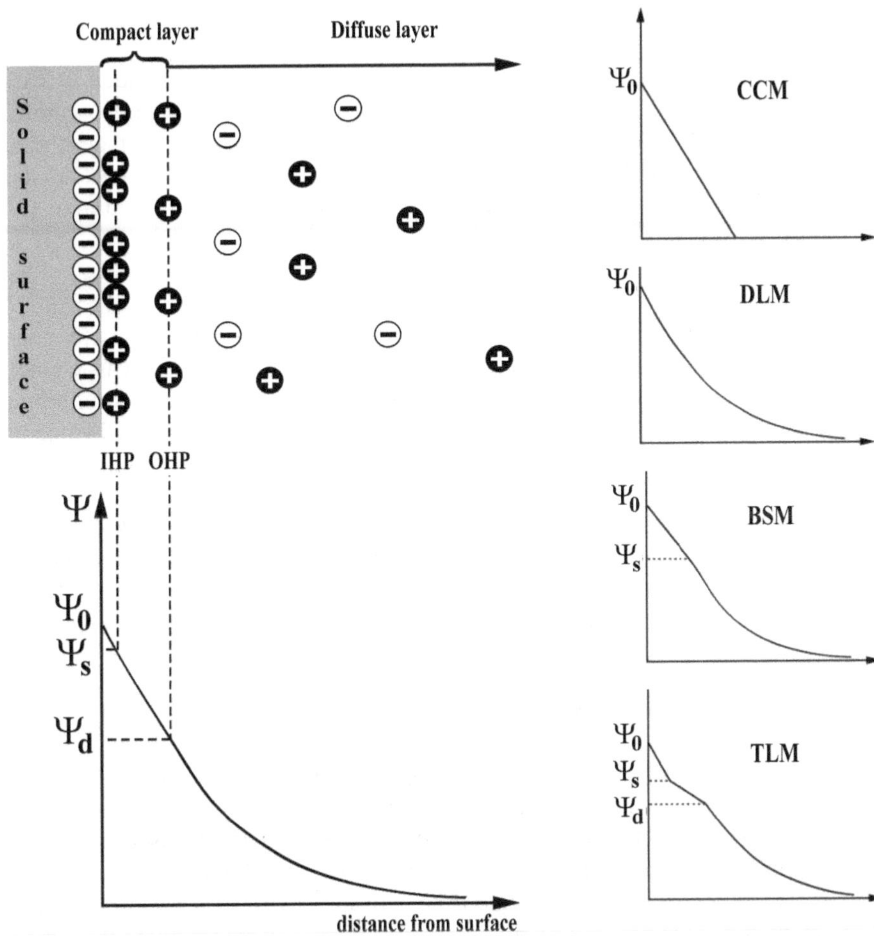

Fig. 4.5 Electric double layer: distribution of the ions in solution in presence of negatively charged surface according to different approximations (*CCM* constant capacitance model, *DLM* diffuse layer model, *BSM* basic Stern model, *TLM* triple layer model)

The potential can be evaluated solving the Poisson-Boltzmann equation, which is obtained introducing a Boltzmann distribution for the charge density in the Poisson equation. Different approximate solutions can be obtained according to simplifications and model assumptions that differ for the nature of surface charge, the number and the disposition of the potential planes and the position of the adsorbed species (Westall and Hohl 1980).

According to the type of biosorbent particles and the operating conditions in the systems different approximations of the potential profile can be then included inside the hypothesised set of reactions between active sites and ionic species in solution.

Multi-Layer Models

The double layer and potential profile at the interface can be described by different approximations assuming a series of charged planes (Fig. 4.5). Helmholtz model considers a layer of counter-ions parallel to the charged surface plane, thus generating a rigid double layer. This schematisation is analogous to a plane capacitor with a linear dependence between surface charge and potential at the interface

$$\sigma_0 = C\psi_0 \tag{4.45}$$

where σ_0 is the surface charge determined by taking in consideration only the ions specifically adsorbed on active sites, and C is the capacitance (constant capacitance model, CCM) (Fig. 4.5) whose value is regressed along with the other equilibrium parameters. Because no relation with ionic strength is included, regressed value of C is referred only to the particular level of ionic strength in which experiments were performed.

CCM completely neglects the effects of thermal energy which are taken in consideration in the Gouy-Chapman model in which counter-ion charge (σ_d) balancing charged sites is diffuse in the interface solution layer (diffuse layer model, DLM) (Fig. 4.5). In this case surface charge and potential are related by an analytical equation depending by ionic strength

$$\sigma_0 = -\sigma_d = \sqrt{8\bar{\varepsilon}\,\bar{\varepsilon}_0 \mathrm{RTI}}\,\sinh(\mathrm{zF}\psi_0/2\mathrm{RT}) \tag{4.46}$$

where $\bar{\varepsilon}$ and $\bar{\varepsilon}_0$ are dielectric constants, R is the gas constant, T is the absolute temperature, z is the ion charge and F the Faraday number.

Both Helmholtz and Gouy-Chapman schematisations do not represent adequately EDL: the first overestimates its rigidity, while the second neglects its structure. Stern model proposes an intermediate schematisation with ions near the surface laying in a rigid Helmholtz plane, and outer ions dispersed according to the Gouy-Chapman diffuse layer model. This model, also known as basic Stern model (BSM) (Fig. 4.5), assumes that solid surface has σ_0 charge and ψ_0 potential. In the inner Helmholtz plane (IHP) ions are localised which directly interact with surface charged sites, while the outer Helmholtz plane (OHP) is where the diffuse layer starts. According to this description a potential difference exists between IHP and OHP which are characterised by ψ_β and ψ_d values, respectively. Nevertheless in the BSM, Stern assumed that $\psi_\beta = \psi_d$ avoiding the regression of another adjustable parameter, namely a second capacitance, C_2. This approximation falls in the triple layer model (TLM) (4.5) where ions can interact with surface on both planes (IHP and OHP) with different mechanisms and strength.

According to the BSM the integral capacity of the EDL is given by the series of two capacitors, the Helmholtz compact and the Gouy-Chapman diffuse capacitors

$$\frac{1}{C_T} = \frac{1}{C_{compact}} + \frac{1}{C_{diffuse}} \tag{4.47}$$

For low ionic strength the capacity of the diffuse layer is much lower than that of the compact layer, then the total capacity of the system can be approximated by the diffuse capacity. On the other hand for high ionic strength diffuse layer capacity is larger than that of the compact layer and then the total capacity is approximated by the compact layer capacity. According to these observations CCM can be considered a good approximation of BSM for high ionic strength, while DLM is the limit case for low ionic strength.

It is noteworthy that for high ionic strength activities rather than concentrations should be used for each species involved both in solution and surface reactions. Activity coefficients for species in solution can be evaluated according to the Davies equation, whereas for surface species activity coefficients are generally lumped into an exponential potential term in the surface complexation constants (Volesky 2003; Yun and Volesky 2003).

Mechanistic models accounting for the effect of the potential profile on surface concentration of ionic species are generally known as surface complexation models (SCM) and have been widely applied for representing metal ions binding onto clay minerals and soils in general (Stumm 1987). Surface complexation models were also used for heavy metal biosorption modelling (Hetzer et al. 2006; Ravat et al. 2000; Hiemstra and Van Riemsdijk 1999; Daughney and Fein 1998; Chen et al. 1997; Chen and Yiacoumi 1997; Fein et al. 1997; Schiewer and Volesky 1997). As an example of SCM application for biosorption modelling, the work of Chen et al. (1997) is reported investigating the effect of pH and ionic strength on copper removal by calcium alginate. These authors used a basic Stern layer approximation and assumed that an anphoteric site is present on the biological matrix used for metal removal (two-pK-basic Stern model). Four reactions are then assumed according to the following scheme

$$SOH + H^+ \overset{K_{H1}}{\leftrightarrow} SOH_2^+ \tag{4.48}$$

$$SOH \overset{K_{H2}}{\leftrightarrow} SO^- + H^+ \tag{4.49}$$

$$SOH + Cu^{2+} \overset{K_{Cu}}{\leftrightarrow} SO^-Cu^{2+} + H^+ \tag{4.50}$$

$$SOH + Cu^{2+} + H_2O \overset{K_{CuOH}}{\leftrightarrow} SO^-CuOH^+ + H^+ \tag{4.51}$$

where

$$K_{H1} = \frac{[SOH_2^+]}{[SOH][H^+]\exp(-y_\beta)} \tag{4.52}$$

$$K_{H2} = \frac{[SO^-][H^+]\exp(-y_\beta)}{[SOH]} \tag{4.53}$$

$$K_{Cu} = \frac{[SO^{-}Cu^{2+}][H^{+}]\exp(-y_{\beta})}{[SOH][Cu^{2+}]\exp(-2y_{d})} \quad (4.54)$$

$$K_{CuOH} = \frac{[SO^{-}CuOH^{+}][H^{+}]^{2}\exp(-y_{\beta})}{[SOH][Cu^{2+}]\exp(-y_{d})} \quad (4.55)$$

$$y_{\beta} = \frac{e\psi_{\beta}}{kT} \quad (4.56)$$

$$y_{d} = \frac{e\psi_{d}}{kT} \quad (4.57)$$

with ψ_{β} and ψ_{d} the potentials at the inner and outer layers of the EDL, respectively. Surface parameters including the surface charge σ_0 and the capacitance C were determined by potentiometric titration modelling according to reactions reported in Eqs. 4.48 and 4.49.

As for copper biosorption the general procedure of mechanistic models is used to obtain a relation between metal concentration in solid phase ($[SO^{-}M^{2+}]$ + $[SO^{-}MOH^{+}]$) and bulk solution concentrations of metals and protons. Then starting from the site mass balance ($[S]_{Tot} = [SOH] + [SOH_2^{+}] + [SO^{-}] + [SO^{-}M^{2+}] + [SO^{-}MOH^{+}]$) expressions of [SOH], $[SOH_2^{+}]$, and $[SO^{-}]$ derived from the previous equilibrium constants were obtained and substituted.

Solution reactions for the formation of copper hydroxo-complexes in solution were also included in the model scheme.

The authors conclude that the SCM they used successfully represent the effect of pH on both surface charge and copper biosorption, suggesting that copper removal may result from the binding of copper aquo- and hydroxo-complexes.

It should be noted that different values of surface complexation constants were obtained for the two investigated level of ionic strength, while according to the basic stern approximation the effect of this factor is included and then experimental data collected at different ionic strength should have been regressed together obtaining unique values complexation constants. The authors argument the fact that different set of parameters were obtained hypothesising that other reactions occurred than those included in the selected reaction scheme. Basic stern layer differ from constant capacitance model approximation because this last one cannot take in consideration the effect of variable ionic strength, and then different capacitance values were obtained for different ionic strength. If Basic stern approximation is used it means that the models should be able to describe also ionic strength effect, otherwise if the aim is only including electrostatic effect for a selected ionic strength the constant capacitance model should be more adequate. If a model architecture is complicated by introducing a phenomenon it means that the model have to represent the effect of such phenomenon otherwise only additional parameters were added making the model more flexible but reducing parameter identifiability.

Daughney and Fein (1998) tried to identified the best approximation for including ionic strength effect on proton, cadmium, lead and copper biosorption onto two bacteria. They observed that the surface characteristics of these biomasses were affected by ionic strength variation because different set of stability constants and sites concentrations were obtained by regressing titrations performed at different ionic strength. They compared CCM and BSM for representing both titration curves at constant ionic strength and titrations obtained at different ionic strength.

They found that CCM was more effective than BSM in describing ionic strength-dependent acid-base behaviour using a three pK model. It should be noted that the better performance of CCM were ascertained by statistical analysis and not by considering its predicting ability. On the other side regressing titration data obtained at different ionic strength with the CCM means that the developed model cannot predict the effect of ionic strength for the intrinsic characteristic of the CCM approximation.

The discussion above denoted the main lack of multi-layers electrostatic models, i.e. the quite complete arbitrary choice of the EDL approximation. If experimental data cannot be represented by neglecting electrostatic corrections due to EDL, the only way to bypass this problem seem to be performing a wide experimental investigation including both potentiometric titrations and metal removal at different levels of ionic strength and developing a mechanistic model assuming an EDL approximation able to represent all data by a unique set of complexation constants (Milne et al.1995a, b). Some independent experimental investigations for the determination of specific parameters of the EDL (such as surface charge and zeta potential) can be also useful for obtaining initial guess values of regressed parameters.

The heterogeneity of biological matrices used for biosorption could limit the reliability of electrostatic mechanistic models and the interpretation of the parameters for the difficulties in identifying and quantifying surface functional groups and determining model parameters for the electrostatic field. For these reasons mechanistic models with no electrostatic correction should be preferred especially for strongly sorbed ions when the chemical contribution to the adsorption free energy dominates over the electrostatic term (Güçlü and Apak 2003; Davis and Leckie 1980).

The Donnan Model

Another way of introducing electrostatic effects in biosorption modelling is the Donnan approach hypothesising that for certain kinds of solid sorbents (permeable particles with dissociated groups distributed in the whole particle volume and completely neutralised by electrolyte ions) the potential is constant in the particle and drop to zero out of it. By this way the adsorbent particles can be considered as a concentrated aqueous solution of the analogous monomer of the active groups separated from the external one by an interface characterised by a potential difference, the Donnan potential, due to the different mobility of the ions in the system. Typical Donnan-type gels are humic substances (Marinsky and Ephraim 1986), synthetic resins (Biesuz et al. 1998) and algal biomasses (Schiewer and Wong 2000; Schiewer 1999; Schiewer and Volesky 1997).

The Donnan equation (Donnan and Allmand 1914), based on the equality of the chemical potential of the different ionic species in the two phases, allows relating the ion concentrations in solid phase to those in liquid phase

$$\left[\overline{G}^{+}\right]\left[\overline{G}^{-}\right] = [G^{+}][G^{-}] \tag{4.58}$$

where G^+G^- is the general electrolyte and $[\bar{G}^+]$ and $[\bar{G}^-]$ refer to the ionic concentrations in biophase. By this way, when a heterogeneous reaction is considered as in the case of acid-base titration of solid adsorbents, the intrinsic reaction constant (that is the constant in solid phase) can be related to the apparent heterogeneous constant containing concentrations both in solid and liquid phase.

The apparent constant (K_{appH}) of a generic heterogeneous equilibrium can be calculated as a function of both the intrinsic constant (K_{intH}) and the counter-ion concentration in liquid and solid phase on the base of the Donnan equation.

$$K_{appH} = \frac{\left[\overline{HS}\right]}{\left[\overline{S}^{-}\right][H^{+}]} = \frac{\left[\overline{HS}\right]}{\left[\overline{S}^{-}\right]\left[\overline{H}^{+}\right]}\frac{\left[\overline{G}^{+}\right]}{[G^{+}]} = K_{intH}\frac{\left[\overline{G}^{+}\right]}{[G^{+}]} = K_{intH}\frac{[G^{-}]}{\left[\overline{G}^{-}\right]} \tag{4.59}$$

It was generally observed that the apparent heterogeneous constants are influenced by the specific operating conditions of pH and ionic strength, while the estimates of the intrinsic constants from experimental data according to Donnan equation, are independent and very near to the equilibrium constants reported in the literature for aqueous solutions of the relative monomer of the functional group present in the solid (i.e. acetic acid is the equivalent monomer for a resin with carboxylic groups) (Pagnanelli et al. 2004; Pesavento et al. 1994).

Schiewer and Wong (2000) use the Donnan model to include electrostatic corrections in an equilibrium model for heavy metal biosorption onto a marine algae. The model reaction scheme include protonation of a monoprotic weakly acidic site, and complexation with bivalent metal to form neutral surface species. A concentration factor (λ) is introduced based on the Donnan equilibrium to correct bulk solution concentration of ionic species and obtain intraparticle concentration. Starting from electro-neutrality condition an analytical equation is obtained relating λ to the ionic strength in solution. The introduction of the Donnan equilibrium into this mechanistic model allowed to account for the ionic strength effect on potentiometric titrations and in metal binding. The authors observed that model performances were improved by accounting for the swelling behaviour of algae and then introducing an empirical correction for the volume of biophase as a function of ionic strength and concentration of binding sites of the biomasses.

4.2.2.3 Continuous Models

Langmuir-type models generally lack in representing heavy metal biosorption mainly because surface heterogeneity and non ideal competition among ionic species in solution are neglected. Surface heterogeneity indicates the different chemical properties of functional groups present on the biomass that even if have the same chemical nature (i.e. carboxylic groups) present a wide variety of affinity for ionic species in solutions depending on the chemical and physical surrounding of the specific group. In this case not a single value of affinity constant nor a sum of different values can be considered, but a continuous approach assuming a distribution function of the affinity constants.

Non ideal competition among ionic species relates to an adsorption behaviour which cannot be described according to the classical Langmuir isotherm assuming that species adsorb independently and without any effect of the degree of coverage.

Surface Heterogeneity

Sorption capacity of a biomass is strictly related to the nature and concentration of its functional groups, whose heterogeneity is, in general, unknown. Heterogeneity characterisation of the functional groups of a biosorbent can be obtained from proton binding curves determined by potentiometric titrations of biomass suspensions. In particular, the continuous approach relates the total fraction of protonated sites on the adsorbent ($\theta_{T,H}$) to the bulk pH by the following general integral:

$$\theta_{T,H} = \frac{[SH]}{[S^-] + [SH]} = \int_{\Delta logK_H} \theta_{L,H}\,(K_H, H) f\,(logK_H) dlogK_H \qquad (4.60)$$

where S is the generic free active site with an affinity constant K_H whose logarithm is distributed according to a certain affinity distribution $f(logK_H)$ over a specified range $\Delta logK_H$. The integral can be solved to obtain an analytical expression for $\theta_{T,H}$ choosing the proper local isotherm $\theta_{L,H}$ and $f(logK_H)$ distribution.

A Langmuir-type equation is commonly used to describe the local equilibrium among protons and active sites:

$$\theta_{L,H} = \frac{K_H\,[H]}{1 + K_H\,[H]} \qquad (4.61)$$

The heterogeneity analysis of experimental data can be used to have preliminary information about the number of active site types and the shape of $f(logK_H)$ (Nederlof et al. 1990). This analysis is based on different Linear Isotherm Approximation (LIA) methods: the ideal Langmuir-type local isotherm is approximated by different simple functions (such as step or ramp functions) allowing the integral inversion and the determination of the affinity distribution from the derivative of the experimental data. The more the local isotherm approximation is precise, the more the order of the derivative of experimental data rises requiring the use of smoothing techniques to eliminate the effect of experimental random errors and avoid spurious peaks in $f(\log K_H)$.

In the literature there are different analytical equations which were obtained inverting Eq. 4.60 by using Langmuir model as local isotherm and different kinds of distributions for $f(\log K_H)$ (De Wit et al. 1993). One of the possible distribution functions for $f(\log K_H)$ is the Sips distribution (Sips 1948):

$$f(\log K_H) = \frac{\ln(10)\sin(m\pi)}{\pi\left[\left(\frac{K_H}{\tilde{K}_H}\right)^{-m} + 2\cos(m\pi) + \left(\frac{K_H}{\tilde{K}_H}\right)^{m}\right]} \tag{4.62}$$

where $\tilde{K}_H$ is the median value of the distribution, and m ($0<m<1$) is a parameter related to the shape of the distribution and in particular as m decreases the distribution becomes larger (for $m=1$ it is a Dirac distribution meaning non heterogeneous site or better different sites but with specific value of the affinity constants).

The relation among fraction of protonated sites $\theta_{T,H}$ and negative charge on the adsorbent Q_H can be expressed as

$$Q_H = Q_{max}(1 - \theta_{T,H}) \tag{4.63}$$

where Q_{max} is the absolute value of the maximum negative charge on solid ($Q_{max} = \sum_k [S_k]_{Tot}$ according to the previous development in the discrete approach).

The analytical solution of the integral for Langmuir local isotherm and Sips distribution can be expressed as the general Langmuir-Freundlich isotherm relating the negative charge on the adsorbent Q_H with the hydrogen ion concentration in solution (De Wit et al. 1993):

$$Q_H = \frac{Q_{max}}{1 + \left(\tilde{K}_H\,[H]\right)^m} \tag{4.64}$$

On the base of heterogeneity analysis (which identify the number of peaks and their shapes in the affinity distribution) the previous expression can be easily extended considering the presence of different affinity distributions, each one characteristic of a specific functional group.

$$Q_H = \sum_{k=1}^{M} \frac{Q_{max,k}}{1 + \left(\tilde{K}_{H,k}\,[H^+]\right)^{m_k}} \tag{4.65}$$

A non-linear regression of the experimental data was then performed to determine the adjustable parameters characteristic of each site, which define concentration $Q_{max,k}$, position ($\tilde{K}_{H,k}$) and shape (m_k) of the related affinity distribution.

Figure 4.4 showed the effect of the parameter m_k (k = 1…2) on the acid-base properties of a biomass with two kinds of heterogeneous weakly acidic sites ($Q_{max,1}$ = 2 mmol/g, $\log \widetilde{K}_{H,1} = 4$, $Q_{max,2}$ = 5 mmol/g, $\log \widetilde{K}_{H,2} = 10$). Distinct flexs points observed for $m_1=m_2=1$ gradually become less evident as m_k values decreases due to the enlargement and then superimposition of the distribution functions of the two acidic sites.

Continuous models have been widely applied in the literature to represent the heterogeneous acid-base properties of different natural matrices such as humic acids (De Wit et al. 1993; Nederlof et al. 1993), bacteria (Plette et al. 1995), algae (Vilar et al. 2007) and agro-industrial wastes (Martin-Lara et al. 2009; Pagnanelli et al. 2003a, 2004, 2008)

Non Ideal Competition

Sorption data of multi-component systems can be represented according to a continuous approach by assuming a competitive Langmuir extension as the local isotherm:

$$\theta_{L,i} = \frac{K_i C_{e,i}}{1 + \sum_{a=1}^{N} K_a C_{e,a}} = \frac{K_L \left(k_i C_{e,i}\right)}{1 + K_L \sum_{a=1}^{N} \left(k_a C_{e,a}\right)} \tag{4.66}$$

where the equilibrium constant of the reaction among the generic site and the ith component (K_i) is assumed to be made up of two independent contributes related to the site heterogeneity (K_L) and to the sorption characteristic of each component (k_i):

$$K_i = K_L k_i \tag{4.67}$$

Introducing this local isotherm (Eq. 4.66) and the Sips distribution (Eq. 4.62) in the integral function for the adsorption of the ith component (Eq. 4.68)

$$\theta_{T,i} = \int_{\Delta \log K_i} \theta_{L,i} \left(K_i, C_{e,i}\right) f\left(\log K_i\right) d\log K_i$$

the general Langmuir-Freundlich isotherm for multi-component systems can be then obtained:

$$\theta_{T,i} = \frac{\widetilde{K}_i C_{e,i}}{\left\{1 + \left(\sum_{a=1}^{N} \widetilde{K}_a C_{e,a}\right)^m\right\} \left(\sum_{a=1}^{N} \widetilde{K}_a C_{e,a}\right)^{1-m}} \tag{4.68}$$

where $\widetilde{K}_i = \widetilde{K}_L k_i$, then the shape of the distribution function of each component is the same (just one m value for all the component) but shifted for the different component on the $\log K_i$ axis (congruent distributions).

If the monocomponent sorption of different species on the same surface presents different shapes of the distribution function, namely different m_k in Eq. 4.65 (non congruent distribution case) (Benedetti et al. 1995), a modified competitive Langmuir isotherm can be considered as local isotherm to account for specific non ideal sorption characteristics of each component:

$$\theta_{L,i} = \frac{K_L g(k_i C_{e,i})}{1 + K_L \sum_{a=1}^{N} g(k_i C_{e,i})} \tag{4.69}$$

The simplified case assuming the same mathematical form of $g(k_i c_i)$ for all the com ponents was considered (in theory $g(k_i C_{e,i})$ can be different for each component):

$$g(k_i C_{e,i}) = (k_i C_{e,i})^{n_i} \tag{4.70}$$

The total fraction of active sites occupied by the i-th species is then obtained by the inversion of the integral function by assuming this modified local isotherm (non ideal competitive adsorption model). In particular assuming a Sips distribution the general Langmuir-Freundlich isotherm is obtained

$$\theta_{T,i} = \frac{\left(\tilde{K}_i C_{e,i}\right)^{n_i}}{\sum_{a=1}^{N}\left(\tilde{K}_a C_{e,a}\right)^{n_a}} \cdot \frac{\left(\sum_{a=1}^{N}\left(\tilde{K}_a C_{e,a}\right)^{n_a}\right)^{p}}{1 + \left(\sum_{a=1}^{N}\left(\tilde{K}_a C_{e,a}\right)^{n_a}\right)^{p}} \tag{4.71}$$

where

$$\tilde{K}_i = \tilde{K}_L^{\frac{1}{n_i}} k_i \tag{4.72}$$

$$m_i = n_i p \tag{4.73}$$

The general procedure for evaluating the parameters for each site is the follow: proton binding data are used to determine m_H describing the combined effect of surface heterogeneity and non ideal sorption (m_H=pn), the median values of the proton affinity $\tilde{K}_H$ and the site concentration Q_{max}. In the second step sorption data of single metal at different pH were used to determine p, n_i and $\tilde{K}_i$.

For other metals only median values of metal affinity constant $\tilde{K}_i$ and non ideal sorption behaviour of the specific metal n_i are determined by single sorption data at different pH (Kinniburgh et al. 1996; Benedetti et al. 1995).

These models generally known as non ideal competition adsorption models (NICA) have been specifically developed to represent ion interactions with humic acids (Koopal et al. 1994, 2001; Kinniburgh et al. 1996; Benedetti et al. 1995), but were also applied for representing site heterogeneity and non ideal adsorption in metal biosorption onto different natural matrices (such as algae, agro-industrial wastes and bacteria) (Vilar et al. 2009; Pagnanelli et al. 2005; Plette et al. 1996).

Electrostatic effects can be also included in the NICA framework by using different EDL approximations (Dupont et al. 2003; Rey-Castro et al. 2003; Kinniburgh et al. 1996, 1999; Benedetti et al. 1996; Milne et al. 1995a, b). A shortcoming of NICA models applied to proton and metal ion binding is that they do not give a correct description of the proton release associated to the metal binding (underestimate of H^+/M^{z+} molar exchange ratio). For this reason NICA models were considered not thermodynamically consistent (Kinniburgh et al. 1999). In order to overcome such problem some modification of NICA models were developed (such as NICCA and CONICA models) taking in consideration different site-metal stoichiometry and binding of hydro-complexes (Plazinski and Rudzinski 2009; Vilar et al. 2009; Herrero et al. 2006; Kinniburgh et al. 1999; Van Riemsdijk et al. 1996).

As a final observation about continuous models it should be noted that isothermal expressions with the same mathematical form of continuous model for single component systems (see Eq. 4.64 for $[H^+]=[M]$) can be obtained assuming an homogeneous surface but cooperative phenomena due to lateral interactions among sorbed species. In particular positive cooperation was represented by m values larger than 1 (Hill equation), while negative cooperation, which typically occur for poly-electrolytes due to electrostatic repulsion beteen free and bound ions, are represented by m values ranging from 0 to 1 (Henderson-Hasselbalch equation) (Koopal et al. 1994, 2001).

4.3 Kinetic Studies in Batch Tests

Kinetic characterisation of metal biosorption is probably the most important factor in adsorption system design being the residence time and the reactor dimensions controlled by the system kinetics. Batch kinetics of metal biosorption are generally characterised by an initial rapid and quantitatively predominant diminution of metal concentration in liquid phase followed by a second slower and insignificant decrease. This trend can be explained by considering the initial abundance of available active sites on the biomass that becoming gradually occupied make biosorption less efficient in the slower stage, but even mass transfer effects due to penetration in the biomass should be taken in consideration.

In fact biosorption of metal ions can consist of various steps (Fig. 4.6a):

- bulk transport of metal ions in solution phase
- film transport involving the diffusion of metals through a hydrodynamic boundary layer around the biosorbent surface
- intraparticle diffusion through the gel phase of the biomass material
- chemical reaction of binding with the active sites

In batch experiments external bulk transport is usually rapid because of mixing and advective flow. Chemical reaction is also too fast to control the sorption rate. The rate determining step in most cases has been established to be either external film diffusion or intraparticle diffusion.

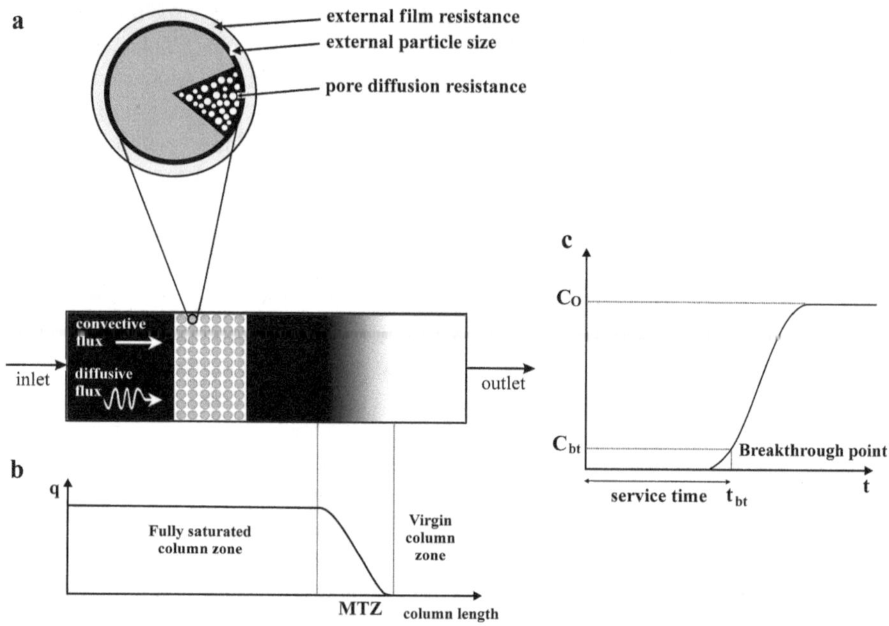

Fig. 4.6 General structure of an adsorbent particle and associated resistances to the solute uptake (**a**), breakthrough curve (**b**), mass transfer zone (MTZ) inside the column (**c**)

As in the case of equilibrium modelling, kinetic models with different degree of complexity can be developed according to the specific aim of models. If the aim of modelling is simply having a mathematical equation resembling experimental data, empirical models neglecting the mass transfer effects can be adopted. Otherwise if kinetic modelling aim to interpret experimental data and predict the effect of influencing factors, phenomenological models should be developed including the description of those mass transport steps limiting the global rate of sorption.

4.3.1 Empirical Kinetic Models

4.3.1.1 Pseudo-First Order Model

Lagergren's first-order rate equation described the adsorption rate on the base of the adsorption capacity

$$\frac{dq}{dt} = k_1 (q_e - q) \tag{4.74}$$

where q is the metal concentration in solid phase at time t, and q_e is the equilibrium metal concentration in solid phase.

In order to distinguish kinetic equations based on concentrations in solution from kinetic equations based on adsorption capacities of solids, Lagergren's first-order rate equation has been called pseudo-first-order.

Integration of Eq. 4.74 (boundary conditions: t=0, q=0 and t=t, q=q) gives

$$q = q_e\left(1 - \exp\left(-k_1 t\right)\right) \tag{4.75}$$

which can be expressed in linear form as

$$\ln(q_e - q) = \ln q_e - k_1 t \tag{4.76}$$

q_e and k_1 parameters can be determined by non linear procedure fitting or, more commonly, by linear regression using plots $\ln(q_e - q)$ versus t obtained for different initial metal concentrations. q_e values are then contrasted with experimental values and if large discrepancies exist, the reaction cannot be classified as pseudo-first order although high regression coefficients were obtained for the linear plots.

Pseudo-first order model have been applied to many biosorption kinetic studies (Febrianto et al. 2009). Generally it does not fit well over the entire contact time range, but applicable only over the initial period of the sorption process.

4.3.1.2 Pseudo-Second Order Model

Pseudo-second order model is derived assuming a second order dependence of the sorption rate on available sites:

$$\frac{dq}{dt} = k_2(q_e - q)^2 \tag{4.77}$$

Integration of Eq. 4.77 (boundary conditions: t=0, q=0 and t=t, q=q) gives

$$q = q_e\left(1 - \frac{1}{1 + q_e k_2 t}\right) \tag{4.78}$$

which can be stated in the linear form

$$\frac{t}{q} = \frac{t}{q_e} + \frac{1}{k_2 q_e^2} \tag{4.79}$$

and pseudo-second order parameters (q_e and k_2) were determined by plotting t/q against t. Even though non linear regression method was proven to be the best way to regress parameters (Ho 2006b) most of biosorption studies utilised the linearised model.

Pseudo-second order model is generally more appropriate than pseudo-first order model to represent kinetic data of heavy metal biosorption in batch reactors. In most systems regression coefficients higher than 0.98 were obtained and calculated q_e values agreed very well with experimental ones (Febrianto et al. 2009).

Table 4.4 Empirical kinetic models for heavy metal biosorption in batch reactors

Model	Equation
Elovich's equation	$\frac{dq}{dt} = \alpha_E \exp(-\beta_E q)$
Ritchie's equation	$\frac{d\theta}{dt} = \alpha_R(1-\theta)^{n_R}$
Third-order model	$\frac{dq}{dt} = k_d C(q_e - q)^2$
First-order reversible model	$\frac{dq}{dt} = -\frac{dC}{dt} = k_1 C - k_2 q$
Langmuir-Hinshelwood model	$\frac{dC}{dt} = \frac{k_1 C}{1+k_2 C}$

θ fraction of occupied sites

Nevertheless due the empirical nature of this model no conclusion about controlling mechanism should be deduced by this good representation of data.

4.3.1.3 Other Empirical Kinetic Models

In the literature other empirical kinetic models were reported to represent the kinetic of heavy metal biosorption in batch reactors without taking in consideration the effect of mass transfer effects or without assessing their negligibility. Some examples were reported in Table 4.4 (Febrianto et al. 2009; Schiewer and Balaria 2009; Bayramoglu and Arica 2008; Sengil and Ozacar 2008; Ho 2006b; Singh et al. 2005; Aksu 2002; Texier et al. 1999).

4.3.2 Assessing Mass Transfer Effects and Controlling Mechanisms

Rate limiting steps in biosorption systems could be identified by performing dedicated tests: external bulk mass transfer and film diffusion resistances can be denoted by changing the agitation of the liquid; intraparticle diffusion effects can be evidenced by performing kinetic tests with different particle size; control by chemical reaction can be denoted by studying the effect of temperature. Such preliminary investigation are often neglected in favour of data elaboration by models developed assuming specific mechanism as controlling biosorption rate. The adequacy or not of these models is then used to infer information about the mechanism controlling the system under study.

4.3.2.1 Weber and Morris Model

Kinetic data of metal biosorption were often analysed by the Weber and Morris model (Eq. 4.80) put in evidence if intraparticle diffusion is the rate controlling step

(Basha et al. 2009; Apiratikul and Pavasant 2008; Sengil and Ozakar 2008; Singh et al. 2005)

$$q = k_{id}\sqrt{t} \tag{4.80}$$

where k_{id} is the intraparticle diffusion rate constant. If experimental data reported as q versus $\sqrt{t}$ give a straight line passing through the origin it is generally assumed that intraparticle diffusion is the rate limiting step. Weber and Morris plots can also present multi-linearity (Basha et al. 2009; Sengil and Ozakar 2008) or intercept different from zero (Apiratikul and Pavasant 2008): in these cases two mechanisms are assumed to be limiting external mass transfer and intraparticle diffusion.

4.3.2.2 Boyd Plots

These plots can be used to distinguish between external-transport-controlled (film diffusion) and intraparticle-transport-controlled rates of biosorption.

Richenberg model (Richenberg 1953) was applied to check that biosorption proceeds via film diffusion or intraparticle diffusion

$$\frac{q}{q_e} = 1 - \frac{6}{\pi^2}\exp\left(-b_B t\right) \tag{4.81}$$

where b_B is a constant.

Previous equation can be written as

$$b_B t = -0.4977 - \ln\left(1 - \frac{q}{q_e}\right) \tag{4.82}$$

Boyd plots were then obtained by plotting $b_B t$ versus time: a straight line passing through the origin is indicative of biosorption processes governed by particle diffusion mechanisms, otherwise they are governed by film diffusion (Basha et al. 2009; Gupta 1998).

4.3.3 Mass Transfer Models

Mass transfer models are phenomenological models taking in consideration mass transfer effects in biosorption kinetic. According to the multi-step nature of biosorption process (Fig. 4.6) these models can include different mechanisms depending on the nature of the biosorbent and the fluido-dynamic conditions adopted inside the reactor. Then in the case of free cell suspensions under effective agitation surface reaction can be the controlling step, while in the case of gel-like biomasses (such as seaweed) or microbial biomasses immobilised in polymeric matrices, intraparticle diffusion is often the rate limiting mechanism. It should be author care performing independent kinetic experiments under different operating conditions to isolate the rate limiting step and then adopt the proper kinetic model to represent data.

In the follow some examples of mass transfer models reported in the literature for biosorption kinetics in batch reactors were reported along with main hypotheses about mass transport mechanism and equilibrium models.

The order of presentation reflects the model complexity and then the number of mechanisms included in kinetic modelling. The simplest case to be represented is that of a free biomass, whose active sites are present only on the surface. For this case, Puranik et al. (1999) developed a mass transfer model considering only external film diffusion and assuming a rapid surface adsorption, negligible intraparticle diffusion and Langmuir or Freundlich isotherm as equilibrium model. The authors found that this simple model was able to represent kinetic biosorption of different metals on three biomasses.

Similarly the McKay model assuming external mass transfer resistance as controlling step and Langmuir isotherm as equilibrium model was applied in biosorption batch kinetic (Singh et al. 2006; Gupta 1998)

$$\ln\left(\frac{C}{C_0} - \frac{1}{1+xK}\right) = \ln\left(\frac{xK}{1+xK}\right) - \left(\frac{1+xK}{xK}\right) k_f a_s t \qquad (4.83)$$

where x is the mass of adsorbent per unit volume, K is the Langmuir constant obtained by multiplying q_{max} and b, k_f is the mass transfer coefficient, and a_s is the outer specific surface of adsorbate per unit volume.

This model has been employed for the determination of the surface mass transfer coefficient by the plot $\ln(C/C_0 - 1/(1+xK))$ which should be linear (Singh et al. 2006; Gupta 1998).

Chu and Hashim (2004) developed a mass transfer model for batch biosorption onto a microalgae considering external mass transfer and intrinsic adsorption kinetics (second order reversible reaction). The authors demonstrated that for their system both these step were necessary to adequately describe the experimental data obtaining physically meaningful rate parameters.

In the case of gel-like matrices and porous biomasses, biosorption reaction is not limited to biosorbent surface then sorbate diffusion should be taken into account. Volesky and coworkers developed a one-dimension intraparticle diffusion model for representing cadmium biosorption rate in a seaweed biomass in batch reactors (Yang and Volesky 1996, 1999b). They excluded the effect of external mass transfer resistance by preliminary tests performed increasing agitation rate until adequate turbulence was created inside the reactor. They assumed that the sorption reaction is much faster than diffusion inside the particle, and then the amount of metal adsorbed inside the biomass particle is in equilibrium with the metal concentration in the liquid phase (Langmuir isotherm is assumed as equilibrium model). Biomass particle is then assumed as a thin plate with the intraparticle diffusion in the thickness direction controlling the diffusion rate of the overall process.

Mass conservation equations for metal ion in the biomass particle and in the bulk solution were

$$\frac{\partial C_r}{\partial t} + \rho_p \frac{\partial q_t}{\partial t} = D_p \frac{\partial^2 C_r}{\partial r^2} \tag{4.84}$$

$$V \frac{dC}{dt} = -D_p S_t \left. \frac{\partial C_r}{\partial r} \right|_{r=R} \tag{4.85}$$

where C and C_r are the metal concentration in the bulk solution and in the biomass gel phase at layer r inside the particle, R is the half-thickness of the particle, ρ_p is the biomass density, V is the volume of the bulk solution, S_t is the total surface area of particles, and D_p is the effective intraparticle diffusion coefficient.

Equilibrium parameters were determined by independent tests and then the only parameter regressed using kinetic batch tests was D_p. The authors found that the rate of cadmium biosorption on Sargassum biomass was properly described by one-dimension intraparticle diffusion model with regresses values of the intraparticle diffusion coefficients smaller than the molecular diffusion coefficient of cadmium in water (Yang and Volesky 1996). Similar modelling approaches assuming intraparticle diffusion as the controlling step were applied by Vilar et al. (2008a) to represent the rate of single and binary metal batch biosorption onto an algal biomass, and by Papageorgiou et al. (2006) for heavy metal batch biosorption by alginate beads.

Models assuming a homogeneous particle structure generally require the regression of a single value of the effective diffusion coefficient for each metal ion. In general, due to tortousity and porosity, the diffusion process inside a porous media is slower than in a liquid phase with the same composition of pore phase. If regressed values of intraparticle diffusion coefficient were larger than the free liquid diffusivity other mechanisms should be taken in consideration. These mechanisms can concern with other limiting steps or with particle matrix characteristics.

In particular the heterogeneity of the matrix of the particle and the simultaneous mechanisms of pore and surface diffusion were described in a wide literature applied to batch kinetic sorption of dye onto activated carbons (Ahmed et al. 1992; McKay and Duri 1991; McKay et al. 1984; Alexander et al. 1978 and references therein), which in theory could be applied also to heavy metal biosorption. As an example, Ahmed et al. (1992) developed a mass transfer model for gold cyanide sorption onto activated carbon assuming both a homogeneous structure of the particle and a modified version with decreased diffusivity in the outer shell of the particle.

For homogeneous matrices whose batch biosorption kinetics cannot be adequately represent by intraparticle diffusion models other mass transfer steps should be considered. Chen and Wang (2004) developed a mass transfer model for batch kinetic of copper biosorption by granular activated carbon considering both external film diffusion and intraparticle pore diffusion as limiting steps and Langmuir isotherm as equilibrium equation

$$\frac{\partial C_p}{\partial t}\left(\varepsilon_p + \rho_p \frac{dq}{dC_p}\right) = \frac{D_p}{r^2}\frac{\partial}{\partial}\left(r^2 \frac{\partial C_p}{\partial r}\right) \tag{4.86}$$

with boundary condition for r=R particle radius

$$\frac{\partial C_p}{\partial r} = \frac{k_f}{D_p}(C - C_s) \tag{4.87}$$

where C is the bulk metal concentration, C_p is the metal concentration within the pores, C_s is the concentration at the solid-liquid interface, D_p is the pore diffusivity, k_f the external mass transfer coefficient, ε_p the porosity of the particle and ρ_p the particle density.

Shrinking core model, originally developed for gas-solid reactions, was extended to heavy metal biosorption (Hui et al. 2003; Cheung et al. 2001; Vegliò et al. 1998) assuming three possible resistances controlling the process: diffusion of the solute through the liquid film surrounding the particle, diffusion of the solute through the particle pores and sorption onto active sites. As an example Vegliò et al. (1998) used the shrinking core model for the kinetic modelling of copper biosorption by immobilised biomass. They did not consider the kinetic control of sorption reaction according to experimental tests with free biomass.

Biosorption extent (χ) in the case of the process controlled by diffusion of metal ions through the liquid film is given by the following expression

$$\chi = \frac{3D_e}{\delta R q_m}\int_0^t C dt \tag{4.88}$$

where D_e is the copper diffusion coefficient in the liquid phase, δ is the liquid film thickness, R is the particle diameter, q_m is the active site concentration in the particle at the beginning of the process.

Consequently if film diffusion is the controlling step, χ versus $\int_0^t C dt$ yields a straight-line relationship.

If the process is controlled by the diffusion through the reacted shell the model is represented by the following expression

$$F(\chi) = 1 - 3(1-\chi)^{\frac{2}{3}} + 2(1-\chi) = \frac{6D_p}{R^2 q_m}\int_0^t C dt \tag{4.89}$$

where D_p is the copper apparent diffusion coefficient in the biomass particle.

Then in the case of particle diffusion control F(χ) versus $\int_0^t C dt$ yields a straight-line relationship, and the apparent diffusivity in the biomass phase could be obtained from the slope.

The authors assessed intraparticle diffusion was the rate controlling step by the straight line obtained reporting data as F(χ) versus $\int_0^t C dt$ finding values of effective intraparticle diffusivities (D_p) that are in agreement with theoretical values of metal ion diffusivity in dilute aqueous solutions.

4.4 Dynamic Behaviour in Continuous Processes

Large scale applications of biosorption processes are strictly related to the development of continuous-flow treatments. Reactor configurations generally used for heavy metal biosorption are fixed bed columns, even though biomasses often require to be immobilised in polymeric matrices to obtain particles with proper mechanical strength and permeability. Alternatively membrane reactors can be used in which continuous stirred tank reactors are integrated with membranes devices able to retain biomass particles within the system. Dynamic models able to represent and predict the system behaviour under changing operating parameters are fundamental process optimisation tools.

4.4.1 Fixed Bed Reactors

In fixed bed columns metal concentration profiles in liquid and solid phase vary in both space and time. The dynamic behaviour of a fixed bed column is described in terms of the effluent concentration profiles versus time, namely the breakthrough curve (Fig. 4.6b). The shape of this curve is determined by the equilibrium isotherm and the transport processes in the column and in the adsorbent. Efficient adsorption performances are obtained when the breakthrough curve is as sharp as possible. For short times the column completely takes up the metal in the feed, but after a while metal breakthrough occurs and effluent concentration gradually increases with time. As for the concentration profile along the column length, the adsorbent close to the column inlet becomes saturated at the prevailing inlet fluid concentration and a concentration gradient develop beyond this saturation zone. The region inside the column in which the solute concentration in the liquid phase change from 90% to 10% of its inlet value is called mass transfer zone (MTZ) (Fig. 4.6c). After complete development, MTZ advances along the length of the column in the direction of the flow at a velocity that depends on solute concentration, the sorbent capacity, and the feed flow rate. When the MTZ reaches the end of the column the effluent concentration starts to increase until the inlet values, then the shape of the breakthrough curve reflects the shape of the MTZ.

Column is operated until a specified breakthrough concentration (C_{bt}) is reached in the effluent at a determinate breakthrough time (t_{bt}) or service time. The position of the breakthrough time is related to the column sorbent capacity due to the amount of metal in the feed, and the operating parameters such as the influent flow rate.

Mechanisms operating in a biosorption process in fixed bed columns are axial dispersion in the direction of the liquid flow, film diffusion resistance, intraparticle diffusion resistance, and sorption kinetic at the adsorbent surface. Rigorous models taking in consideration all these mechanisms present mathematic and numerical difficulties (especially due to the non linearity associated to equilibrium models) and require independent experiments and/or reliable engineering correlations to estimate the numerous equilibrium, transport and kinetic parameters. Otherwise multi-parameter fitting of breakthrough curves may reduce the physical significance of the mechanistic parameters. For these reasons approximate modelling approaches have been widely used which allow to model breakthrough behaviour without the need of numerical solution and with immediate practical benefits in process development and design.

In the follow an overview of approximate and mass transfer model used for biosorption process in fixed bed column reactor is reported.

4.4.1.1 Approximate Models

In this sections were reported models that have been developed and solved under specific theoretical approximations (approximate models) and that due to their simple application are widely used in column performance theory, often without proving the validity of their assumptions.

The Thomas Model

This model was one of the most widely used among approximate models (Gokhale et al. 2009; Apiratikul and Pavasant 2008; Salamatinia et al. 2008; Naddafi et al. 2007; Senthilkumar et al. 2006; Han et al. 2006b; Malkoc and Nuhoglu 2006; Kapoor and Viraraghavan 1998). The Thomas model, assuming Langmuir kinetic of adsorption-desorption and no axial dispersion was developed considering that rate driving force obeys second order reversible reaction kinetic

$$\frac{C}{C_0} = \frac{1}{1 + \exp\left(\frac{k_{Th}}{F}\left(q_m m - C_0 V_{eff}\right)\right)} \tag{4.90}$$

where C is the effluent concentration during time, C_0 is the influent concentration, k_{Th} is the Thomas rate constant, q_m is the maximum solid-phase concentration of the solute, F is the flow rate, V_{eff} is the volume of effluent, m the amount of sorbent in the column.

It can also be used in the linearised form

$$\ln\left(\frac{C_0}{C} - 1\right) = \frac{k_{Th} q_m m}{F} - \frac{k_{Th} C_0}{F} V_{eff} \tag{4.91}$$

The kinetic coefficients k_{Th} and the adsorption capacity of the bed q_m can be determined from the plot of $\ln[(C/C_0)-1]$ against V_{eff} at a given flow rate.

The Adams-Bohart Model

This model, originally developed for gas adsorption, assumes that the adsorption rate is proportional to both the residual capacity of the adsorbent and the concentration of the sorbing species. The solution of the differential equations for mass transfer rates in solid and liquid phases (Eqs. 4.92 and 4.93) gives the Adams-Bohart model (Eq. 4.94) used in different biosorption applications in fixed bed columns (Pakshirajan and Swaminathan 2009; Salamatinia et al. 2008; Texier et al. 2002)

$$\frac{\partial q}{\partial t} = -k_{AB} q C \tag{4.92}$$

$$\frac{\partial C}{\partial h} = -\frac{k_{AB}}{u} q C \tag{4.93}$$

$$\ln \frac{C}{C_0} = k_{AB} C_0 t - k_{AB} q_m \frac{H}{u} \tag{4.94}$$

where q is the metal concentration in the bulk liquid, k_{AB} is the kinetic constant, h is the axial length coordinate, H is the height of the column, and u is the superficial velocity. Equation 4.94 was derived under the assumptions of low concentration field ($C<0.15C_0$, generally valid in the initial part of the breakthrough), and that for $t\rightarrow\infty$, $q\rightarrow q_m$ where q_m is the saturation concentration. Characteristic operational parameters of the column can be determined from the plot of $\ln(C/C_0)$ against t at a given flow rate.

Starting from this model Adams and Bohart, proposed a relationship between bed depth (*H*) and the time taken for breakthrough to occur (service time or breakthrough time) which was the base for Bed Depth Service Time analysis widely used for column design (Gokhale et al. 2009; Hasan et al. 2009; Vijayaraghavan et al. 2005; Texier et al. 2002; Zulfadhly et al. 2001; Gupta 1998).

The Wolborska Model

The Wolborska model (Wolborska 1989) is derived for the description of adsorption dynamics taking in consideration both diffusion and external mass transfer mechanism

$$\frac{\partial C}{\partial t} + u\frac{\partial C}{\partial h} + \frac{\partial q}{\partial t} = D_e \frac{\partial^2 C}{\partial h^2} \tag{4.95}$$

$$\frac{\partial q}{\partial t} = -v\frac{\partial q}{\partial h} = k_f\left(C - C_s\right) \tag{4.96}$$

where C_s is the metal concentration at the solid-liquid interface, D_e is the kinetic coefficient for diffusive transport, k_f the kinetic coefficient for external mass transfer, and ν the migration rate.

Under specific assumptions ($C_s \ll C_b$; $v \ll u$; $D_e \rightarrow 0$) the solution of the previous mass balance equations gives the Wolborska model (Pakshirajan and Swaminathan 2009; Salamatinia et al. 2008)

$$\ln\frac{C}{C_0} = \frac{k_f C_0}{q_m}t - \beta_f\frac{H}{u} \tag{4.97}$$

with

$$\beta_f = \frac{u}{2D_e}\left(\sqrt{1 + \frac{4k_f D_e}{u^2}} - 1\right) \tag{4.98}$$

It is noteworthy that the expression of Wolborska is equivalent to that of Adams-Bohart if $k_{AB}=k_f/q_m$. Then obtaining a straight line in the plot $\ln(C/C_0)$ against t cannot be considered a prove that the mechanism hypotheses under which the two different models were derived, are satisfied.

The Clark Model

This model combines the Freundlich equation and the mass transfer concept (Clark 1987) according to the following equation

$$u\frac{dC}{dh} = k_f\left(C - C_e\right) \tag{4.99}$$

where k_f is the mass transfer coefficient and C_e is the equilibrium concentration in liquid phase.

Solution of the previous equation is the Clark model

$$\frac{C}{C_0} = \left(\frac{1}{1 + Ae^{-rt}}\right)^{1/n-1} \tag{4.100}$$

with

$$A = \left(\frac{C_0^{n-1}}{C_{bt}^{n-1}} - 1\right)e^{rt_{bt}} \tag{4.101}$$

and

$$R\left(n - 1\right) = r \tag{4.102}$$

$$R = \frac{k_{Cl}}{u} v \tag{4.103}$$

Equation 4.100 is a logistic function where n is the Freundlich constant, C_{bt} and t_{bt} are the outlet concentration and the time at breakthrough (or limit effluent concentration). Values of A and r can be determined by Eq. 4.100 by non linear regression thus enabling the prediction of breakthrough curves according to the relation between C/C_0 and t (Pakshirajan and Swaminathan 2009).

The Yoon and Nelson Model

The Yoon and Nelson model (Yoon and Nelson 1984) is less complicated than other models not requiring detailed information about sorbent and solute characteristics, and about physical properties of adsorption bed (Gokhale et al. 2009; Salamatinia et al. 2008; Senthilkumar et al. 2006).

The Yoon and Nelson model for a single component system is

$$\ln \frac{C}{C_0 - C} = k_{YN} t - \tau k_{YN} \tag{4.104}$$

where k_{YN} is the rate constant and τ the time required for 50% adsorbate breakthrough.

If this model adequately describes the breakthrough curve a straight line is obtained by plotting ln $C/(C_0-C)$ versus the sampling time, whose slope and intercept are k_{YN} and τk_{YN}, respectively.

4.4.1.2 Mass Transfer Models

Approximate models such those described before, generally can represent adequately the experimental breakthrough curves (Apiratikul and Pavasant 2008; Naddafi et al. 2007; Aksu and Gonen 2006; Han et al. 2006b; Malkoc and Nuhoglu 2006; Zulfadhly et al. 2001; Kapoor and Viraraghavan 1998). Nevertheless their validity is limited to the range of conditions used during laboratory column tests. In addition due to their theoretical derivation for gas adsorption their application to solid liquid systems can not include the effects of the most influential parameters in heavy metal biosorption such as pH and ionic strength. The development of mass transfer mechanistic models aim to predict the effect of operating conditions on column performance.

Mass transfer models for biosorption processes in fixed bed reactors have originated mainly from research on activated carbon sorption and chromatographic applications (Gadam et al. 1995; Tan and Spinner 1994; Cysewski et al. 1991; Bohart and Adams 1920). The general approach is based on solving the mass balance equations for the solute being transported thought the liquid phase and biosorbed by the solid phase.

The mass balance of a chemical species in the bulk fluid flow was generally written as

$$\frac{\partial C}{\partial t} = D_e \frac{\partial^2 C}{\partial h^2} - u_h \frac{\partial C}{\partial h} - \frac{\rho_p (1-\varepsilon)}{\varepsilon} \left(\frac{\partial q}{\partial t} \right) \tag{4.105}$$

where C is the solute concentration in liquid phase, D_e is the diffusion coefficient in liquid phase, h is the position along the column length, u_h is the fluid phase interstitial velocity, ρ_P is the biosorbent material bulk density, ε the bed porosity, and q solute concentration in sorbent phase.

In the right end side Eq. 4.105 incorporates the diffusive and bulk solute movement in the first two terms, and the sorptive process in the third term.

The main difference in the various approaches reported in the literature is between equilibrium and non equilibrium model. Equilibrium models are the most simplistic approach to model sorption phenomena according to the concept of rapid local equilibrium: microscopic processes of mass transfer to the sorption sites and the sorption reactions are faster than macroscopic process of fluid transport in the column length. Under this assumption the rate of change of metal in sorbed phase at any point in the column instantaneously reflects the rate of change of metal in the liquid phase according to a specified equilibrium equation:

$$\frac{\partial q}{\partial t} = \frac{\partial q}{\partial C} \frac{\partial C}{\partial t} \tag{4.106}$$

Models based on this assumption are known as local equilibrium models (LEM): they neglect mass transfer resistance in the liquid and solid phases and assume instantaneous biosorption reaction (Hatzikioseyian et al. 2001; Chen and Wang 2004; Vilar et al. 2008b).

Non-equilibrium models account for the kinetic of the microscopic phenomena occurring during sorptive process. One widely used expression for sorption rate can be written assuming a linear driving force for the sorption process and a combined film and intraparticle mass transfer resistance (K′) (Naja and Volesky 2006; Kratochvil et al. 1997)

$$\frac{\partial q}{\partial t} = K' \left(q_e (C) - q \right) \tag{4.107}$$

where $q_e(C)$ is the equilibrium metal concentration in solid phase when the liquid concentration is C, and q is the instantaneous metal concentration in solid phase. The relation $q_e(C)$ can be one of the equilibrium models previously described even if simple empirical models (such as Langmuir isotherm) are generally used in dynamic modelling.

Along with the two previously described approaches, in the literature other mass transport models were reported to describe breakthrough curves. These models can be more simplistic than the two approaches described before (for instance neglecting diffusive term (Trujillo et al. 1991) or lumping together intraparticle diffusion and adsorption kinetics (Maiti et al. 2009) or less approximate (i.e. considering intraparticle pore diffusion separately from sorption rate) (Reverberi et al. 2009). In Table 4.5 a synthetic view of the theoretical assumptions and main equations for some kinds of mass transfer dynamic models of biosorption processes in column reactors was reported.

Dynamic mass transfer models generally can describe accurately experimental data of single metal biosorption in column reactors. Nevertheless these models should allow not only to simulate but also to predict the performance of a column under various operating conditions, thus assisting the scale-up process by choosing the conditions under which running the pilot tests. On the other hand the predicting power of such models is sometime doubtful especially when adjustable parameters with clear chemico-physical meaning present unrealistic values and/or different parameter sets are necessary to represent the system behavior in different operating conditions. Diffusion coefficient is a typical example: regressed value of this parameter inside the fixed bed column material can exceed the theoretical values in water, and different values were obtained for representing the same data by different mass transfer models (Reverberi et al. 2009; Naja and Volesky 2006; Kratochvil et al. 1997).

The case of multi-metal system is still scarcely modelled by mass transfer models mainly due to the lack of available ready-to-use and sufficiently sophisticated process modeling software package incorporating all the aspects of biosorption (mechanistic equilibrium equations for multi-metallic systems, mass transfer and fluid flow characteristics). Few examples can be found in the literature in which proprietary computer software were used (Volesky 2003; Kratochvil and Volesky 1998). Nevertheless, a part from the numeric difficulties in model solving, a critical aspect of the description of dynamic multi-metallic biosorption seem to be the accurate description of equilibrium relationship. In fact in continuous flow systems competition among metals gives rise to typical non monotonic trend of the metal concentration in the effluent (overshoot) due to the different affinity between site and metal in the systems, according to the base principle of chromatography technique. Then only if competitive effects have been properly described in the equilibrium equation included in the dynamic model, it is possible to represent these trends.

4.4.2 Membrane Reactors

Continuous biosorption processes in membrane reactors are less diffuse than fixed bed configuration.

Few examples of heavy metal biosorption in membrane reactors were reported in the literature and less those comprehending dynamic models, which in turn are quite all mass transfer models (Ahmady-Asbchin et al. 2009; Beolchini et al. 2005, 2006; Pagnanelli et al. 2003b, c; Koltuniewicz and Bezak 2002; Bayhan et al. 2001; Chang and Chen 1999). Heavy metal biosorption on specially propagated bacterial biomasses in cross flow ultrafiltration membrane reactors was widely investigated both experimentally and by dynamic simulations considering the effect of pH, ionic strength, multi-metallic systems, and multi-stage configurations (Beolchini et al. 2005, 2006;

Table 4.5 Synthetic view of different mass transfer models for biosorption application in fixed-bed columns

Biosorption system	Model assumptions	Equilibrium model	Mass Balance equations	Reference
Cu biosorption by *Sargassum fluidans*	Sorptive process inside the pores	Langmuir model	$\frac{\partial x}{\partial \tau} = \frac{1}{Pe}\frac{d_p}{L}\frac{\partial^2 x}{\partial z^2} - \xi\frac{\partial x}{\partial z} - \frac{1-\varepsilon}{\varepsilon}St(x - x_p)$ $\frac{\partial x_p}{\partial \tau} = -\frac{1-\varepsilon_p}{\varepsilon_p}\frac{\partial}{\partial \tau} + \frac{St}{\varepsilon_p}(x - x_p)$ $\frac{\partial y}{\partial \tau} = Da\left(\sigma\frac{Kx_p}{\frac{1}{\rho_p}+Kx_p} - y\right)$	Reverberi et al. (2009)
Cu biosorption by Ca-preloaded *Sargassum fluidans*	Combined film and intra-particle mass transfer resistance	Competitive ion exchange model	$\frac{\partial x}{\partial \tau} = \frac{1}{Pe}\frac{\partial^2 x}{\partial z^2} - \frac{\partial x}{\partial z} - D_{gM}\left(\frac{\partial y}{\partial \tau}\right)$ $\frac{\partial y}{\partial \tau} = Sh_M(y_e - y)$	Naja and Volesky (2006)
U biosorption by *Sargassum fluidans*, Cd biosorption by *Ascopyllum nodosum*	Rapid local equilibrium	Linear, Langmuir, and Freundlich models	$\frac{\partial x}{\partial \tau} = \frac{1}{Pe}\frac{\partial^2 x}{\partial z^2} - \frac{\partial x}{\partial z} - D_e\left(\frac{\partial y}{\partial \tau}\right)$ $\frac{\partial y}{\partial \tau} = \frac{\partial y}{\partial x}\frac{\partial x}{\partial \tau}$	Hatzikioseyian et al. (2001)
Heavy metal biosorption by *Sphagnum* peat moss immobilised in polysulfone matrix	No axial dispersion	Competitive Langmuir model	$\frac{\partial x}{\partial \tau} = -\frac{\partial x}{\partial z} - P_1(y_e - y)$ $\frac{\partial y}{\partial \tau} = P_2(y_e - y)$	Trujillo et al. (1991)
Cu biosorption by marine algae *Gelidium*	External film diffusion and internal mass transfer resistance described separately by linear driving force approximation	Langmuir model	$\frac{\partial x}{\partial \tau} = \frac{1}{Pe}\frac{\partial^2 x}{\partial z^2} - \frac{\partial x}{\partial z} - \xi N_d(y_e - \langle y \rangle)$ $\frac{\partial \langle y \rangle}{\partial \tau} = N_d(y_e - \langle y \rangle)$ $\frac{\partial \langle y \rangle}{\partial \tau} = \frac{N_f}{\xi}(y - y_f)$	Vilar et al. (2008b)

x: dimensionless concentration in liquid phase, y: dimensionless concentration in solid phase, τ: dimensionless time, z: dimensionless axial coordinate; pedix p refers to concentration inside the pores; pedix f refers to the metal concentration in the film surrounding the particles; <y> is the dimensionless average metal concentration in the solid phase; Pe is the Peclet number; St is the Stanton number; Da is the Damkohler number; Sh_M is the modified Sherwood number. For other symbols see details in references

Pagnanelli et al. 2003b, c). Dynamic simulations were developed assuming perfect mixing inside the reactor, biosorption as an equilibrium process, no transport resistance in the bulk solution and constant reaction volume. Metal retention coefficient changed during time due to the presence of biomass fragments which determined both pore plugging and metal binding by soluble macromolecules (released by biomass) resulting in metal complexes larger than membrane pores and then retained. Metal retention coefficient profiles were represented by an empirical equation regressed on the base of experimental data.

Dynamic simulations were then developed by considering the unsteady mass balance of the metal in the reactor (C) assuming that metal concentration in the reactor and in the retentate are the same:

$$FC_0 - FC - V\frac{d(qx)}{dt} = V\frac{dC}{dt} \tag{4.108}$$

where C is the concentration inside the reactor during time, x is the biomass concentration in the reactor, F the volumetric flow rate and V is the constant suspension volume inside the reactor.

For binary metal systems two distinct equations have to be considered to represent the dynamic behaviour of each component. Also equilibrium relation between metal concentration in the adsorbent phase and in the permeate solution has to be chosen considering the competition among metals for the active sites (competitive Langmuir model).

According to equilibrium hypothesis, metal concentration on the solid, q, is given by equilibrium equations whose parameters were determined by independent equilibrium batch tests. As equilibrium concentration is taken the one in the permeate instead of the concentration in the reactor because not all copper in solution inside the reactor is available for biosorption, since it could be bound to macromolecules coming from biomass degradation.

Under the hypothesis of constant biomass concentration (dx/dt=0) dynamic simulations can adequately represent experimental data on the base of batch equilibrium parameters and empirical models of σ profiles during time.

Koltuniewicz and Bezak (2002) modelled the time-dependent sorption properties inside the membrane reactors by taking into account the different biosorbing capacity of suspended biomass and biofilm formed on membrane: they observed that the number of available sites for biosorption declined during the process with a low similar to that of radio-isotopes decay. Then the sorption properties of biomass particles in the membrane biofilm are related to the age distribution of the sorbent elements evaluated assuming the random renewal of the sorbent at the membrane surface.

Chang and Chen (1999) used a hollow-fiber crossfow microfiltration membrane to retain a biomass of Pseudomonas aeruginosa for continuous biosorption of single and ternary metal systems. These authors used both a rapid equilibrium model and a model accounting for external mass transfer and assuming linear equilibrium relation. The authors found that the rapid equilibrium (using Langmuir and competitive Langmuir for single and multi-component systems, respectively) can represent adequately only single metal biosorption, while the other model can also predict correct trends of selective biosorption in multi-metallic systems.

4.5 Conclusions

The overview reported in this chapter denoted that different degrees of complexity can be adopted in the modelling of biosorption processes. Including all the chemical, physical and electrical phenomena simultaneoulsy occurring during heavy metal biosorption in an utopistic task without practical use. In fact a model of such mathematical complexity could be managed only by specifically trained personnel with a solid mathematical and numerical background.

In addition not all the parameters potentially influencing biosorption could be significantly relevant in the system under study for the range of operating conditions practically used. Then the choice of model complexity to be adopted should be carefully sized to the specificaim. Application of experimental design and statistical analysis of data should be then the first step to isolate the parameters significantly influencing the biosorption process in the selected range of operating conditions. After this, model architecture is developed to include the effect of such parameters according to empiric or mechanistic approaches depending on the aim of modelling.

Model overview ends here with a picture showing how changed during last 15 years the role of models in heavy metal biosorption studies. The picture was "taken" by ISI Web of Knowledge (20th October 2009) by searching in the topic the key words "heavy metal" and "biosorption" and refing for the pubblication type (only articles) and for the language (only English). Figure 4.7 A denoted a rapid increase of publications about this topic especially in the last ten years.

Nevertheless the analysis of the abstracts put in evidence that the use of mathematical models for representing biosorption data is not increased in the same way. In fact the percent of works using models (equilibrium, kinetic and dynamic) remained quite constant during years (54 ± 11 as %) (Fig. 4.7a).

Papers including models were then classified according to the type of models they reported in four hierarchic cathegories:

- empirical equilibrium models (emp eq)
- mechanistic equilibrium models (mecc eq)
- empirical models for equilibrium and kinetics in batch reactor (eq and kin)
- dynamic models for continuous process (dyn)

Each paper was counted one time respecting the hierarchy above (dyn, eq and kin, mecc eq, emp eq) then if a paper includes a dyn model and also empirical equilibrium models it was classified as dyn.

This analysis denoted that the large majority of papers include empirical equilibrium models, while the % values of those using mechanistic approach for equilibrium data (mecc eq) and dynamic models for continuous processes (dyn) remained at the same low level during years. On the other side papers using empirical models

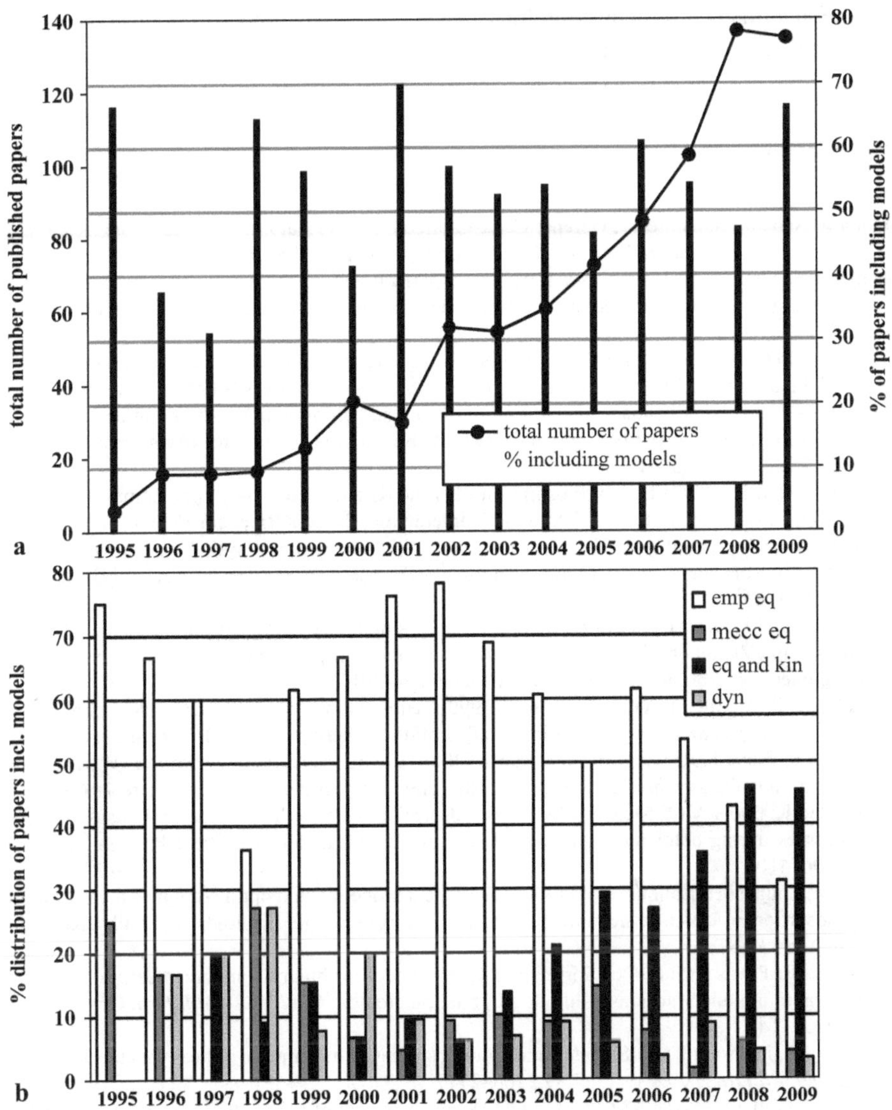

Fig. 4.7 Literature works about biosorption of heavy metals: total number of published papers and % of those including models (**a**), classification of biosorption models (**b**) as those including empirical equilibrium models (emp eq), mechanistic equilibrium models (mecc eq), equilibrium and kinetic empirical models (eq and kin), and dynamic models for continuous processes (dyn)

both for equilibrium and kinetic data in batch reactors present a clear increasing trend during the last ten years.

Then the increasing interest in heavy metal biosorption does not seem to be accompanied by an augmented degree of mathematical and interpretative knowledge about chemico-physical mechanisms operating in metal removal.

Perhaps this could be the step forward that research should do to bring biosorption processes to a large scale application.

References

Ahmady-Asbchin S, Andres Y, Gerente C, Le Cloirec P (2009) Natural seaweed waste as sorbent for heavy metal removal from solution. Environ Technol 30:755–762

Ahmed FE, Young BD, Bryson AW (1992) Comparison and modelling of the adsorption kinetics of gold cyanide onto activated carbon and resin in a silica slurry. Hydrometallurgy 30: 257–275

Aksu Z (2002) Determination of the equilibrium, kinetic and thermodynamic parameters of the batch biosorption of nickel(II) ions onto *Chlorella vulgaris*. Process Biochem 38:89–99

Aksu Z, Gonen F (2006) Binary biosorption of phenol and chromium(VI) onto immobilized activated sludge in a packed bed: prediction of kinetic parameters and breakthrough curves. Sep Purif Technol 49:205–216

Aksu Z, Gulen H (2002) Binary biosorption of iron(III) and iron(III)-cyanide complex ions on *Rhizopus arrhizus*: modeling of synergistic interaction. Process Biochem 38:161–173

Aksu Z, Acikel U, Kutsal T (1997) Application of multicomponent adsorption isotherms to simultaneous biosorption of iron(III) and chromium (VI) on *C. vulgaris*. J Chem Technol Biotechnol 70:368–378

Aksu Z, Açikel U, Kutsal T (1999a) Investigation of simultaneous biosorption of copper (II) and chromium (VI) on dried *Chlorella vulgaris* from binary metal mixtures: application of multicomponent adsorbtion isoterms. Sep Sci Technol 34:501–524

Aksu Z, Calik A, Dursun AY, Demircan Z (1999b) Biosorption of iron(III)-cyanide complex anions to *Rhizopus arrhizus*: application of adsorption isotherms. Process Biochem 34:483–491

Al-Asheh S, Banat F, Al-Omari R, Duvnjak Z (2000) Predictions of binary sorption isotherms for the sorption of heavy metals by pine bark using single isotherm data. Chemosphere 41:659–665

Alexander F, Poots VJP, McKay G (1978) Adsorption kinetics and diffusional mass transfer processes during color removal from effluent using silica. Ind Eng Chem Process Des Dev 17:406–410

Amini M, Younesi H, Bahramifar N (2009) Statistical modeling and optimization of the cadmium biososprtion process in aqueous solution using *Aspergillus niger*. Colloids Surf A Physicochem Eng Asp 337:67–73

Apiratikul R, Pavasant P (2006) Sorption isotherm model for binary component sorption of copper, cadmium and lead ions using dried green macroalga, *Caulerpa lentillifera*. Chem Eng J 119:135–145

Apiratikul R, Pavasant P (2008) Batch and column studies of biosorption of heavy metals by *Caulerpa lentillifera*. Biores Technol 99:2766–2777

Bailey SE, Olin TJ, Bricka RM, Adrian DD (1999) A review of potentially low-cost sorbents for heavy metals. Water Res 33:2469–2479

Basha S, Murthy ZVP, Jha B (2008) Isotherm modeling for biosorption of Cu(II) and Ni(II) from wastewater onto brown seaweed, *Cystoseira indica*. Environ Energy Eng 54:3291–3302

Basha S, Murthy ZVP, Jha B (2009) Removal of Cu(II) and Ni(II) from industrial effluents by brown seaweed, *Cystoseira indica*. Ind Eng Chem Res 48:961–975

Bayhan YK, Keskinler B, Cakici A, Levent M, Akay G (2001) Removal of divalent heavy metal mixtures from water by *Saccharomyces cerevisiae* using crossflow microfiltration. Water Res 35:2191–2200

Bayramoglu G, Arica MY (2008) Removal of heavy mercury(II), cadmium(II) and zinc(II) metal ions by live and heat inactivated *Lentinus edodes pellets*. Chem Eng J 143:133–140

Benedetti MF, Milne CJ, Kinniburgh DG, van Riemsdijl WH, Koopal LK (1995) Metal ion binding by humic substances: application of the non-ideal competitive adsorption model. Environ Sci Technol 29:446–457

Benedetti MF, van Riemsdijk WH, Koopal LK (1996) Humic substances considered as a heterogeneous Donnan gel phase. Environ Sci Technol 30:1805–1813

Beolchini F, Pagnanelli F, Toro L, Vegliò F (2005) Continuous biosorption of copper and lead in single and binary systems using *Sphaerotilus natans* cells confined by a UF/MF membrane reactor: experimental validation of dynamic models. Hydrometallurgy 76:73–85

Beolchini F, Pagnanelli F, Toro L, Vegliò F (2006) Ionic strength effect on copper biosorption by *Sphaerotilus natans*: equilibrium study and dynamic modelling in membrane reactor. Water Res 40:144–152

Biesuz R, Pesavento M, Gonzalo A, Valiente M (1998) Sorption of proton and heavy metal ions on a macroporous chelating resin with an iminodiacetate active group as a function of temperature. Talanta 47:127–136

Bohart GS, Adams EQ (1920) Some aspects of the behavior of charcoal with respect to chlorine. J Am Chem Soc 42:423–444

Carmona MER, Da Silva MAP, Ferreira Leite SG (2005) Biosorption of chromium using factorial experimental design. Process Biochem 40:779–788

Chang JS, Chen CC (1998) Quantitative analysis and equilibrium models of selective adsorption in multimetal systems using a bacterial biosorbent. Sep Sci Technol 33:611–632

Chang JS, Chen CC (1999) Biosorption of lead, copper and cadmium with continuous hollow-fiber microfiltration processes. Sep Sci Technol 34:1607–1627

Chen J, Yiacoumi S (1997) Biosorption of metal ions from aqueous solutions. Sep Sci Technol 32:51–69

Chen JP, Wang L (2004) Characterization of metal adsorption kinetic properties in batch and fixed bed reactors. Chemosphere 54:397–404

Chen JP, Yang L (2006) Study of heavy metal biosorption onto raw and chemically modified *Sargassum* sp. via spectroscopic and modelling analysis. Langmuir 22:8906–8914

Chen H, Tendeyong F, Yiacoumi S (1997) Equilibrium and kinetic studies of copper ion uptake by calcium alginate. Environ Sci Technol 31:1433–1439

Chen Z, Ma W, Han M (2008) Biosorption of nickel and copper onto treated alga (*Undaria pinnatifida*): application of isotherm and kinetic models. J Hazard Mater 155:327–333

Cheung CW, Chan CK, Porter JF, McKay G (2001) Combined diffusion model for the sorption of cadmium, copper and zinc ions onto bone char. Environ Sci Technol 35:1511–1522

Cheung CW, Ko DCK, Porter JP, McKay G (2003) Binary metal sorption on bone char mass transport model using IAST. Langmuir 19:4144–4153

Chojnacka K, Chojnacki A, Gorecka H (2005) Biosorption of Cr^{3+}, Cd^{2+} and Cu^{2+} ions by blue-green algae *Spirulina* sp.: kinetics, equilibrium and the mechanism of the process. Chemosphere 59:75–84

Chong KH, Volesky B (1995) Description of two-metal biosorption equilibria by Langmuir-type models. Biotechnol Bioeng 47:451–460

Chong KJ, Volesky B (1996) Metal biosorption equilibria in a ternary system. Biotechnol Bioeng 49:629–638

Chu KH, Hashim MA (2004) Quantitative analysis of copper biosorption by the microalga *Chlorella vulgaris*. Environ Eng Sci 21:139–147

Clark RM (1987) Evaluating the cost and the performance of field scale granular activated carbon systems. Environ Sci Technol 16:101–114

Cox JS, Smith DS, Warren LA, Ferris FG (1999) Characterizing heterogeneous bacterial surface functional groups using discrete affinity spectra for proton binding. Environ Sci Technol 33:4514–4521

Crist RH, Martin JR, Carr D, Watson JR, Clarke HJ, Crist DR (1994) Interaction of metals and protons with algae. Environ Sci Technol 28:1859–1866

Cysewski P, Jaulmes A, Lemque R, Sebille B, Vidal-Madjar C, Jilge G (1991) Multivalent ion-exchange model of biopolymer chromatography for mass overload conditions. J Chromatogr 548:61–79

Dang VBH, Doan HD, Dang-Vu T, Lohi A (2009) Equilibrium and kinetics of biosorption of cadmium(II) and copper(II) ions by wheat straw. Biores Technol 100:211–219

Daughney CJ, Fein JB (1998) The effect of ionic strength on the adsorption of H^+, Cd^{2+}, Pb^{2+} and Cu^{2+} by *Bacillus subtilis* and *Bacillus licheniformis*: a surface complexation model. J Colloid Interface Sci 198:53–77

Davis JA, Leckie JO (1980) Surface ionization and complexation at the oxide/water interface. 3. Adsorption of anions. J Colloid Interface Sci 74:32–43

Davis TA, Volesky B, Vieira RHSF (2000) Sargassum seaweed as biosorbent for heavy metals. Water Res 34:4270–4278

De Wit JCM, Van Riemsdijk WH, Koopal LK (1993) Proton binding to humic substances. 2. Chemical heterogeneity and adsorption models. Environ Sci Technol 27:2015–2022

Demirbas A (2008) Heavy metal biosorption onto agro-based waste material: a review. J Hazard Mater 157:220–229

Donnan FG, Allmand AJ (1914) Ionic equilibria across semi-permeable membranes. Proc Chem Soc 30:180–181

Dupont L, Bouanda J, Dumonceau J, Aplincourt M (2003) Metal ion binding onto a lignocellulosic substrate extracted from wheat bran: a NICA-Donnan approach. J Colloid Interface Sci 263:35–41

Esposito A, Pagnanelli F, Vegliò F (2002) pH related equilibria models for biosorption in single metal systems. Chem Eng Sci 57:307–313

Febrianto J, Kosasih AN, Sunarso J, Ju YH, Indraswati N, Ismadji S (2009) Equilibrium and kinetic studies in adsorption of heavy metals using biosorbent: a summary of recent studies. J Hazard Mater 162:616–645

Fein JB, Daughney CJ, Yee N, Davis TA (1997) A chemical equilibrium model for metal adsorption onto bacterial surfaces. Geochim Cosmochim Acta 61:3319–3328

Figueira MM, Volesky B, Ciminelli VST (1997) Assessment of interference in biosorption of a heavy metal. Biotechnol Bioeng 54:344–350

Figueira MM, Volesky B, Mathieu HJ (1999) Instrumental analysis study of iron species biosorption by *Sargassum* biomass. Environ Sci Technol 33:1840–1846

Figueira MM, Volesky B, Ciminelli VST, Roddick FA (2000) Biosorption of metals in Brown seaweed biomass. Water Res 34196–34204

Fiol N, Villaescusa I, Matinez M, Miralles N, Poch J, Serarols J (2006) Sorption of Pb(II), Ni(II), Cu(II) and Cd(II) from aqueous solutions by olive stone waste. Sep Purif Technol 50: 132–140

Fourest E, Volesky B (1996) Contribution of sulphonate groups and alginate to heavy metal biosorption by the dry biomass of *Sargassum fluidans*. Environ Sci Technol 30:277–282

Fowle DA, Fein JB (1999) Competitive adsorption of metal cations onto two gram positive bacteria: testing the chemical equilibrium model. Geochim Cosmochim Acta 63:3059–3067

Fowle DA, Fein JB (2000) Experimental measurements of the reversibility of metal-bacteria adsorption reactions. Chem Geol 168:27–36

Gadam SD, Gallant SR, Gramer SM (1995) Transient profiles in ion exchange displacement chromatography. AIChE J 41:1667–1686

Gadd GM (2008) Biosorption: critical review of scientific rational, environmental importance and significance for pollution treatment. J Chem Technol Biotechnol 84:13–28

Gavrilescu M (2004) Removal of heavy metal from the environment by biosorption. Eng Life Sci 4:219–232

Ghorbani F, Younesi H, Ghasempouri SM, Zinatizadeh AA, Amini M, Daneshi A (2008) Application of response surface methodology for optimization of cadmium biosorption in an aqueous solution by *Saccharomyces cerevisiae*. Chem Eng J 145:267–275

Gokhale SV, Jyoti KK, Lele SS (2009) Modeling of chromium(VI) biosorption by immobilized *Spirulina platensis* in packed column. J Hazard Mater 170:735–743

Güçlü K, Apak RJ (2003) Modeling the adsorption of free and heavy metal complex-bound EDTA onto red mud by a nonelectrostatic surface complexation model. J Colloid Interface Sci 260:280–290

Gupta VK (1998) Equilibrium uptake, sorption dynamics, process development and column operations for the removal of copper and nickel from aqueous solutions and wastewater using activated slag, a low cost adsorbent. Ind Eng Chem Res 37:192–202

Han X, Wong YS, Tam NFY (2006a) Surface complexation mechanism and modelling in Cr(III) biosorption by a microalgal isolate, *Chlorella miniata*. J. Colloid Interface Sci 303:365–371

Han R, Zhang J, Zou W, Xiao H, Shi J, Liu H (2006b) Biosorption of copper(II) and lead(II) from aqueous solution by chaff in a fixed bed column. J Hazard Mater B133:262–268

Hasan SH, Srivastava P, Ranjan D, Talat M (2009) Biosorption of Cr(VI) from aqueous solution using *A. hydrophila* in up-flow column: optimization of process variables. Appl Microbiol Biotechnol 83:567–577

Hatzikioseyian A, Tsezos M and Mavituna F (2001) Application of simplified rapid equilibrium models in simulating experimental breackthrough curves from fixed bed biosorption reactors. Hydrometallurgy 59:395–406

Herrero R, Lodeiro P, Rey-Castro C, Vilarino T, de Vicente MES (2005) Removal of inorganic mercury from aqueous solutions by biomass of the marine macroalga *Cystoseira baccata*. Water Res 39:3199–3210

Herrero R, Cordero B, Lodeiro P, Rey-Castro C, de Vicente MES (2006) Interactions of cadmium(II) and protons with dead biomass of marine algae *Fucus* sp. Marine Chem 99:106–116

Hetzer A, Daughney CJ, Morgan HW (2006) Cadmium ion biosorption by the thermophilic bacteria *Geobacillus stearothermophilus* and *G. thermocatenulatus*. Appl Environ Microbiol 72:4020–4027

Hiemstra T, Van Riemsdijk WH (1999) Surface structural ion adsorption modeling of competitive binding of oxyanions by metal (hydr)oxides. J Colloid Interface Sci 210:182–193

Himmelblau DM (1978) Process analysis by statistical methods. Wiley, New York, pp 176–207

Ho Y (2006a) Isotherms for the adsorption of lead onto peat: comparison of linear and non linear methods. Polish J Environ Stud 15:81–86

Ho Y (2006b) Second-order kinetic model for the sorption of cadmium onto three fern: a comparison of linear and non linear methods. Water Res 40:119–125

Ho YS, McKay G (2000) Correlative biosorption equilibria model for a binary batch system. Chem Eng Sci 55:817–825

Hui CW, Chen B, McKay G (2003) Pore-surface diffusion model for batch adsorption processes. Langmuir 19:4188–4196

Jain JS, Snoeyink VL (1973) Adsorption from bisolute systems on active carbon. J Water Pollut Control Fed 45:2463–2479

Juang RS, Shao HJ (2002) A simplified equilibrium model for sorption of heavy metal ions from aqueous solution on chitosan. Water Res 36:2999–3008

Kapoor A, Viraraghavan T (1998) Removal of heavy metals from aqueous solutions using immobilized fungal biomass in continuous mode. Water Res 32:1968–1977

Kim YH, Park JY, Yoo YJ (1998) Modeling of biosorption by marine brown *Undaria pinnatifida* based on surface complexation mechanism. Korean J Chem Eng 15:157–163

Kinniburgh DG, Milne CJ, Benedetti MF, Pinheiro JP, Filius J, Koopal LK, Van Riemsdijk WH (1996) Metal ion binding by humic acid: application of the NICA-Donnan model. Environ Sci Technol 30:1687–1698

Kinniburgh DG, Van Riemsdijk WH, Koopal LK, Borkovec M, Benedetti MF, Avena MJ (1999) Ion binding to natural organic matter: competition, heterogeneity, stoichiometry and thermodynamic consistency. Colloid Surf 151:147–166

Ko DCK, Cheung CW, Choy KKH, Porter JF, McKay G (2004) Soprtion equilibria of metal ions on bome char. Chemosphere 54:273–281

Koltuniewicz A, Bezak K (2002) Engineering of membrane biosorption. Desalination 144:219–226

Koopal LK, Van Riemsdijk WH, De Wit JCM, Benedetti MF (1994) Analytical isotherm equations for multicomponent adsorption to heterogeneous surfaces. J Colloid Interface Sci 166:51–60

Koopal LK, Van Riemsdijk WH, Kinniburgh DG (2001) Humic matter and contaminants. General aspects and modeling metal ion binding. Pure App Chem 73:2005–2016

Kratochvil D, Volesky B (1998) Advances in the biosorption of heavy metals. Trends Biotechnol 16:291–300

Kratochvil D, Volesky B, Demopoulos G (1997) Optimizing Cu removal/recovery in a biosorption column. Water Res 31:2327–2339

Langmuir I (1918) The adsorption of gases on plane surfaces of glass, mica and platinum. J Am Chem Soc 40:1361–1402

Lodeiro P, Fuentes A, Herrero R, de Vicente MES (2008) Cr(III) binding by surface polymers in natural biomass: the role of carboxylic groups. Environ Chem 5:355–365

Maiti SK, Bera D, Chattopadhyay P, Ray L (2009) Determination of kinetic prameters in the bosorption of Cr (VI) on imobilized *Bacillus cereus* M-16(1) in a continuous packed bed column reactor. Appl Biochem Biotechnol 159:488–504

Malkoc E, Nuhoglu Y (2006) Fixed bed studies for the sorption of chromium(VI) onto tea factory waste. Chem Eng Sci 61:4363–4372

Marinsky JA, Ephraim J (1986) A unified physicochemical description of the protonation and metal ion complexation equilibria of natural organic acids (humic and fulvic acids). 1. Analysis of the influence of polyelectrolyte properties on protonation equilibria in ionic media: fundamental concepts. Environ Sci Technol 20:349–354

Martin-Lara MA, Hernainz F, Calero M, Blazquez G, Tenorio G (2009) Surface chemistry evaluation of some solid wastes from olive-oil industry used for lead removal from aqueous solutions. Biochem Eng J 44:151–159

Matos MDN, Diniz VG, de Abreu CAM, Knoechelmann A, da Silva VL (2009) Bioadsorption and ion exchange of Cr^{3+} and Pb^{2+} solutions with algae. Adsorption 170:735–743

McKay G, Duri BA (1991) Multicomponent dye adsorption onto carbon using a solid diffusion mass-transfer model. Ind Eng Chem Res 30:385–395

McKay G, Allen SJ, McConvey IF, Walters JHR (1984) External mass-transfer and homogeneous solid-phase diffusion effects during the adsorption of dyestuff. Ind Eng Chem Process Des Dev 23:221–226

Milne CJ, Kinniburgh DG, De Wit JCM, Van Riemsdijk WH, Koopal LK (1995a). Analysis of proton binding by a peat humic acid using a simple electrostatic model. Geochim Cosmochim Acta 59:1101–1112

Milne CJ, Kinniburgh DG, De Wit JCM, Van Riemsdijk WH, Koopal LK (1995b) Analysis of metal-ion binding by a peat humic acid using a simple electrostatic model. J Colloid Interface Sci 175:448–460

Mukhopadhyay M (2008) Role of surface properties during biosorption of copper by pretreated *Aspergillus niger* biomass. Colloid Surf A—Physicochem Eng Asp 329:95–99

Naddafi K, Nabizadeh R, Saeedi R, Mahvi AH, Vaezi F, Yaghmaeian K, Ghasri A, Nazmara S (2007) Biosorption of lead(II) and cadmium(II) by protonated *Sargassum glaucescens* biomass in a continuous packed bed column. J Hazard Mater 147:785–791

Naja G, Voleski B (2006) Behaviour of the mass transfer zone in a biosorption column. Environ Sci Technol 40:3996–4003

Namasivayam C, Sureshkumar MV (2008) Removal of chromium(VI) from water and wastewater using surfactant modified coconut coir pit as a biosorbent. Biores Technol 99:2218–2225

Nederlof MN, Van Riemsdijk WH, Koopal LK (1990) Determination of adsorption affinity distributions: a general framework for methods related to local isotherm approximations. J Colloid Interface Sci 135:410–426

Nederlof MN, de Wit JC, van Riemsdijk WH, Koopal LK (1993) Determination of proton affinity distributions for humic substances. Environ Sci Technol 27:846–856

Pagnanelli F, Petrangeli Papini M, Trifoni M, Toro L, Vegliò F (2000) Biosorption of metals ions on *Arthrobacter* sp.: biomass characterization and biosorption modeling. Environ Sci Technol 34:2773–2778

Pagnanelli F, Trifoni M, Beolchini F, Toro L, Vegliò F (2001) Equilibrium biosorption studies in single and multi-metal systems. Process Biochem 37:115–124

Pagnanelli F, Esposito A, Vegliò F (2002) Multi-metallic modeling for biosorption of binary systems. Water Res 36:4095–4105

Pagnanelli F, Mainelli S, Vegliò F, Toro L (2003a) Heavy metal removal by olive pomace: adsorbent characterisation and equilibrium modelling. Chem Eng Sci 58:4709–4717

Pagnanelli F, Beolchini F, Esposito A, Toro L, Vegliò F (2003b) Mechanistic modeling of heavy metal biosorption in batch and membrane reactor systems. Hydrometallurgy 71:201–208

Pagnanelli F, Di Biase A, Beolchini F, Vegliò F (2003c) Effect of equilibrium models on biosorption of heavy metals in single and two stage UF/MF membrane reactor systems. Biochem Eng J 15:27–35

Pagnanelli F, Vegliò F, Toro L (2004) Modeling of the acid-base properties of natural and synthetic adsorbent materials used for heavy metal removal from aqueous solutions. Chemosphere 54:905–915

Pagnanelli F, Mainelli S, De Angelis S, Toro L (2005) Biosorption of protons and heavy metals onto olive pomace: modelling of competition effects. Water Res 39:1639–1651

Pagnanelli F, Mainelli S, Toro L (2008) New biosorbent materials for heavy metal removal: product development guided by active site characterisation Water Res 42:2953–2962

Pakshirajan K, Swaminathan T (2009) Biosorption of copper and cadmium in packed bed columns with live immobilized fungal biomass of *Phanerochaete chrysosporium*. Appl Biochem Biotechnol 157:159–173

Papageorgiou SK, Katsaros FK, Kouvelos EP, Nolan JW, Deit HL, Kanellopoulos NK (2006) Heavy metal sorption by calcium alginate beads from *Laminata digitata*. J Hazard Mater B137:1765–1772

Papageorgiou SK, Katsaros FK, Kouvelos EP, Kanellopoulos NK (2009) Prediction of binary adsorption isotherms of Cu^{2+}, Cd^{2+} and Pb^{2+} on calcium alginate beads from single adsorption data. J Hazard Mater 162:1347–1354

Perez-Marin AB, Ballester A, Gonzales F, Blazquez ML, Munoz JA, Saez J, Meseguer Zapata V (2008) Study of cadmium, zinc and lead biosorption by orange wastes using the subsequent addition method. Biores Technol 99:8101–8106

Pesavento M, Biesuz R, Cortina JL (1994) Sorption of metal ions on a weak acid cation-exchange resin containing carboxylic groups. Anal Chim Acta 298:225–232

Plazinski W, Rudzinski W (2009) Modeling the effect of surface heterogeneity in equilibrium of heavy metal ion biosorption by using the ion exchange model. Environ Sci Technol 43:7465–7471

Plette ACC, van Riemsdijk WH, Benedetti MF, van der Wal A (1995) pH dependent charging behavior of isolated cell walls of a Gram-positive soil bacterium. J Colloid Interface Sci 173:354–363

Plette ACC, Benedetti MF, van Riemsdijk WH (1996) Competitive binding of protons, calcium, cadmium and zinc to isolated cell walls of a Gram-positive soil bacterium. Environ Sci Technol 30:1902–1910

Preetha B, Viruthagiri T (2007) Application of response surface methodology for the biosorption of copper using *Rhizopus arrhizus*. J Hazard Mater 143:506–510

Puranik PR, Modak JM, Paknikar KM (1999) A comparative study of the mass transfer kinetics of metal biosorption by microbial biomass. Hydrometallurgy 52:189–197

Quintelas C, Rocha Z, Silva B, Fonseca B, Figueiredo H, Tavares T (2009) Biosorptive performance of an *Escherichia coli* biofilm supported on zeolite NaY for the removal of Cr(VI), Cd(II), Fe(III) and Ni(II). Chem Eng Sci 152:110–115

Radke CJ, Prausnitz JM (1972) Thermodynamics of multi-solute adsorption for dilute liquid solutions. AIChe J 18:761–768

Ravat C, Dumonceau J, Monteil-Rivera F (2000) Acid/base and Cu(II) binding properties of natural organic matter extracted from wheat bran: modelling by surface complexationn model. Water Res 34:1327–1339

Reverberi AP, Dovì VG, Fabiano B, Maga L (2009) Using non equilibrium biosorption column modelling for improving process efficiency. J Clean Prod 17:963–968

Rey-Castro C, Lodeiro P, Herrero R, de Vincente MES (2003) Acid-base properties of brown seaweed biomass considered as Donnan Gel. A model reflecting electrostatic effects and chemical heterogeneity. Environ Sci Technol 37:5159–5167

Richenberg D (1953) Properties of ion exchange resins in relation to their structure. III. kinetic of exchange. J Am Chem Soc 75:589–597

Sag Y, Kutsal T (1996) Fully competitive biosorption of chromium(VI) and iron(III) ions from binary metal mixture by *R. arrhizus*: use of the competitive Langmuir model. Process Biochem 31:573–585

Sag Y, Kaya A, Kutsal T (1998) The simultaneous biosorption of Cu (II) and Zn on *Rhizopus arrhizus*: application of the sorption models. Hydrometallurgy 50:297–314

Sag Y, Akcael B, Kautsal T (2001) Evaluation, interpretation and representation of three-metal biosorption equilibria using a fungal biosorbent. Process Biochem 37:35–50

Salamatinia B, Kamaruddin AK, Abdullah AZ (2008) Modeling of the continuous copper and zinc removal by sorption onto sodium hydroxide-modified oil palm frond in a fixed-bed column. Chem Eng J 145:259–266

Sanchez A, Ballester A, Blazquez ML, Gonalez F, Munoz J, Hammaini A (1999) Biosorption of copper and zinc by *Cymodocea nodosa*. FEMS Microbiol Rev 23:527–536

Sari A, Mendil D, Tuzen M, Soylak M (2008) Biosorption of Cd(II) and Cr(III) from aqueous solution by moss (*Hylocomium splendens*) biomass: equilibrium, kinetic and thermodynamic studies. Chem Eng Sci 144:1–9

Schiewer S (1999) Modelling complexation and electrostatic attraction in heavy metal biosorption by *Sargassum* biomass. J Appl Phycol 11:79–87

Schiewer S, Balaria A (2009) Biosorption of Pb2+ by original and protonated citrus peels: equilibrium, kinetics, and mechanism. Chem Eng J 146:211–219

Schiewer S, Volesky B (1995) Modeling of the proton-metal ion exchange in biosorption. Environ Sci Technol 29:3049–3058

Schiewer S, Volesky B (1996) Modeling multi metal ion exchange in biosorption. Environ Sci Technol 30:2921–2927

Schiewer S, Volesky B (1997) Ionic strength and electrostatic effects in biosorption of divalent metal ions and protons. Environ Sci Technol 31:2478–2485

Schiewer S, Wong MH (2000) Ionic strength effect in biosorption of metals by marine algae. Chemosphere 41:271–282

Schiewer S, Fourest E, Chong KH, Volesky B (1995) Ion exchange in biosorption by dried seaweed: experiments and model predictions. In: Jerez CA, Vargas T, Toledo H, Wiertz JV (eds) Biohydrometallurgical processing. University of Chile, Santiago, pp 219–228

Seki H, Suzuki A (1998) Biosorption of heavy metal ions to brown algae, *Macrocystis pyrifera*, *Kjellmaniella crassiforia*, and *Undaria pinnatifida*. J Colloid Interface Sci 206:297–301

Seki H, Suzuki A (2002) Adsorption of heavy metal ions to floc-type biosorbents. J Colloid Interface Sci 249:295–300

Seki H, Suzuki A, Mitsueda SI (1998) Biosorption of heavy metal ions on *Rhodobacter sphaeroides* and *Alcaligenes eutrophus* H16. J Colloid Interface Sci 197:185–190

Sengil IA, Ozacar M (2008) Biosorption of Cu(II) from aqueous solutions by mimosa tannin gel. J Hazard Mater 157:277–285

Senthilkumar R, Vijayaraghavan K, Thilakavathi M, Iyer PVR, Velan M (2006) Seaweeds for the remediation of wastewaters contaminated with Zn(II) ions. J Hazard Mater 136:791–799

Singh KK, Rastogi R, Hasan SH (2005) Removal of Cr(VI) from waste waters using rice bran. J Colloid Interface Sci 290:61–68

Singh KK, Singh AK, Hasan SH (2006) Low cost biosorbent wheat bran for the removal of cadmium from wastewater: kinetic and equilibrium study. Biores Technol 97:994–1001

Sips R (1948) Structure of a catalyst surface. J Chem Phys 16:490–495

Stumm W (1987) Aquatic surface chemistry. Wiley, New York

Su H, Chen S, Tan T (2007) Surface active site model for Ni^{2+} adsorption of the surface imprinted adsorbent. Process Biochem 42:612–619

Tan HKS, Spinner IH (1994) Multicomponent ion exchange column dynamics. Can J Chem Eng 72:330–341

Texier AC, Andrès Y, Le Cloirec P (1999) Selective Biosorption of Lanthanide (La, Eu, Yb) ions by *Pseudomonas aeruginosa*. Environ Sci Technol 33:489–495

Texier AC, Andres Y, Faur-Brasquet C, Le Cloirec P (2002) Fixed-bed study for lanthanide (La, Eu, Yb) ions removal from aqueous solutions by immobilized *Pseudomonas aeruginosa*: experimental data and modelization. Chemosphere 47:333–342

Trujillo EM, Jeffers TH, Ferguson C, Stevenson HQ (1991) Mathematically modeling the removal of heavy metals from wastewater using immobilized biomass. Environ Sci Technol 25:1559–1565

Vaghetti JCP, Lima EC, Royer B, da Cunha BM, Cardoso NF, Brasil JL, Dias SLP (2009) Pecan nutshell as biosorbent to remove Cu(II), Mn(II) and Pb(II) from aqueous solutions. J Hazard Mater 162:270–280

Van Riemsdijk WH, De Wit JCM, Mous SLJ, Koopal LK, Kinniburgh DG (1996) An analytical isotherm equation (CONICA) for non ideal mono- and bidentate competitive ion adsorption to heterogeneous surfaces. J Colloid Interface Sci 183:35–50

Vegliò F, Beolchini F, Toro L (1998) Kinetic modeling of copper biosorption by immobilized biomass. Ind Eng Chem Res 37:1107–1111

Vijayaraghavan K, Jegan J, Palanivelu K, Velan M (2005) Biosorption of cobalt(II) and nickel(II) by seaweed: batch and column studies. Sep Purif Technol 44:53–59

Vilar VJP, Botelho CMS, Boaventura RAR (2007) Modeling equilibrium and kinetics of metal uptake by algal biomass in continuous stirred and packed bed adsorbers. Adsorption 13:587–601

Vilar VJP, Botelho CMS, Boaventura RAR (2008a) Kinetics modeling of biosorption by algal biomass from binary metal solutions using batch contactors. Biochem Eng J 38:319–325

Vilar VJP, Botelho CMS, Loureiro JM, Boaventura RAR (2008b) Biosorption of copper by marine algae *Gelidium* and algal composite material in a packed bed column. Biores Technol 99:5830–5838

Vilar VJP, Botelho CMS, Pinheiro JPS, Domingos RF, Boaventura RAR (2009) Copper removal by algal biomass: biosorbent characterization and equilibrium modeling. J Hazard Mater 163:1113–1122

Volesky B (2003) Biosorption process simulation tools. Hydrometallurgy 71:179–190

Volesky B (2007) Biosorption and me. Water Res 41:4017–4029

Wang J, Chen C (2009) Biosorbents for heavy metals and their future. Biotechnol Adv 27:195–226

Warren LA, Ferris FG (1998) Continuum between sorption and precipitation of Fe(III) on microbial surfaces. Environ Sci Technol 32:2331–2337

Webster EA, Murphy AJ, Chudek JA, Gadd GM (1997) Metabolism-independent binding of toxic metals by *Ulva lactuca*: cadmium binds to oxygen-containing groups, as determined by NMR. Biometals 10:105–117

Westall J, Hohl H (1980) A comparison of electrostatic models for the oxide/solution interface. Adv Colloid Interface 12:265–294

Williams CJ, Aderhold D, Edyvean RGJ (1998) Comparison between biosorbents for the removal of metal ions from aqueous solutions. Water Res 32:216–224

Wolborska A (1989) Adsorption on activated carbon of p-nitrophenol from aqueous solution. Water Res 23:85–91

Yang J, Volesky B (1996) Intraparticle diffusivity of Cd ions in a new biosorbent material. J Chem Tech Biotechnol 66:355–364

Yang J, Volesky B (1999a) Modeling uranium-proton ion exchange in biosorption. Environ Sci Technol 33:4079–4085

Yang J, Volesky B (1999b) Cadmium biosorption rate in protonated *Sargassum* biomass. Environ Sci Technol 33:751–757

Yoon YH, Nelson JH (1984) Application of gas adsorption kinetics. I. A theoretical model for respirator cartridge service time. Am Ind Hyg Assoc J 45:509–516

Yun YS, Volesky B (2003) Modeling of lithium interference in cadmium biosorption. Environ Sci Technol 37:3601–3608

Yun YS, Park D, Park JM, Volesky B (2001) Biosorption of trivalent chromium on the brown seaweed biomass. Environ Sci Technol 35:4353–4358

Zulfadhly Z, Mashitah MD, Bhatia S (2001) Heavy metal removal in fixed-bed column by the macro fungus *Pycnoporus sanguineus*. Environ Pollut 112:463–470

Zulkali MMD, Ahmad AL, Norulakmal NH (2006) *Oryza sativa* L. husk as heavy metal adsorbent: optimization with lead as model solution. Biores Technol 97:21–25

Chapter 5
Bacterial Biosorption and Biosorbents

Yeong-Sang Yun, Kuppusamy Vijayaraghavan and Sung Wook Won

Abstract Water pollution due to presence of metals has become one of the most serious environmental problems today. Biosorption, using inactive/dead biomaterials such as bacteria, fungi, algae and industrial/agricultural wastes, is regarded as cost-effective technology for the treatment of metal-bearing wastewaters. In recent years, several biosorbents have been investigated, but the bacterial biomass has since proven to be the most effective and promising biosorbent for wide variety of metals. The state of the art in the field of biosorption of metals by bacterial biomass is discussed in this article. It is their basic cell wall constituents that are responsible for this high metal uptake. The properties of the cell wall constituents, such as peptidoglycan, and the role of functional groups, such as carboxyl, amine and phosphonate, are discussed on the basis of their biosorption potential. A systematic comparison of the literature, based on the metal binding capacity of bacterial biomass under different conditions, is also provided. To enhance biosorption capacity, biomass can be chemically modified, via various techniques, or genetically engineered. The problems associated with microbial biosorption are analysed, with suitable remedies discussed. Thus, this chapter presents the achievements and current status of bacterial biosorption technology, and hopes to provide insights into this research frontier.

Keywords Biosorption • Wastewater treatment • Bacteria • Metals • Immobilization

5.1 Introduction

The early 1980s witnessed the capability of some microorganisms for the accumulation of metallic elements. Numerous research reports have been published from toxicological points of view, but these have been concerned with the ac-

---

Y.-S. Yun (✉)
Division of Environmental and Chemical Engineering, Research Institute of Industrial Technology, Chonbuk National University, Jeonjubuk 561-756 Jeonju, South Korea
e-mail: ysyun@chonbuk.ac.kr

P. Kotrba et al. (eds.), *Microbial Biosorption of Metals*,
DOI 10.1007/978-94-007-0443-5_5, © Springer Science+Business Media B.V. 2011

cumulation due to the active metabolism of living cells, the effects of metal on the metabolic activities of the microbial cell and the consequences of accumulation on the food chain (Volesky 1987). However, further research has revealed that inactive/dead microbial biomass can passively bind metal ions via various physicochemical mechanisms. The biomass can chemically interact and sequester metallic ions from solutions. With this new finding, research on biosorption has become active, with numerous biosorbents of different origins having been proposed for the removal of metals. Researchers have understood and explained that biosorption depends not only on the type or chemical composition of the biomass, but also on the external physicochemical factors and solution chemistry. Many investigators have been able to explain the mechanisms responsible for biosorption, which may be one or combination of ion exchange, complexation, coordination, adsorption, chelation and microprecipitation, etc. (Volesky and Schiewer 1999; Vegliò and Beolchini 1997).

5.2 Bacterial Structure and Components Responsible for Biosorption Ability

Bacteria represent a major group of unicellular living organisms belonging to the prokaryotes, which are ubiquitous in soil and water, and as symbionts of other organisms. Bacteria can be found in a wide variety of shapes, which include cocci (such as *Streptococcus*), rods (such as *Bacillus*), spiral (such as *Vibrio cholerae*) and filamentous (such as *Actinomyces*). All bacteria have a relatively simple cell structure, which lack cell nuclei, but possess cell walls. The bacterial cell wall provides structural integrity to the cell, but differs from that of all other organisms due to the presence of peptidoglycan (poly-*N*-acetylglucosamine and *N*-acetylmuramic acid), which is located immediately outside of the cytoplasmatic membrane. Peptidoglycan is responsible for the rigidity of the bacterial cell wall, and determines the cell shape. It is also relatively porous and considered as an impermeable barrier to small substrates.

The cell walls of all bacteria are not identical. In fact, the cell wall composition is one of the most important factors in the analysis and differentiation of bacterial species. Accordingly, two general types of bacteria exist, of which Gram-positive bacteria are comprised of a thick peptidoglycan layer connected by amino acid bridges. Imbedded in the Gram-positive cell wall are polyalcohols, known as teichoic acids, some of which are lipid-linked to form lipoteichoic acids. Because lipoteichoic acids are covalently linked to lipids within the cytoplasmic membrane, they are responsible for linking peptidoglycan to the cytoplasmic membrane. The cross-linked peptidoglycan molecules form a network, which covers the cell like a grid. Teichoic acids give the Gram-positive cell wall an overall negative charge, due to the presence of phosphodiester bonds between the teichoic acid monomers. In general, 90% of the Gram-positive cell wall is comprised of peptidoglycan.

On the contrary, the cell wall of Gram-negative bacteria is much thinner, and composed of only 20% peptidoglycan. In addition, the cell wall contains an additional outer membrane composed of phospholipids and lipopolysaccharides. The highly charged nature of lipopolysaccharides confers an overall negative charge on the Gram-negative cell wall. Except in unusual cases, teichioic acid, teichuronic acid and lipoteichoic acid are not found in Gram-negative cell walls (Yee and Fein 2001).

Sherbert (1978) showed that the anionic functional groups present in the peptidoglycan, teichoic acids and teichuronic acids of Gram-positive bacteria, and the peptidoglycan, phospholipids, and lipopolysaccharides of Gram-negative bacteria were the components primarily responsible for the anionic character and metal binding capability of the cell wall. Extracellular polysaccharides are also capable of binding metals (McLean et al. 1992). However, their availability depends on the bacterial species and growth conditions; they can be easily removed by simple mechanical disruption or chemical washing (Yee and Fein 2001).

5.2.1 Mechanism of Bacterial Biosorption

The bacterial cell wall is the first component that comes into contact with metal ions, where the solutes can be deposited on the surface or within the cell wall structure. Since the mode of solute uptake by dead/inactive cells is extracellular, the chemical functional groups of the cell wall play vital roles in biosorption. Due to the nature of the cellular components, several functional groups are present on the bacterial cell wall, including carboxyl, phosphonate, amine and hydroxyl groups.

Since negatively charged and abundantly available, carboxyl groups actively participate in binding of metal cations. Golab and Breitenbach (1995) indicated that the carboxyl groups of the cell wall peptidoglycan of *Streptomyces pilosus* were responsible for the binding of copper. Yee and Fein (2001) confirmed that carboxyl groups were responsible for the binding of cadmium via Cd-carboxyl complexation on the bacterial surface. Conversely, amino groups are very effective at removing metal ions, as it not only chelates cationic metal ions, but also adsorbs anionic metal species via electrostatic interactions or hydrogen bonds Kang et al. (2007) observed that amino groups protonated at pH 3 attracted negatively charged chromate ions via electrostatic interaction. In general, increasing pH increases the overall negative charge on the surface of cells until all the relevant functional groups are deprotonated, which favours the electrochemical attraction and adsorption of cations. Anions would be expected to interact more strongly with cells with increasing concentration of positive charges, due to the protonation of functional groups at lower pH values.

The solution chemistry affects not only the bacterial surface chemistry, but the metal speciation as well. Metal ions in solution undergo hydrolysis with increasing pH, the extent of which differs at different pH values and with each metal, but the usual sequence of hydrolysis is the formation of hydroxylated monomeric species, followed by the formation of polymeric species, and then the formation of crystalline oxide precipitates after aging (Baes and Mesmer 1976). For example, in the case of nickel solution, López et al. (2000) indicated that within the pH range 1–7, nickel existed in solution as Ni^{2+} ions (90%); whereas at pH 9, Ni^{2+} (68%), $Ni_4OH_4^{4+}$ (10%) and $Ni(OH)^+$ (8.6%) co-existed. The different chemical species of a metal occurring with pH changes will have variable charges and ability to absorb at solid-liquid interfaces. In many instances, biosorption experiments conducted at high alkaline pH values have been reported to complicate evaluation of the biosorbent potential as a result of metal precipitation (Iqbal and Saeed 2007; Selatnia et al. 2004b).

5.3 Characterization of Bacterial Surface

Characterization of bacterial biomass and the biosorption mechanisms can be elucidated using different methods, including potentiometric titrations (Won et al. 2005; Yee and Fein 2001; Texier et al. 2000), Fourier transform infrared spectroscopy (Çabuk et al. 2006; Vannela and Verma 2006), X-ray diffraction (Kazy et al. 2006; López et al. 2000), X-ray photoelectron spectroscopy (Won et al. 2010; Deng and Ting 2005a), Scanning electron microscopy (Vijayaraghavan et al. 2007; Tunali et al. 2006; Lu et al. 2006), Atomic force microscopy (Yin et al. 2008; Pan et al. 2006), Transmission electron microscopy (Kazy et al. 2006; Vannela and Verma 2006; Mullen et al. 1989) and Energy dispersive X-ray microanalysis (Tunali et al. 2006; Kazy et al. 2006).

Potentiometric titrations have aided several researchers in the determination of the nature and number of binding sites. Yee and Fein (2001) titrated two Gram-negative and seven Gram-positive bacteria, and determined the pK_a values and number of available binding sites. Davis et al. (2000) successfully correlated the amount of acidic groups, determined via potentiometric titrations, with the metal uptake capacity. The nature of the binding sites and their involvement during biosorption can be approximately evaluated using FT-IR. Vannela and Verma (2006) analyzed the FT-IR spectra of virgin and Cu^{2+} exposed *Spirulina platensis*. Several band transformations allowed the authors to predict the possible involvement of amide, amino and carboxyl groups in the biosorption of Cu^{2+}. Won et al. (2005) used FT-IR spectra to confirm the presence of carboxyl, amine and phosphonate groups in *C. glutamicum* biomass. XPS are widely used to distinguish different forms of the same element and identify the existence of a particular element in a material. Won et al. (2010) developed PEI-modified *Escherichia coli* biomass through cross-linking polyethyleneimine (PEI) onto the surface of biomass and analyzed the PEI-modified *E. coli* biomass before and after Pt biosorption using XPS (Fig. 5.1), and they reported that the Pt

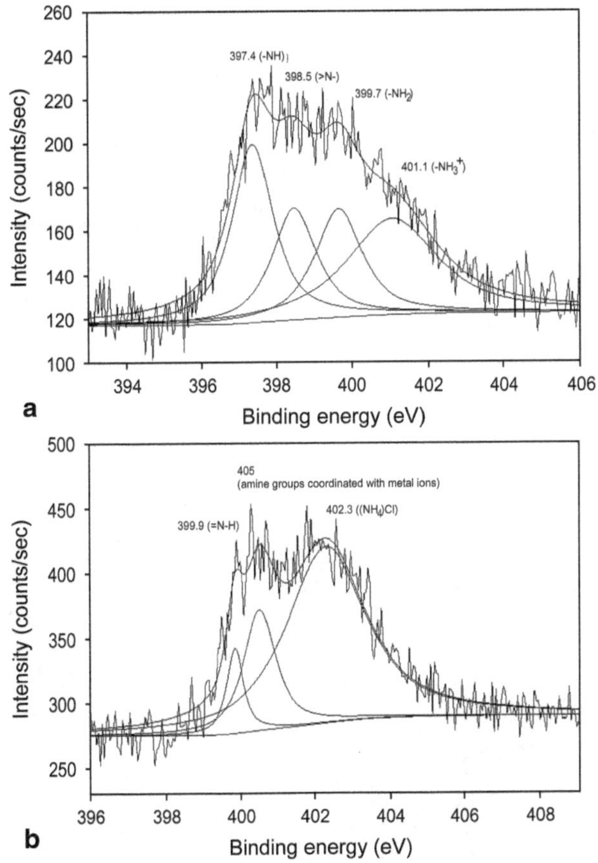

Fig. 5.1 XPS N 1s spectra for the PEI-modified *E. coli* biomass before (**a**) and after (**b**) biosorption. (According to Won et al. 2010)

biosorption on the PEI-modified biomass can be explained by the electrostatic attraction of anionic Pt-chloride complexes to protonated amino groups on the biomass surface.

EDX can provide information regarding the chemical and elemental characteristics of a biomass. Tunali et al. (2006) analyzed both Pb^{2+} and Cu^{2+} loaded *Bacillus* sp. using EDX, and confirmed the involvement of an ion exchange mechanism during their biosorption. In order to elucidate the chemical nature of bacterial cell-bound lanthanum, Kazy et al. (2006) employed XRD analysis, and confirmed the involvement of cellular carboxyl and phosphate groups in the binding of lanthanum by *Pseudomonas* biomass. To analyze the morphology of the cell surface before and after biosorption, SEM micrographs are often used. With the aid of SEM photographs, Lu et al. (2006) visualized the surface of metal-loaded *Enterobacter* sp., which appeared to be vague and damaged by the heavy-metal ions. Vijayaraghavan et al. (2007) used SEM photographs to show the pattern of *C. glutamicum* immobilization within a polysulfone matrix. AFM is an ideal tool for determining changes in surface morphology. For example, AFM was used to investigate the cell surface morphology of raw and polyallyamine hydrochloride (PAH)-modified *E. coli* biomass. The surface morphology of biomass was prone to change when it was modified by cross-linking reaction with polymer. As shown in Fig. 5.2, the surface of the PAH-modified biomass was clearly differentiated from that of the raw biomass. This superficial change was attributed to the addition of polymer to the biomass surface. Pan et al. (2006) showed AFM images of *Bacillus cereus* cells under the exposure to different amounts of Pb^{2+} ions solutions, and the biomass shape has been changed from a rod-like structure to a spindle-like structure after Pd^{2+} biosorption. They mentioned that these morphological changes of the sample can be attributed to the interactions between heavy metal and the surface of *B. cereus* biomass. Methods for analyzing the biomass surface and possible biosorption mechanism; therefore, are well established.

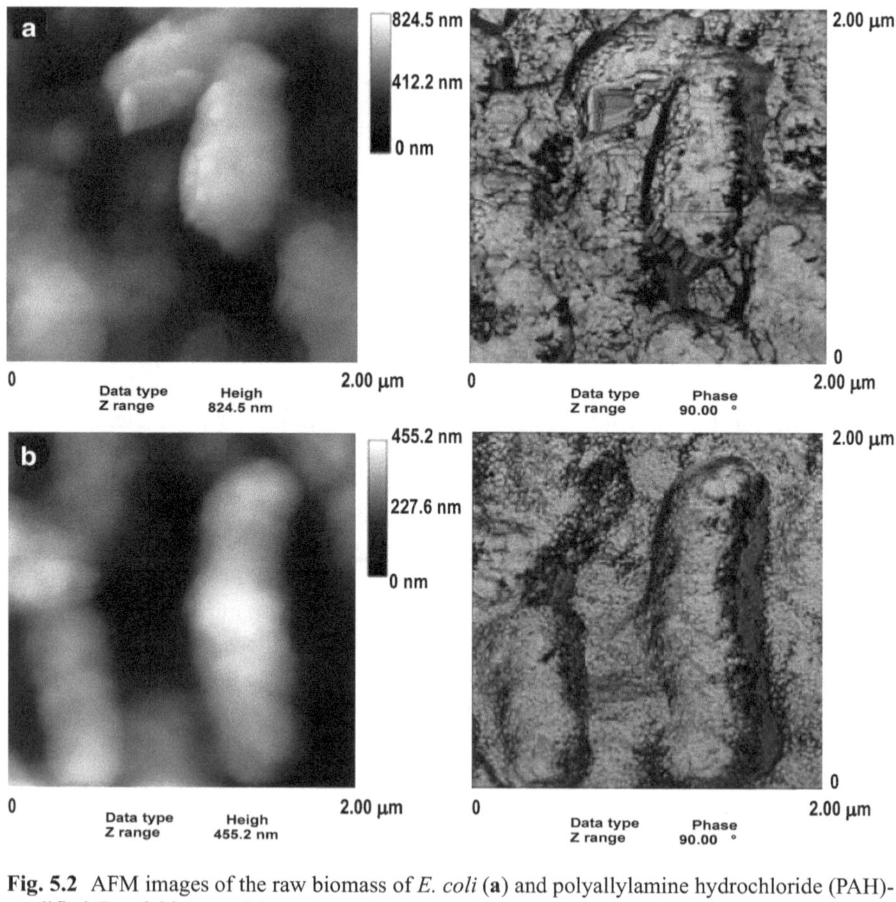

Fig. 5.2 AFM images of the raw biomass of *E. coli* (**a**) and polyallylamine hydrochloride (PAH)-modified *E. coli* biomass (**b**)

5.4 Preparation of Bacterial Biosorbents

In recent years, interest has focused on increasing the sorption capacity of the biomass. Several types of biomass, regarded as industrial wastes following certain processes, possess low biosorption capacities. As sorption mainly takes place on the biomass surface, increasing/activating the binding sites on the surface would be an effective approach for enhancing the biosorption capacity.

5.4.1 Chemically Modified Biosorbents

Chemical modification procedures include pre-treatment, binding site enhancement, binding site modification and polymerization. Common chemical pre-treatment's include acid, alkaline, ethanol and acetone treatments of the biomass (Vijayaraghavan and Yun 2007b; Göksungur et al. 2005; Selatnia et al. 2004a; Bai and Abraham 2002). The success of a chemical pre-treatment strongly depends on the cellular components of the biomass itself. In many instances, acidic pre-treatment has proved to be successful; this is because some of the impurities and ions blocking the binding sites can be easily eliminated. However, utmost care and careful screening methods must be employed for selecting appropriate chemical agents for pre-treatment. Sar et al. (1999) observed that the metal (Cu^{2+} and Ni^{2+}) uptake capacity of lyophilized *Pseudomonas aeruginosa* cells was enhanced when pre-treated with NaOH, NH_4OH or toluene; whereas, oven heating (80°C), autoclaving, acid, detergent and acetone treatments were inhibitory. Even though these chemical pre-treatments were almost essential for most of the biosorbents, especially industrial wastes, vast improvements in the biosorption capacities cannot always be expected.

Conversely, enhancement or modification of the binding sites on a biomass seems to enhance the biosorption capacities by many multiples. Carboxyl, amine, phosphonate, sulfonate and hydroxyl groups have become well established as being responsible for metal binding. As the density of these groups is low, most biosorbents show low sorption capacities. Various procedures are available for the enhancement of these functional groups on the biomass. General, futile/less important functional groups can be converted into active binding groups via several chemical treatment methods. Jeon and Höll (2003) used chloroacetic acid to introduce carboxyl in the place of hydroxyl groups. The carboxylated biomass was then treated with ethylenediamine, followed by carbodiimide, to form an aminated biomass, and observed increase in the amount of amine groups increased the uptake of mercury by 47% compared to the control. Li et al. (2007) employed citric acid to modify an alkali-saponified biomass, which resulted in an increase in the total acidic sites, but a decrease of basic sites. In particular, they reported that biomass modified using 0.6 mol/l citric acid at 80°C for 2 h exhibited cadmium uptake capacity twice than that of the raw biomass.

Another efficient way for the introduction of functional groups onto the biomass surface is the grafting of long polymer chains onto the biomass surface via direct grafting or polymerization of a monomer. However, very little research has focused specifically on this aspect. Deng and Ting (2005a, b, c, 2007) worked extensively with polyethylenimine, composed of a large number of primary and secondary amine groups, which when cross-linked with biomass exhibited good biosorption abilities towards chromium(VI), copper, lead, nickel and arsenic. In 2005, Deng and Ting copolymerized acrylic acid onto the biomass surface to enhance the carboxyl groups, which resulted in five and seven fold enhancements for the uptakes of copper and cadmium, respectively, compared with the pristine biomass. Poly(amic acid), from reaction of pyromellitic dianhydride and thiourea, which comprises a large number of carboxyl and secondary amine groups in a molecule, when grafted on the biosorbent surface, exhibited 15- and 11-fold increases in the uptakes of cadmium and lead compared to the pristine biomass (Yu et al. 2007).

5.4.2 Genetically Modified Biosorbents

Genetic engineering has the potential to improve or redesign microorganisms, where biological metal-sequestering systems will have a higher intrinsic capability as well as specificity and greater resistance to ambient conditions. It is well known that virgin biosorbents usually lack specificity in metal binding, which may cause difficulties in the recovery and recycling of the desired metal(s). Genetic modification is a potential alternative to enhance the selectivity as well as the accumulating properties of the cells.

Genetic modification would be feasible especially when the microbial biomass is produced by fermentation processes where genetically engineered microorganisms are used. Nowadays, many kinds of amino acids and nucleic acids are being produced in an industrial scale by using genetically engineered microbial cells.

Higher organisms respond to the presence of metals, with the production of cysteine-rich peptides, such as glutathione (GSH) (Singhal et al. 1997), phytochelatins (PCs) and metallothioneins (MTs) (Mehra and Winge 1991), which can bind and sequester metal ions in biologically inactive forms (Bae et al. 2000; Hamer 1986). The overexpression of MTs in bacterial cells will result in enhanced metal accumulation and; thus, it offers a promising strategy for the development of microbial-based biosorbents for the remediation of metal contamination (Pazirandeh et al. 1995). In addition to the high selectivity and accumulation capacity, Pazirandeh et al. (1995) demonstrated that the uptake by recombinant *E. coli* (expressing the *Neurospora crassa* metallothionein gene within the periplasmic space) was rapid. Greater than 75% Cd uptake occurred in the first 20 min with maximum uptake achieved in less than 1 h. However, the expression of such cysteine-rich proteins is not devoid of problems, due to the predicted interference with redox pathways in the cytosol. More importantly, the intracellular expression of MTs may prevent the recycling of the biosorbents, as the accumulated metals cannot be easily released (Gadd and White 1993). Chen and Georgiou (2002) suggested a solution to bypass this transport problem by expressing MTs on the cell surface. Sousa et al. (1996) demonstrated the possibility of inserting MTs into the permissive site 153 of the LamB sequence. The expression of the hybrid proteins on the cell surface dramatically increases the whole-cell accumulation of cadmium. Also, the expression of proteins on the surface offers an inexpensive alternative for the preparation of affinity adsorbents (Georgiou et al. 1993).

The use of PCs in a similar manner to MTs has also been suggested (Bae et al. 2000). PCs are short, cysteine-rich peptides, with the general structure (γGlu-Cys)$_n$Gly ($n=2-11$) (Zenk 1996). PCs offer many advantages over MTs, due to their unique structural characteristics, particularly the continuously repeating γGlu-Cys units. Also, PCs have been found to exhibit higher metal-binding capacity (on a per cysteine basis) than MTs (Mehra and Mulchandani 1995). However, the development of organisms overexpressing PCs requires a thorough knowledge of the mechanisms involved in the synthesis and chain elongation of these peptides.

Several biosorbents, displaying metal-binding peptides on the cell surface, have been successfully engineered. A typical example includes creating a repetitive metal-binding motif, consisting of (Glu-Cys)$_n$Gly (Bae et al. 2000). These peptides simulate the structure of PCs; however, differ in the fact the peptide bond between the glutamic acid and cysteine is a standard α peptide bond. Phytochelatin analogs were present on the bacterial surface, which enhanced the accumulation of Cd^{2+} and Hg^{2+} by 12- (Bae et al. 2000) and 20-fold (Bae et al. 2001), respectively.

Attempts to create recombinant bacteria with improved metal-binding capacity have so far been restricted to mostly *Escherichia coli*, as *E. coli* greatly facilitates genetic engineering experiments, and has more surface area per unit of cell mass, which should potentially give higher rates of metal removal from solution (Chen and Wilson 1997). Nevertheless, a Gram-positive surface display system also possesses its own merits compared to Gram-negative bacteria (Samuelson et al. 2000; Malik et al. 1998): (a) translocation through only one membrane is required, and (b) Gram-positive bacteria have been shown to be more rigid and; therefore, less sensitive to shear forces (Kelemen et al. 1979) due to the thick cell wall surrounding the cells, which potentially make them more suitable for field applications, such as biosorption. Samuelson et al. (2000) generated recombinant *Staphylococcus xylosus* and *Staphylococcus carnosus* strains, with surface-exposed chimeric proteins containing polyhistidyl peptides. Both strains of staphylococci gained improved nickel-binding capacities due to the introduction of the H1 or H2 peptide into their surface proteins.

Owing to their high selectivity, genetically engineered biosorbents may prove very competitive for the separation of toxins and other pollutants from dilute contaminated solutions.

5.4.3 *Immobilized Biosorbents*

Bacterial biosorbents are basically small particles, with low density, poor mechanical strength and little rigidity. Even though they have merits, such as high biosorption capacity, rapid equilibrium attainment, less process cost and good particle mass transfer. They often suffer several drawbacks, the most important include solid-liquid separation problems, possible biomass swelling, impossible to regenerate/reuse and develop a high pressure drop when used in the column mode (Vijayaraghavan and Yun 2007a; Vegliò and Beolchini 1997).

Several established techniques are available to make biosorbents suitable for process applications, including immobilization techniques, such as entrapment and cross linking, which have been found to be practical for biosorption (Volesky 2001; Vegliò and Beolchini 1997). Immobilization of microorganisms within a polymeric matrix has exhibited greater potential, especially in packed or fluidized bed reactors, with benefits including the control of particle size, regeneration and reuse of the biomass, easy separation of biomass and effluent, high biomass loading and minimal clogging under continuous flow conditions (Hu and Reeves 1997). Various synthetic (acrylamide, polyurethane, polyvinyl alcohol, resins) and natural polymer derivatives of algal polysaccharides (alginate, carrageenan, agar, agarose), and chitosan, an amino polysaccharide derived from chitin, has been experimentally used. Several researchers successfully immobilized bacterial biomass for metal biosorption (Bai and Abraham 2003; Beolchini et al. 2003; Khoo and Ting, 2001; Yan and Viraraghavan 2001; Prakasham et al. 1999; Hu and Reeves 1997). Important immobilization matrices used in biosorbent immobilization include sodium alginate (Xiangliang et al. 2005; Bai and Abraham 2003), polysulfone (Vijayaraghavan et al. 2007; Beolchini et al. 2003), polyacrylamide (Bai and Abraham 2003) and polyurethane (Hu and Reeves 1997). The choice of immobilization matrix is a key factor in the environmental application of immobilized biomass. The polymeric matrix determines the mechanical strength and chemical resistance of the final biosorbent particle to be utilized for successive sorption-desorption cycles (Bai and Abraham 2003).

However, care must be taken to avoid the practical problems generated during the immobilization process; in particular, the mass transfer limitations and additional process costs. After immobilization, the biomass will usually be retained within the interior of the matrix used for the immobilization; hence, mass transfer resistance will play a vital role in deciding the rate of biosorption. The presence of mass transfer resistance usually slows the attainment of equilibrium; however, a successful immobilization matrix should allow all the active binding sites to have access to the solute, even at a slower rate. Vijayaraghavan et al. (2008) reported the immobilization of *C. glutamicum* within a polysulfone matrix, which delayed the attainment of equilibrium; however, the nickel uptake was almost comparable to that of the free biomass. Next, immobilizing the biomass usually enhances the process costs. Biosorption is usually portrayed as a cost effective process, which is often highlighted as the main attraction of biosorption compared to that of other proven technologies. While immobilizing the biomass for the sole purpose of biosorption will generally enhance the process costs, it is also often necessary for practical implementation of biosorption to real applications. Many can argue, what is the need for using microbial biomass for biosorption when highly rigid and efficient biosorbents, such as seaweeds, are available? Microbial biomaterials, such as bacteria and fungi, exhibit high metal uptakes. Also, the microbial wastes generated by many fermentation/food industries cause a nuisance, and their disposal is of great concern. For instance, *Corynebacterium glutamicum*, a Gram-positive bacterium, is widely used for the biotechnological production of amino acids. Currently, the production of amino acids from fermentative processes using *C. glutamicum* amounts to 1,500,000 and 550,000 t per year of L-glutamate and L-lysine, respectively (Hermann 2003). Hence the waste *C. glutamicum* generated after fermentation is usually high, but the potential utilization of this waste is of interest. Yun and co-workers examined the biosorption potential of *C. glutamicum* and identified its excellent metal and dye binding capacity. However, this biomass is associated with problems during desorption, as strong acidic/alkaline environment affect the physical structure of *C. glutamicum*. Vijayaraghavan et al. (2008) immobilized *C. glutamicum* using polysulfone, and the biosorbent showed unaltered nickel biosorption potential over three successive sorption-desorption cycles.

A review of the literature relating to biosorption revealed that several microbial biomasses have been cultivated and explored for their biosorption potential. The cost for producing biomass for the sole purpose of its transformation into biosorbents has been shown to be too expensive (Tsezos 2001). Furthermore, the continuous supply of biomass cannot be assured, which will have a huge impact on its successful application in industrial biosorption applications.

5.5 Some Case Studies of Bacterial Biosorbents

A wide variety of bacterial species were used for the biosorption of a broad range of metal ions. Table 5.1 summarizes some of the important results of metal biosorption using bacterial biomass. A direct comparison of experimental data is not possible, due to different systematic experimental conditions employed (pH, pH control, temperature, equilibrium time and biomass dosage). However, Table 5.1 provides basic information to evaluate the possibility of using bacterial biomass for the uptake of metal ions. Also, it should be indicated that Table 5.1 is only comprised of biosorption studies that employed either inactive or dead bacterial biomasses. Some variability has been observed in the results when the same bacterium is employed for the same metal, but under different instances. Apart from the different experimental conditions, this may be due to the fact that the biomass may have be pre-treated or immobilized to improve the biosorbent characteristics, as highlighted in Table 5.1. Also, for most metal ions, acidic pH conditions resulted in maximum biosorption, due to the involvements in the carboxyl and other acidic functional groups responsible for the binding metal cations via various mechanisms. However, the mechanisms for the biosorption have not always been confirmed or discussed in most studies; therefore, generalizations are not possible for these cases. The extent of biosorption depends not only on the type of metal ions, but also on the bacterial genus, due to variations in the cellular constituents. Very short contact times were generally sufficient to attain metal-bacterial biomass equilibrium, which is because

Table 5.1 Important results from the literature on metal biosorption by various bacterial species

Metal	Organism	Operating conditions			Uptake (mg/g)	References
		pH	Temp (°C)	Other information		
Aluminum	*Chryseomonas luteola*	5.0	NA	$M = 1$ g/l, $t_{eq} = 1$ h	55.2 (L)	Ozdemir and Baysal (2004)
Chromium(VI)	*Aeromonas caviae*	2.5	20	$M = 0.5$ g/l; $t_{eq} = 2$ h	284.4 (L)	Loukidou et al. (2004a)
	Bacillus coagulans	2.5	28 ± 3	$M = 2$ g/l, $t_{eq} = 1$ h	39.9 (E)	Srinath et al. (2002)
	Bacillus licheniformis	2.5	50	$M = 1$ g/l, $t_{eq} = 2$ h	69.4 (L)	Zhou et al. (2007)
	Bacillus megaterium	2.5	28 ± 3	$M = 2$ g/l, $t_{eq} = 1$ h	30.7 (E)	Srinath et al. (2002)
	Bacillus thuringiensis	2.0	25	$M = 1$ g/l	83.3 (L)	Şahin and Öztürk (2005)
	Chryseomonas luteola	4.0	NA	$M = 1$ g/l, $t_{eq} = 1$ h	3.0 (L)	Ozdemir and Baysal (2004)
	Pseudomonas sp.	4.0	NA	$M = 1$ g/l, $t_{eq} = 1.5$ h	95.0 (L)	Ziagova et al. (2007)
	Pseudomonas fluorescens TEM08	2	27	$M = 1$ g/l	40.8 (L)	Uzel and Ozdemir (2009)
	Staphylococcus xylosus	1.0	NA	$M = 1$ g/l, $t_{eq} = 1.5$ h	143.0 (L)	Ziagova et al. (2007)
	Zoogloea ramigera	2.0	25	NA	27.5 (L)	Sağ and Kutsal (1989)
Copper	*Arthrobacter* sp.	5	30	–	175.9 (L)	Hasan and Srivastava (2009)
	Bacillus sp. (ATS-1)	5.0	25	$M = 2$ g/l, $t_{eq} = 2$ h	16.3 (E)	Tunali et al. (2006)
	Bacillus subtilis IAM 1026	5	25	$M = 0.5$ g/l, $t_{eq} = 1$ h	20.8 (L)	Nakajima et al. (2001)
	Enterobacter sp. J1	5.0	25	$t_{eq} = 24$ h	32.5 (L)	Lu et al. (2006)
	Geobacillus thermoleovorans sub.sp. *stromboliensis*	5	60	$M = 2.5$ g/l, $t_{eq} = 1$ h	41.5 (L)	Özdemir et al. (2009)
	Geobacillus toebii sub.sp. *decanicus*	4	60	$M = 2.5$ g/l, $t_{eq} = 1$ h	48.5 (L)	Özdemir et al. (2009)
	Micrococcus luteus IAM 1056	5	25	$M = 0.5$ g/l, $t_{eq} = 1$ h	33.5 (L)	Nakajima et al. (2001)
	Pseudomonas aeruginosa PU21	5.0	NA	$M = 1–2$ g/l; $t_{eq} = 24$ h	23.1 (L)	Chang et al. (1997)
	Pseudomonas cepacia	7	30	NA	65.3 (L)	Savvaidis et al. (2003)
	Pseudomonas putida	6.0	NA	NA	6.6 (L)	Pardo et al. (2003)
	Pseudomonas putida	5.5	30	$M = 1$ g/l, $t_{eq} = 24$ h	96.9 (L)	Uslu and Tanyol (2006)
	Pseudomonas putida CZ1	4.5	30	$M = 1$ g/l; $t_{eq} = 24$ h	15.8 (L)	Chen et al. (2005)
	Pseudomonas stutzeri IAM 12097	5	25	$M = 0.5$ g/l, $t_{eq} = 1$ h	22.9 (L)	Nakajima et al. (2001)
	Sphaerotilus natans	6	NA	$M = 3$ g/l; $t_{eq} = 0.5$ h	60 (E)	Beolchini et al. (2006)
	Sphaerotilus natans[b]	5.5	30	NA	5.4 (L)	Beolchini et al. (2006)

Table 5.1 (continued)

Metal	Organism	Operating conditions			Uptake (mg/g)	References
		pH	Temp (°C)	Other information		
	Streptomyces coelicolor	5.0	25	M = 1 g/l; t_{eq} = 8 h	66.7 (L)	Öztürk et al. (2004)
	Thiobacillus ferrooxidans[a]	6.0	37	M = 0.2 g/l; t_{eq} = 2 h	198.5 (L)	Ruiz-Manriquez et al. (1997)
	Thiobacillus ferrooxidans[a]	5.0	40	M = 300 g/l; t_{eq} = 2 h	39.84 (L)	Liu et al. (2004)
Cadmium	*Aeromonas caviae*	7.0	20	M = 1 g/l; t_{eq} = 2 h	155.3 (L)	Loukidou et al. (2004b)
	Bacillus circulans	7.0	30	M = 0.5 g/l; t_{eq} = 2 h	26.5 (E)	Yilmaz and Ensari (2005)
	Enterobacter sp. J1	6.0	25	t_{eq} = 24 h	46.2 (L)	Lu et al. (2006)
	Geobacillus thermoleovorans sub.sp. *stromboliensis*	4	70	M = 2.5 g/l, t_{eq} = 1 h	38.8 (L)	Özdemir et al. (2009)
	Geobacillus toebii sub.sp. *decanicus*	6	70	M = 2.5 g/l, t_{eq} = 1 h	29.2 (L)	Özdemir et al. (2009)
	Pseudomonas aeruginosa PU21	6.0	NA	M = 1–2 g/l; t_{eq} = 24 h	42.4 (L)	Chang et al. (1997)
	Pseudomonas putida	6.0	NA	NA	8.0 (L)	Pardo et al. (2003)
	Pseudomonas sp.	7.0	NA	M = 1 g/l, t_{eq} = 1.5 h	278.0 (L)	Ziagova et al. (2007)
	Staphylococcus xylosus	6.0	NA	M = 1 g/l, t_{eq} = 1.5 h	250.0 (L)	Ziagova et al. (2007)
	Streptomyces pimprina[a]	5.0	NA	t_{eq} = 1 h	30.4 (L)	Puranik et al. (1995)
	Streptomyces rimosus[a]	8.0	NA	M = 3 g/l	64.9 (L)	Selatnia et al. (2004a)
Iron (III)	*Streptomyces rimosus*[a]	NA	NA	M = 3 g/l; t_{eq} = 4 h	122.0 (L)	Selatnia et al. (2004c)
Lead	*Bacillus* sp. (ATS-1)	3.0	25	M = 2 g/l, t_{eq} = 2 h	92.3 (E)	Tunali et al. (2006)
	Corynebacterium glutamicum	5.0	20 ± 2	M = 5 g/l, t_{eq} = 2 h	567.7 (E)	Choi and Yun (2004)
	Enterobacter sp. J1	5.0	25	t_{eq} = 24 h	50.9 (L)	Lu et al. (2006)
	Pseudomonas aeruginosa ASU 6a	6	30 ± 2	M = 1 g/l, t_{eq} = 0.5 h	123 (L)	Gabr et al. (2008)
	Pseudomonas aeruginosa PU21	5.5	NA	M = 1–2 g/l; t_{eq} = 24 h	79.5 (L)	Chang et al. (1997)
	Pseudomonas aeruginosa PU21[b]	5	50	M = 200 g/l	0.7 (L)	Lin and Lai (2006)
	Pseudomonas putida	5.5	25	M = 1 g/l, t_{eq} = 24 h	270.4 (L)	Uslu and Tanyol (2006)
	Pseudomonas putida	6.5	NA	–	56.2 (L)	Pardo et al. (2003)
	Streptomyces rimosus[a]	–	NA	M = 3 g/l; t_{eq} = 3 h	135.0 (L)	Selatnia et al. (2004b)
	Streptoverticillium cinnamoneum[a]	4.0	28 ± 3	M = 2 g/l, t_{eq} = 0.5 h	57.7 (E)	Puranik and Paknikar (1997)
	Symphortcarpus albus	5.5	45	M = 1 g/l	62.1 (L)	Akar et al. (2009)

Table 5.1 (continued)

Metal	Organism	Operating conditions			Uptake (mg/g)	References
		pH	Temp (°C)	Other information		
Manganese	*Geobacillus thermoleovorans* sub.sp. *strombolienis*	6	70	M=2.5 g/l, t_{eq}=1 h	23.2 (L)	Özdemir et al. (2009)
	Geobacillus toebii sub.sp. *decanicus*	6	70	M=2.5 g/l, t_{eq}=1 h	13.9 (L)	Özdemir et al. (2009)
Mercury	*Bacillus* sp.	6	25	M=2 g/l, t_{eq}=2 h	7.9 (L)	Green-Ruiz (2006)
Nickel	*Bacillus thuringiensis*	6	35	M=1 g/l, t_{eq}=8 h	45.9 (L)	Öztürk (2007)
	Geobacillus thermoleovorans sub.sp. *strombolienis*	4	70	M=2.5 g/l, t_{eq}=1 h	42.0 (L)	Özdemir et al. (2009)
	Geobacillus toebii sub.sp. *decanicus*	4	70	M=2.5 g/l, t_{eq}=1 h	21.0 (L)	Özdemir et al. (2009)
	Pseudomonas aeruginosa ASU 6a	7	30±2	M=1 g/l, t_{eq}=0.5 h	113.6 (L)	Gabr et al. (2008)
	Pseudomonas fluorescens TEM08	5	27	M=1 g/l	12.4 (L)	Uzel and Ozdemir (2009)
	Streptomyces rimosus[a]	6.5	NA	M=3 g/l, t_{eq}= 2 h	32.6 (L)	Selatnia et al. (2004d)
Palladium	*Desulfovibrio desulfuricans*	2.0	30	M=0.15 g/l, t_{eq}=4 d	128.2 (L)	de Vargas et al. (2004)
	Desulfovibrio fructosivorans	2.0	30	M=0.15 g/l, t_{eq}=4 d	119.8 (L)	de Vargas et al. (2004)
	Desulfovibrio vulgaris	2.0	30	M=0.15 g/l, t_{eq}=4 d	106.3 (L)	de Vargas et al. (2004)
Platinum	*Desulfovibrio desulfuricans*	2.0	30	M=0.15 g/l, t_{eq}=4 d	62.5 (L)	de Vargas et al. (2004)
	Desulfovibrio fructosivorans	2.0	30	M=0.15 g/l, t_{eq}=4 d	32.3 (L)	de Vargas et al. (2004)
	Desulfovibrio vulgaris	2.0	30	M=0.15 g/l, t_{eq}=4 d	40.1 (L)	de Vargas et al. (2004)
Thorium	*Arthrobacter nicotianae* IAM 12342	3.5	30	M=0.15 g/l, t_{eq}=1 h	75.9 (E)	Nakajima and Tsuruta (2004)
	Bacillus licheniformis IAM 111054	3.5	30	M=0.15 g/l, t_{eq}=1 h	66.1 (E)	Nakajima and Tsuruta (2004)
	Bacillus megaterium IAM 1166	3.5	30	M=0.15 g/l, t_{eq}=1 h	74.0 (E)	Nakajima and Tsuruta (2004)
	Bacillus subtilis IAM 1026	3.5	30	M=0.15 g/l, t_{eq}=1 h	71.9 (E)	Nakajima and Tsuruta (2004)
	Corynebacterium equi IAM 1038	3.5	30	M=0.15 g/l, t_{eq}=1 h	46.9 (E)	Nakajima and Tsuruta (2004)
	Corynebacterium glutamicum IAM 12435	3.5	30	M=0.15 g/l, t_{eq}=1 h	36.2 (E)	Nakajima and Tsuruta (2004)
	Micrococcus luteus IAM 1056	3.5	30	M=0.15 g/l, t_{eq}=1 h	77.0 (E)	Nakajima and Tsuruta (2004)
	Nocardia erythropolis IAM 1399	3.5	30	M=0.15 g/l, t_{eq}=1 h	73.8 (E)	Nakajima and Tsuruta (2004)
	Zooglea ramigera IAM 12136	3.5	30	M=0.15 g/l, t_{eq}=1 h	67.8 (E)	Nakajima and Tsuruta (2004)

Table 5.1 (continued)

Metal	Organism	Operating conditions			Uptake (mg/g)	References
		pH	Temp (°C)	Other information		
Uranium	*Arthrobacter nicotianae* IAM 12342	3.5	30	M=0.15 g/l, t_{eq}=1 h	68.8 (E)	Nakajima and Tsuruta (2004)
	Bacillus licheniformis IAM 111054	3.5	30	M=0.15 g/l, t_{eq}=1 h	45.9 (E)	Nakajima and Tsuruta (2004)
	Bacillus megaterium IAM 1166	3.5	30	M=0.15 g/l, t_{eq}=1 h	37.8 (E)	Nakajima and Tsuruta (2004)
	Bacillus subtilis IAM 1026	3.5	30	M=0.15 g/l, t_{eq}=1 h	52.4 (E)	Nakajima and Tsuruta (2004)
	Citrobacter freundii	–	55	–	48.0 (L)	Xie et al. (2008)
	Corynebacterium equi IAM 1038	3.5	30	M=0.15 g/l, t_{eq}=1 h	21.4 (E)	Nakajima and Tsuruta (2004)
	Corynebacterium glutamicum IAM 12435	3.5	30	M=0.15 g/l, t_{eq}=1 h	5.9 (E)	Nakajima and Tsuruta (2004)
	Micrococcus luteus IAM 1056	3.5	30	M=0.15 g/l, t_{eq}=1 h	38.8 (E)	Nakajima and Tsuruta (2004)
	Nocardia erythropolis IAM 1399	3.5	30	M=0.15 g/l, t_{eq}=1 h	51.2 (E)	Nakajima and Tsuruta (2004)
	Zooglea ramigera IAM 12136	3.5	30	M=0.15 g/l, t_{eq}=1 h	49.7 (E)	Nakajima and Tsuruta (2004)
Zinc	*Aphanothece halophytica*	6.5	30	M=0.2 g/l, t_{eq}=1 h	133.0 (L)	Incharoensakdi and Kitjaharn (2002)
	Pseudomonas putida	7.0	NA	NA	6.9 (L)	Pardo et al. (2003)
	Geobacillus thermoleovorans sub.sp. *stromboliensis*	4	70	M=2.5 g/l, t_{eq}=1 h	29.0 (L)	Özdemir et al. (2009)
	Geobacillus toebii sub.sp. *decanicus*	5	80	M=2.5 g/l, t_{eq}=1 h	21.1 (L)	Özdemir et al. (2009)
	Pseudomonas putida CZ1	5.0	30	M=1 g/l; t_{eq}=24 h	17.7 (L)	Chen et al. (2005)
	Streptomyces rimosus	7.5	20	M=3 g/l	30.0 (L)	Mameri et al. (1999)
	Streptomyces rimosus[a]	7.5	20	M=3 g/l	80.0 (L)	Mameri et al. (1999)
	Streptoverticillium cinnamoneum[a]	5.5	28±3	M=2 g/l, t_{eq}=0.5 h	21.3 (E)	Puranik and Paknikar (1997)
	Thiobacillus ferrooxidans[a]	6.0	25	M=0.2 g/l, t_{eq}=2 h	82.6 (L)	Celaya et al. (2000)
	Thiobacillus ferrooxidans[a]	6.0	40	M=300 g/l; t_{eq}=2 h	172.4 (L)	Liu et al. (2004)

E experimental uptake, *L* uptake predicted by the Langmuir model, *M* biomass dosage, t_{eq} equilibrium time, *NA* not available

[a] Chemically modified and "b" in Table 5.1

[b] Immobilized

the biomass was either used in the form of a fine powder or as wet cells; where mass transfer resistances are usually negligible. The very fast kinetics observed with bacterial biomass represents an advantageous aspect for the design of waste water treatment systems.

Recently, several bacterial biosorbents showed potential for the recovery of precious metals such as gold, silver and platinum group metals (Ru, Rh, Pd, Os, Ir, and Pt) from wastewater. Due to high market prices, recovery of precious metals from effluents is interesting. Zhang et al. (2005) observed that the strain *Corynebacterium* SH09 biosorbed as well as bioreduced diamine silver complex. de Vargas et al. (2004) evaluated the biosorption capacities of palladium and platinum using three different species of *Desulfovivbrio*: *Desulfovibrio desulfuricans, Desulfovibrio fructosivorans* and *Desulfovibrio vulgaris*. They reported that the most promising Pd and Pt biosorption results were obtained using *D. desulfuricans* with a maximum uptake of 128.2 mg/g and 62.5 mg/g for Pd and Pt accumulation respectively, at pH 2. Won et al. (2010) applied their developed biosorbent, PEI-modified *E. coli* biomass, to ICP wastewater containing Pt ions and successfully recovered metallic form of platinum with a recovery efficiency of over 98.7% by a combined method of biosorption and incineration.

5.6 Conclusions

Bacterial biomass represents an efficient and potential class of biosorbents for the removal of metal ions from solutions. Several researchers identified superior biosorption performance of bacterial biomass compared to other biomaterials. Unfortunately, the difficulties in reusing the microbial biomass, as well as the poor selectivity, hinder their applications under real conditions. However, this hurdle can easily be overcome by modification and immobilization techniques. Thus, biosorption technology using bacterial biomass possesses all characteristics to be employed commercially. Although some attempts have been made at the commercialization of biosorption for wastewater treatment, the progress is very modest considering that there has been more than a decade of fundamental research. However, it is no small feat to replace well established conventional techniques. But through continued research, especially on pilot and full-scale biosorption process, the situation is likely to change in the near future, with biosorption technology becoming more beneficial and attractive than currently used technologies.

References

Akar ST, Gorgulu A, Anilan B, Kaynak Z, Akar T (2009) Investigation of the biosorption characteristics of lead(II) ions onto *Symphoricarpus albus*: batch and dynamic flow studies. J Hazard Mater 165:126–133

Bae W, Chen W, Mulchandani A, Mehra RK (2000) Enhanced bioaccumulation of heavy metals by bacterial cells displaying synthetic phytochelatins. Biotechnol Bioeng 70:518–524

Bae W, Mehra R, Mulchandani A, Chen W (2001) Genetic engineering of *Escherichia coli* for enhanced uptake and bioaccumulation of mercury. Appl Environ Microbiol 67: 5335–5338

Baes CF, Mesmer RE (1976) The hydrolysis of cations. Wiley, New York, pp 241–310

Bai RS, Abraham TE (2002) Studies on enhancement of Cr(VI) biosorption by chemically modified biomass of *Rhizopus nigricans*. Water Res 36:1224–1236

Bai RS, Abraham TE (2003) Studies on chromium(VI) adsorption-desorption using immobilized fungal biomass. Biores Technol 87:17–26

Beolchini F, Pagnanelli F, Toro L, Vegliò F (2003) Biosorption of copper by *Sphaerotilus natans* immobilised in polysulfone matrix: equilibrium and kinetic analysis. Hydrometallurgy 70:101–112

Beolchini F, Pagnanelli F, Toro L, Vegliò F (2006) Ionic strength effect on copper biosorption by *Sphaerotilus natans*: equilibrium study and dynamic modelling in membrane reactor. Water Res 40:144–152

Çabuk A, Akar T, Tunali S, Tabak Ö (2006) Biosorption characteristics of *Bacillus* sp. ATS-2 immobilized in silica gel for removal of Pb(II). J Hazard Mater 136:317–323

Celaya RJ, Noriega JA, Yeomans JH, Ortega LJ, Ruiz-Manriquez A (2000) Biosorption of Zn by *Thiobacillus ferrooxidans*. Bioproc Eng 22:539–542

Chang J-S, Law R, Chang C-C (1997) Biosorption of lead, copper and cadmium by biomass of *Pseudomonas aeruginosa* PU21. Water Res 31:1651–1658

Chen W, Georgiou G (2002) Cell-surface display of heterologous proteins: from high-throughput screening to environmental applications. Biotechnol Bioeng 79:496–503

Chen S, Wilson DB (1997) Genetic engineering of bacteria and their potential for Hg^{2+} bioremediation. Biodegradation 8:97–103

Chen XC, Wang YP, Lin Q, Shi JY, Wu WX, Chen YX (2005) Biosorption of copper(II) and zinc(II) from aqueous solution by *Pseudomonas putida* CZ1. Colloids Surf B: Biointerfaces 46:101–107

Choi SB, Yun Y-S (2004) Lead biosorption by waste biomass of *Corynebacterium glutamicum* generated from lysine fermentation process. Biotechnol Lett 26:331–336

Davis TA, Volesky B, Vieira RHSF (2000) *Sargassum* seaweed as biosorbent for heavy metals. Water Res 34:4270–4278

de Vargas I, Macaskie LE, Guibal E (2004) Biosorption of palladium and platinum by sulfate-reducing bacteria. J Chem Technol Biotechnol 79:49–56

Deng S, Ting Y-P (2005a) Characterization of PEI-modified biomass and biosorption of Cu(II), Pb(II) and Ni(II). Water Res 39:2167–2177

Deng S, Ting YP (2005b) Fungal biomass with grafted poly(acrylic acid) for enhancement of Cu(II) and Cd(II) biosorption. Langmuir 21:5940–5948

Deng S, Ting YP (2005c) Polyethylenimine-modified fungal biomass as a high-capacity biosorbent for Cr(VI) anions: sorption capacity and uptake mechanisms. Environ Sci Technol 39:8490–8496

Deng S, Ting YP (2007) Removal of As(V) and As(III) from water with a PEI-modified fungal biomass. Water Sci Technol 55:177–185

Gabr RM, Hassan SHA, Shoreit AAM (2008) Biosorption of lead and nickel by living and non-living cells of *Pseudomonas aeruginosa* ASU 6a. Int Biodeterior Biodegrad 62:195–203

Gadd GM, White C (1993) Microbial treatment of metal pollution: a working biotechnology? Trends Biotechnol 11:353–359

Georgiou G, Poetschke HL, Stathopoulos C, Francisco JA (1993) Practical applications of engineering Gram-negative bacterial cell surfaces. Trends Biotechnol 11:6–10

Göksungur Y, Üren S, Güvenç U (2005) Biosorption of cadmium and lead ions by ethanol treated waste baker's yeast biomass. Biores Technol 96:103–109

Golab Z, Breitenbach M (1995) Sites of copper binding in *Streptomyces pilosus*. Water Air Soil Poll 82:713–721

Green-Ruiz C (2006) Mercury(II) removal from aqueous solutions by nonviable *Bacillus* sp. from a tropical estuary. Biores Technol 97:1907–1911

Hamer DH (1986) Metallothionein. Annu Rev Biochem 55:913–951

Hasan SH, Srivastava P (2009) Batch and continuous biosorption of Cu^{2+} by immobilized biomass of *Arthrobacter* sp. J Environ Manag 90:3313–3321

Hermann T (2003) Industrial production of amino acids by coryneform bacteria. J Biotechnol 104:155–172

Hu MZC, Reeves M (1997) Biosorption of uranium by *Pseudomonas aeruginosa* strain CSU immobilized in a novel matrix. Biotechnol Prog 13:60–70

Incharoensakdi A, Kitjaharn P (2002) Zinc biosorption from aqueous solution by a halotolerant cyanobacterium *Aphanothece halophytica*. Curr Microbiol 45:261–264

Iqbal M, Saeed A (2007) Production of an immobilized hybrid biosorbent for the sorption of Ni(II) from aqueous solution. Proc Biochem 42:148–157

Jeon C, Höll WH (2003) Chemical modification of chitosan and equilibrium study for mercury ion removal. Water Res 37:4770–4780

Kang S-Y, Lee J-U, Kim K-W (2007) Biosorption of Cr(III) and Cr(VI) onto the cell surface of *Pseudomonas aeruginosa*. Biochem Eng J 36:54–58

Kazy SK, Das SK, Sar P (2006) Lanthanum biosorption by a *Pseudomonas* sp.: equilibrium studies and chemical characterization. J Ind Microbiol Biotechnol 33:773–783

Kelemen MV, Sharpe JE (1979) Controlled cell disruption: a comparison of the forces required to disrupt different micro-organisms. J Cellul Sci 35:431–441

Khoo K-M, Ting Y-P (2001) Biosorption of gold by immobilized fungal biomass. Biochem Eng J 8:51–59

Li X, Tang Y, Xuan Z, Liu Y, Luo F (2007) Study on the preparation of orange peel cellulose adsorbents and biosorption of Cd^{2+} from aqueous solution. Sep Purif Technol 55:69–75

Lin C-C, Lai Y-T (2006) Adsorption and recovery of lead(II) from aqueous solutions by immobilized *Pseudomonas aeruginosa* PU21 beads. J Hazard Mater 137:99–105

Liu H-L, Chen B-Y, Lan Y-W, Cheng YC (2004) Biosorption of Zn(II) and Cu(II) by the indigenous *Thiobacillus thiooxidans*. Chem Eng J 97:195–201

López A, Lázaro N, Priego JM, Marquès AM (2000) Effect of pH on the biosorption of nickel and other heavy metals by *Pseudomonas fluorescens* 4F39. J Ind Microbiol Biotechnol 24: 146–151

Loukidou MX, Karapantsios TD, Zouboulis AI, Matis KA (2004a) Diffusion kinetic study of chromium(VI) biosorption by *Aeromonas caviae*. Ind Eng Chem Res 43:1748–1755

Loukidou MX, Karapantsios TD, Zouboulis AI, Matis KA (2004b) Diffusion kinetic study of cadmium(II) biosorption by *Aeromonas caviae*. J Chem Technol Biotechnol 79:711–719

Lu W-B, Shi J-J, Wang C-H, Chang J-S (2006) Biosorption of lead, copper and cadmium by an indigenous isolate *Enterobacter* sp. J1 possessing high heavy-metal resistance. J Hazard Mater 134:80–86

Malik P, Terry TD, Bellintani F, Perham RN (1998) Factors limiting display of foreign peptides on the major coat protein of filamentous bacteriophage capsids and a potential role for leader peptidase. FEBS Lett 436:263–266

Mameri N, Boudries N, Addour L, Belhocine D, Lounici H, Grib H, Pauss A (1999) Batch zinc biosorption by a bacterial nonliving *Streptomyces rimosus* biomass. Water Res 33:1347–1354

McLean RJC, Beauchemin D, Beveridge TJ (1992) Influence of oxidation-state on iron-binding by *Bacillus licheniformis* capsule. Appl Environ Microbiol 55:3143–3149

Mehra RK, Mulchandani P (1995) Glutathione-mediated transfer of Cu(I) into phytochelatins. Biochem J 307:697–705

Mehra RK, Winge DR (1991) Metal ion resistance in fungi-molecular mechanisms and their regulated expression. J Cellul Biochem 45:30–40

Mullen MD, Wolf DC, Ferris FG, Beveridge F, Flemming CA, Bailey GW (1989) Bacterial sorption of heavy metals. Appl Environ Microbiol 55:3143–3149

Nakajima A, Tsuruta T (2004) Competitive biosorption of thorium and uranium by *Micrococcus luteus*. J Radioanal Nucl Chem 260:13–18

Nakajima A, Yasuda M, Yokoyama H, Ohya-Nishiguchi H, Kamada H (2001) Copper biosorption by chemically treated *Micrococcus luteus* cells. World J Microbiol Biotechnol 17:343–347

Ozdemir G, Baysal SH (2004) Chromium and aluminium biosorption on *Chryseomonas luteola* TEM05. Appl Microbiol Biotechnol 64:599–603

Özdemir S, Kilinc E, Poli A, Nicolaus B, Güven K (2009) Biosorption of Cd, Cu, Ni, Mn and Zn from aqueous solutions by thermophilic bacteria, *Geobacillus toebii* subsp. *decanicus* and *Geobacillus thermoleovorans* subsp. *stromboliensis*: equilibrium, kinetic and thermodynamic studies. Chem Eng J 152:195–206

Öztürk A (2007) Removal of nickel from aqueous solution by the bacterium *Bacillus thuringiensis*. J Hazard Mater 147:518–523

Öztürk A, Artan T, Ayar A (2004) Biosorption of nickel(II) and copper(II) ions from aqueous solution by *Streptomyces coelicolor* A3(2). Colloids Surf B Biointerface 34:105–111

Pan J, Ge X, Liu R, Tang H (2006) Characteristic features of *Bacillus cereus* cell surfaces with biosorption of Pb(II) ions by AFM and FT-IR. Colloids Surf B Biointerface 52:89–95

Pardo R, Herguedas M, Barrado E, Vega M (2003) Biosorption of cadmium, copper, lead and zinc by inactive biomass of *Pseudomonas putida*. Anal Bioanal Chem 376:26–32

Pazirandeh M, Chrisey LA, Mauro JM, Campbell JR, Gaber BP (1995) Expression of the *Neurospora crassa* metallothionein gene in *Escherichia coli* and its effects on heavy-metal uptake. Appl Microbiol Biotechnol 43:1112–1117

Prakasham RS, Merrie JS, Sheela R, Saswathi N, Ramakrishna SV (1999) Biosorption of chromium(VI) by free and immobilized *Rhizopus arrhizus*. Environ Poll 104:421–427

Puranik PR, Paknikar KM (1997) Biosorption of lead and zinc from solutions using *Streptoverticillium cinnamoneum* waste biomass. J Biotechnol 55:113–124

Puranik PR, Chabukswar NS, Paknikar KM (1995) Cadmium biosorption by *Streptomyces pimprina* waste biomass. Appl Microbiol Biotechnol 43:1118–1121

Ruiz-Manriquez A, Magaña PI, López V, Guzmán R (1997) Biosorption of Cu by *Thiobacillus ferrooxidans*. Bioproc Eng 18:113–118

Sağ Y, Kutsal T (1989) Application of adsorption isotherms to chromium adsorption on *Z. ramigera*. Biotechnol Lett 11:141–144

Şahin Y, Öztürk A (2005) Biosorption of chromium(VI) ions from aqueous solution by the bacterium *Bacillus thuringiensis*. Proc Biochem 40:1895–1901

Samuelson P, Wernérus H, Svedberg M, Ståhl S (2000) Staphylococcal surface display of metal-binding polyhistidyl peptides. Appl Environ Microbiol 66:1243–1248

Sar P, Kazy SK, Asthana RK, Singh SP (1999) Metal adsorption and desorption by lyophilized *Pseudomonas aeruginosa*. Int Biodeterior Biodegrad 44:101–110

Savvaidis I, Hughes MN, Poole RK (2003) Copper biosorption by *Pseudomonas cepacia* and other strains. World J Microbiol Biotechnol 19:117–121

Selatnia A, Bakhti MZ, Madani A, Kertous L, Mansouri Y (2004a) Biosorption of Cd^{2+} from aqueous solution by a NaOH treated bacterial dead *Streptomyces rimosus* biomass. Hydrometallurgy 75:11–24

Selatnia A, Boukazoula A, Kechid N, Bakhti MZ, Chergui A, Kerchich Y (2004b) Biosorption of lead(II) from aqueous solution by a bacterial dead *Streptomyces rimosus* biomass. Biochem Eng J 19:127–135

Selatnia A, Boukazoula A, Kechid N, Bakhti MZ, Chergui A (2004c) Biosorption of Fe^{3+} from aqueous solution by a bacterial dead *Streptomyces rimosus* biomass. Proc Biochem 39:1643–1651

Selatnia A, Madani A, Bakhti MZ, Kertous L, Mansouri Y, Yous R (2004d) Biosorption of Ni^{2+} from aqueous solution by a NaOH-treated bacterial dead *Streptomyces rimosus* biomass. Min Eng 17:903–911

Sherbert GV (1978) The biophysical characterization of the cell surface. Academic Press, London

Singhal RK, Andersen ME, Meister A (1997) Glutathione, a first line of defense against cadmium toxicity. FASEB J 1:220–223

Sousa C, Cebolla A, de Lorenzo V (1996) Enhanced metalloadsorption of bacterial cells displaying poly-His peptides. Nat Biotechnol 14:1017–1020

Srinath T, Verma T, Ramteke PW, Garg SK (2002) Chromium(VI) biosorption and bioaccumulation by chromate resistant bacteria. Chemosphere 48:427–435
Texier A-C, Andrès Y, Illemassene M, Le Cloirec P (2000) Characterization of lanthanide ions binding sites in the cell wall of *Pseudomonas aeruginosa*. Environ Sci Technol 34:610–615
Tsezos M (2001) Biosorption of metals. The experience accumulated and the outlook for technology development. Hydrometallurgy 59:241–243
Tunali S, Çabuk A, Akar T (2006) Removal of lead and copper ions from aqueous solutions by bacterial strain isolated from soil. Chem Eng J 115:203–211
Uslu G, Tanyol M (2006) Equilibrium and thermodynamic parameters of single and binary mixture biosorption of lead(II) and copper(II) ions onto *Pseudomonas putida*: effect of temperature. J Hazard Mater 135:87–93
Uzel A, Ozdemir G (2009) Metal biosorption capacity of the organic solvent tolerant *Pseudomonas fluorescens* TEM08. Biores Technol 100:542–548
Vannela R, Verma SK (2006) Cu^{2+} removal and recovery by SpiSORB: batch stirred and up-flow packed bed columnar reactor systems. Bioproc Biosys Eng 29:7–17
Vegliò F, Beolchini F (1997) Removal of metals by biosorption: a review. Hydrometallurgy 44:301–316
Vijayaraghavan K, Yun Y-S (2007a) Chemical modification and immobilization of *Corynebacterium glutamicum* for biosorption of reactive black 5 from aqueous solution. Ind Eng Chem Res 46:608–617
Vijayaraghavan K, Yun Y-S (2007b) Utilization of fermentation waste (*Corynebacterium glutamicum*) for biosorption of reactive black 5 from aqueous solution. J Hazard Mater 141:45–52
Vijayaraghavan K, Han MH, Choi SB, Yun Y-S (2007) Biosorption of Reactive black 5 by *Corynebacterium glutamicum* biomass immobilized in alginate and polysulfone matrices. Chemosphere 68:1838–1845
Vijayaraghavan K, Lee MW, Yun Y-S (2008) Evaluation of fermentation waste (*Corynebacterium glutamicum*) as a biosorbent for the treatment of nickel(II)-bearing solutions. Biochem Eng J 41:228–233
Volesky B (1987) Biosorbents for metal recovery. Trends Biotechnol 5:96–101
Volesky B (2001) Detoxification of metal-bearing effluents: biosorption for the next century. Hydrometallurgy 59:203–216
Volesky B, Schiewer S (1999) Biosorption of metals. In: Flickinger M, Drew SW (eds) Encyclopedia of bioprocess technology, pp 433–453. Wiley, New York
Won SW, Choi SB, Yun Y-S (2005) Interaction between protonated waste biomass of *Corynebacterium glutamicum* and anionic dye Reactive Red 4. Colloid Surf A Physicochem Eng Asp 262:175–180
Won SW, Mao J, Kwak I-S, Sathishkumar M, Yun Y-S (2010) Platinum recovery from ICP wastewater by a combined method of biosorption and incineration. Biores Technol 101:1135–1140
Xiangliang P, Jianlong W, Daoyong Z (2005) Biosorption of Pb(II) by *Pleurotus ostreatus* immobilized in calcium alginate gel. Proc Biochem 40:2799–2803
Xie S, Yang J, Chen C, Zhang X, Wang Q, Zhang C (2008) Study on biosorption kinetics and thermodynamics of uranium by *Citrobacter freundii*. J Environ Radioactiv 99:126–133
Yan G, Viraraghavan T (2001) Heavy metal removal in a biosorption column by immobilized *M. rouxii* biomass. Biores Technol 78:243–249
Yee N, Fein J (2001) Cd adsorption onto bacterial surfaces: a universal adsorption edge? Geochim Cosmochim Acta 65:2037–2042
Yilmaz EI, Ensari NY (2005) Cadmium biosorption by *Bacillus circulans* strain EB1. World J Microbiol Biotechnol 21:777–779
Yin H, He B, Peng H, Ye J, Yang F, Zhang N (2008) Removal of Cr(VI) and Ni(II) from aqueous solution by fused yeast: study of cations release and biosorption mechanism. J Hazard Mater 158:568–576
Yu J, Tong M, Sun X, Li B (2007) A simple method to prepare poly(amic acid)-modified biomass for enhancement of lead and cadmium adsorption. Biochem Eng J 33:126–133
Zenk MH (1996) Heavy metal detoxification in higher plants: a review. Gene 179:21–30

Zhang H, Li Q, Lu Y, Sun D, Lin X, Deng X, He N, Zheng S (2005) Biosorption and bioreduction of diamine silver complex by *Corynebacterium*. J Chem Technol Biotechnol 80:285–290

Zhou M, Liu Y, Zeng G, Li X, Xu W, Fan T (2007) Kinetic and equilibrium studies of Cr(VI) biosorption by dead *Bacillus licheniformis* biomass. World J Microbiol Biotechnol 23:43–48

Ziagova M, Dimitriadis G, Aslanidou D, Papaioannou X, Tzannetaki EL, Liakopoulou-Kyriakides M (2007) Comparative study of Cd(II) and Cr(VI) biosorption on *Staphylococcus xylosus* and *Pseudomonas* sp. in single and binary mixtures. Biores Technol 98:2859–2865

Chapter 6
Fungal Biosorption and Biosorbents

Thiruvenkatachari Viraraghavan and Asha Srinivasan

Abstract The common filamentous fungi can sorb heavy metals from aqueous solutions. Fungal biosorption largely depends on parameters such as pH, metal ion and biomass concentration, physical or chemical pretreatment of biomass, presence of various ligands in solution, and to a limited extent on temperature. The cell-wall fraction of biomass plays an important role in the sorption of heavy metals. The fungal biosorbents widely used for heavy metal removal are reviewed in this chapter, mainly focusing on their performance, pretreatment, regeneration/reuse, modeling of biosorption, their potential application and future. The potential of fungal biomass as sorbents has been established by the available data, and more research and development of the fungal biosorption technology is recommended. The mismatch between strong scientific progress in the field of biosorption and lack of commercialization of research is evident.

Keywords Biosorption • Adsorption • Heavy metals • Fungi • Regeneration • Immobilization

6.1 Introduction

Fungi and yeasts accumulate micronutrients, such as Cu, Zn and Mn, and non-nutrient metals, like U, Ni, Cd, Sn, Hg, in amounts higher than the nutritional requirement. Bacteria, fungi, yeast and algae can remove heavy metals and radio nuclides from aqueous solutions in substantial quantities. The potential of fungal biomass as adsorbents for the removal of heavy metals and radionuclides from polluted waters has been widely recognized. Fungi and yeast can be easily grown in substantial amounts using relatively unsophisticated fermentation techniques and inexpensive growth media. Thus, fungal biosorption can serve as an economical means for removal/recovery of metal ions from aqueous solutions.

T. Viraraghavan (✉)
Faculty of Engineering and Applied Science, University of Regina, S4S0A2 Regina, Canada
e-mail: t.viraraghavan@uregina.ca

P. Kotrba et al. (eds.), *Microbial Biosorption of Metals*,
DOI 10.1007/978-94-007-0443-5_6,

6.2 Metal Ion Uptake and Biosorption Equilibria

Biosorption utilizes both living and dead biological cells as well as chemically pretreated microorganisms. The metal ion uptake by living and dead cells can consist of two different modes. The first uptake mode is independent of cell metabolic activity and involves surface binding of metal ions to cell walls and extracellular material. This is referred to biosorption or passive uptake. The second mode of metal uptake into the cell across the cell membrane is dependent on the cell metabolism, and is referred to intracellular uptake, active uptake or bioaccumulation. Intracellular uptake of the metal ions occur by the cells' metabolism using only living cells, while cell surface sorption allows interaction between toxic metal ions and functional groups such as carboxylate, hydroxyl, sulfate, phosphate and amino groups present on the cell surface. These interactions occur through ion-exchange, complexation and physical adsorption. The use of dead biomass seems to be preferred due to absence of toxicity limitations, absence of nutrient requirements in the feed solution and reuse of regenerated biomass. Live and dead cells of fungi and yeast are capable of removing heavy metals from waste streams. The toxicity of heavy metals on the growth of fungi is well known and will not be the part of this review. The models used to describe the sorption phenomena and removal of various metals by fungi are reviewed in the following sections.

The available literature indicates that equilibria of biosorption of heavy metals and organic compounds follow an adsorption-type isotherm. The degree of biosorption of a metal ion on a biosorbent has been found to be a function of the equilibrium metal-ion concentration in solution at constant pH and temperature conditions. The single-solute adsorption isotherm models of Langmuir, Freundlich and Brunauer-Emmett-Teller (BET) have been shown to describe the biosorption equilibrium. The Langmuir model can be described as (Weber 1972) (Eq. 6.1):

$$\mathrm{x/m} = abCe/1 + b\mathrm{Ce} \tag{6.1}$$

where x/m = amount of metal ion biosorbed per unit weight of biomass; a = constant related to the energy or net enthalpy of biosorption; b = amount of metal ion biosorbed per unit weight of biomass; Ce = equilibrium concentration of metal ion in solution after biosorption.

The Langmuir model is based on the assumption that maximum adsorption occurs when a saturated monolayer of solute molecules is present on the adsorbent surface, and the energy of adsorption is constant and that there is no migration of adsorbing molecules in the surface plane. The Freundlich isotherm has the form (Weber 1972) (Eq. 6.2)

$$\mathrm{Q} = K\mathrm{Ce}^{1/\mathrm{n}} \tag{6.2}$$

where Q = metal uptake capacity of biomass; K = biosorption equilibrium constant indicative of biosorptive uptake capacity; n = biosorption equilibrium constant; Ce = equilibrium metal-ion concentration. The Freundlich model is basically empirical, and was developed for heterogeneous surfaces. The model is a useful means of data description. The BET isotherm represents the multi-layer adsorption at the adsorbent surface and assumes that a Langmuir isotherm applies to each layer (Weber 1972) (Eq. 6.3):

$$Ce/(Cs - Ce) = 1/BQ^\circ + ((B - 1)/BQ^\circ)(Ce/Cs) \quad (6.3)$$

Cs = saturation concentration of the metal ion; Q° = amount adsorbed per unit weight of biomass for monolayer biosorption; B = constant relating to the energy of interaction with the surface.

The adsorption models described above were developed for gas adsorption on surfaces. The application of these models to complex biological system may not be able to explain the biosorption behavior. The applicability of the models should be considered as a mathematical representation of the biosorption equilibrium over a given metal-ion concentration range. The mechanistic conclusions from the good fit of the models alone should be avoided. In spite of the above limitations, these models can provide information on metal-uptake capacities and differences in metal uptake between various species.

The biosorption of metals by cell surface binding can take place for both living and dead cells, and is of particular interest in the removal and recovery of metal ions.

Table 6.1 gives the biosorptive capacities of various fungi for different metals.

Table 6.1 The biosorptive capacity of various fungal organisms

Biomass type	Metal	Biosorption capacity (mg/g)	Reference
Aspergillus sydoni	Cr(VI)	1.76	Kumar et al. (2008)
Aspergillus niger	Cr(VI)	3.1	Mungasavalli et al. (2007)
Aspergillus niger	Pb	32.60	Dursun (2006)
Aspergillus niger	Cu	15.6	Dursun et al. (2003a)
Aspergillus niger	Pb	34.4	Dursun et al. (2003a)
Aspergillus niger	Pb	93	Spanelova et al. (2003)
Aspergillus niger	Cu	9.53	Dursun et al. (2003b)
Aspergillus niger	Pb	2.25	Kapoor et al. (1999)
Aspergillus niger	Cd	1.31	Kapoor et al. (1999)
Aspergillus niger	Cu	0.75	Kapoor et al. (1999)
Aspergillus niger	Ni	1.75	Kapoor et al. (1999)
Aspergillus niger	Cu	23.62	Mukhopadhyay et al. (2007)
Aspergillus niger	Cu	28.7	Dursun (2006)
Aspergillus niger	Cr(VI)	117.33	Khambhaty et al. (2009)
Aspergillus terreus	Pb	201.1	Kogej and Pavko (2001)
Aspergillus terreus	Cu	160–180	Gulati et al. (2002)
Candida sp.	Cu	4.8	Donmez and Aksu (1999)
Kluyveromyces marxianus	Cu	6.44	Donmez and Aksu (1999)
Lentinus sajor caju	Cr(VI)	191.24	Bayramoglu et al. (2005)
Mucor hiemalis	Cr(VI)	53.5	Tewari et al. (2005)
Mucor rouxii	Cd	20.31	Yan and Viraraghavan (2003)
Mucor rouxii	Zn	53.85	Yan and Viraraghavan (2003)
Mucor rouxii	Pb	53.75	Yan and Viraraghavan (2003)

Table 6.1 (continued)

Biomass type	Metal	Biosorption capacity (mg/g)	Reference
Mucor rouxii	Ni	20.49	Yan and Viraraghavan (2003)
Neurospora crassa	Pb	49.1	Kiran et al. (2005)
Neurospora crassa	Cu	12.3	Kiran et al. (2005)
Penicillium simplicissium	Cd	52.50	Fan et al. (2008)
Penicillium simplicissium	Zn	65.60	Fan et al. (2008)
Penicillium simplicissium	Pb	76.90	Fan et al. (2008)
Penicillium canescens	Cd	102.7	Say et al. (2003b)
Penicillium canescens	Pb	213.2	Say et al. (2003b)
Penicillium canescens	Hg	54.8	Say et al. (2003b)
Penicillium chrysogenum	Ni	82.5	Su et al. (2006)
Penicillium chrysogenum	Cd	210.2	Deng and Ting (2005b)
Penicillium chrysogenum	Cu	108.3	Deng and Ting (2005b)
Penicillium chrysogenum	Cu	92	Deng and Ting (2005a)
Penicillium chrysogenum	Pb	204	Deng and Ting (2005a)
Penicillium chrysogenum	Ni	55	Deng and Ting (2005a)
Penicillium chrysogenum	Ni	260	Tan et al. (2004)
Penicillium chrysogenum	Cr(III)	18.6	Tan and Cheng (2003)
Penicillium chrysogenum	Ni	13.2	Tan and Cheng (2003)
Penicillium chrysogenum	Zn	6.8	Tan and Cheng (2003)
Penicillium chrysogenum	Pb	96	Skowronski et al. (2001)
Penicillium chrysogenum	Cd	21.5	Skowronski et al. (2001)
Penicillium chrysogenum	Zn	13	Skowronski et al. (2001)
Penicillium chrysogenum	Cu	11.7	Skowronski et al. (2001)
Penicillium italicum	Cu	0.4–2	Ahluwalia and Goyal (2007)
Penicillium italicum	Zn	0.2	Ahluwalia and Goyal (2007)
Penicillium purpurogenum	Cr(VI)	36.5	Say et al. (2004)
Penicillium purpurogenum	Cd	110.4	Say et al. (2003a)
Penicillium purpurogenum	Pb	252.8	Say et al. (2003a)
Penicillium purpurogenum	Hg	70.4	Say et al. (2003a)
Penicillium purpurogenum	As	35.6	Say et al. (2003a)
Phanerochaete chrysosporium	Pb	419.4	Kogej and Pavko (2001)
Phanerochaete chrysosporium	Cu	20.23	Say et al. (2001)
Rhizopus nigricans	Pb	403.2	Say et al. (2001)
Rhizopus arrhizus	Cu	10.8	Dursun et al. (2003b)
Rhizopus arrhizus	Cr(VI)	78	Aksu and Balibek (2007)
Saccharomyces cerevisiae	Pb	211.2	Say et al. (2001)
Saccharomyces cerevisiae	Cu	7.11	Donmez and Aksu (1999)
Saccharomyces cerevisiae	Pb	79.2	Al-Saraj et al. (1999)
Saccharomyces cerevisiae	Cu	6.4	Al-Saraj et al. (1999)
Saccharomyces cerevisiae	Zn	23.4	Al-Saraj et al. (1999)
Saccharomyces cerevisiae	Hg	64.2	Al-Saraj et al. (1999)
Saccharomyces cerevisiae	Co	9.9	Al-Saraj et al. (1999)
Saccharomyces cerevisiae	Ni	8	Al-Saraj et al. (1999)
Saccharomyces cerevisiae	Cd	35.5–58.4	Park et al. (2003)
Saccharomyces cerevisiae	Cr(VI)	32.6	Ozer and Ozer (2003)
Saccharomyces cerevisiae	Ni	46.3	Ozer and Ozer (2003)
Saccharomyces cerevisiae	Pb	270.3	Ozer and Ozer (2003)

6.3 Metal Uptake by Fungal Biomass

The living cells of *Penicillium, Aspergillus, Trichoderma, Rhizopus, Mucor, Saccharomyces* and *Fusarium* have been shown to biosorb metal ions (Singh 2006; Kapoor and Viraraghavan 1995). Among the *Penicillium* sp., *Phanerochaete chrysogenum* was studied most. *P. chrysogenum* can extract gold from a cyanide solution. *P. spinulosum* was reported to be capable of removing Cu, Au, Zn, Cd, Mn. *P. chrysogenum* exhibited preferential sorption in the following order: Pb>Cu>Zn>Cd>Ni>Co. The species of *Penicillium* was observed to sorb uranium and lead well (Wang and Chen 2009).

The living fungal cells of *A. niger, M. rouxii* and *R. arrhizus* have been shown to take up precious metals such as gold and silver (Mullen et al. 1992). Living cells of *R. arrhizus* and *A. niger* could remove Cu with maximum specific uptake capacity of 10.76 and 9.53 mg/g at 75 mg/L of initial Cu concentration (Dursun et al. 2003b).

The metal ion uptake by living cells is a function of cell age, composition of growth media, contact time, pH of metal solution, and temperature. The kinetics of biosorption of metals is usually biphasic in nature, consisting of an initial rapid phase, contributing up to 90% of biosorption, and lasting for 10 min. The second phase is slower and has been found to last for up to 4 h. Kinetic studies revealed that maximum biosorption of *P. chrysosporium* for Ni and Pb were achieved generally in the first 30 min of contact. Equilibrium was reached in a contact time of 3 h (Ceribasi and Yetis 2001). Cell age has also been found to affect biosorption of metal ions. Increased biosorption has been observed during the lag period or early stages of growth and declined as cultures reached stationary phase. *A. niger, P. spinulosum* and *T. viridae* showed a similar uptake pattern (Kapoor and Viraraghavan 1995). The growth medium controls the composition and structure of cell wall, which in turn affects biosorption. Research has indicated that it is possible to manipulate fungal cells to increase the metal biosorptive capacities and, therefore, produce efficient fungal biosorbents. Mapolelo and Torto (2004) reported that using 10–20 mM glucose increased removal efficiency for Cd, Cr(III), Cu, Pb and Zn by 30–40% using *S. cerevisiae.* Cd adsorption by live *M. rouxii* biomass was observed to be higher for the yeast-malt medium compared to yeast, peptone and glucose or dextrose and peptone media (Yan and Viraraghavan 2003).

Biosorption of metal ions strongly depends on pH. Biosorption of Ni, Cd, and Pb by *P. digitatum* was observed to be inhibited below pH 3.0 (Kapoor and Viraraghavan 1995). Cadmium biosorption was found to increase with pH from acidic (2.0) to basic range (7.0) (Mullen et al. 1992). The degree of biosorption of cadmium was explained using proton competitive adsorption reactions. The surface charge of the fungal cells is predominantly negative over the pH range of 3–10. The magnitude of negative surface charge for various fungal cells follows the order *Fusarium solani*>*T. viridae*>*A. nidullans*>*A. niger*+*A. oryzae*>*P. notatum.* Cadmium removal capacity followed the order *F. solani*>*T. viridae*>*A. nidullans*+*A. oryzae*>*P. notatum*>*A. niger.* Thus, biosorption appeared to be proportional to negative surface charge, the only exception being *A. niger.* Optimum operating pH was 5.0 for biosorption of Ni and Pb by *Phanerochaete chrysosporium* (Ceribasi and Yetis 2001).

The metal uptake discussed so far was mainly due to the passive mode. The biosorption of heavy metal ions on the cell surface occurs by ion-exchange and complexation reactions with functional groups. The various functional groups believed to be involved in metal binding include carboxyl, amine, amides, hydroxyl, phosphate and sulfhydryl groups. Metals such as silver, gold and uranium have also been removed as a result of precipitation/crystallization on fungal and algal surfaces. Biosorption contributed by precipitation and crystallization can be possibly explained by BET isotherms.

The metal uptake can also take place by active mode, which is dependent on the cell metabolic cycle, and metal ions are transported into the cell material across the cell wall. The active mode can contribute significantly to metal removal for yeast. The metal uptake by active mode has been observed for metals such as Cu, Cd, Ni, Zn, Co, Mn and Ca. Research has indicated that at high metal concentrations active mode may not contribute significantly to metal uptake, especially for filamentous fungi. The mechanism of intracellular uptake is complex and not fully understood.

The biosorptive capacity of dead fungal cells has been studied extensively in comparison to living cells. The biosorptive capacity of dead cells may be greater, equivalent to or less than that of living cells. Use of dead biomass in industrial applications offers certain advantages over living cells. Systems using living cells are likely to be more sensitive to metal-ion concentration (toxicity effects) and adverse operating conditions (pH and temperature). Furthermore, constant nutrient supply is required for systems using living cells (increased operating cost for waste streams devoid of nutrients) and recovery of metals and regeneration of biosorbent is more complicated for living cells. The dead biomass can be procured from industrial sources as a waste product from established fermentation processes.

The cells can be killed for biosorption by physical and chemical methods. Physical methods include vacuum and freeze drying, boiling, autoclaving and mechanical disruption. The chemical methods include contacting the biomass with various organic and inorganic compounds. *S. cerevisiae* were modified by methanol, formaldehyde and glutaraldehyde to remove Cu from aqueous solutions. Pretreatment of *Mucor rouxii* biomass with detergent and alkali chemicals such as NaOH, Na_2CO_3, and $NaHCO_3$ were investigated for biosorption of Pb, Cd, Ni and Zn (Yan and Viraraghavan 2000). The effect of pretreatment of *A. niger* biomass on biosorption of lead, cadmium, copper and nickel has also been studied (Kapoor and Viraraghavan 1998a). Table 6.2 provides the methods employed to pretreat biomass and their effect on metal removal. The alkali treatment of fungal biomass has been shown to increase significantly the metal sorption capacity of *Aspergillus, Mucor* and *Penicillium*. The increased removal capacity results from deacetylation of chitin in the cell wall to form chitosan—glucan complexes with a higher affinity to metal ions.

Biosorption of metal ions is a rapid process and often reaches equilibrium within four to six hours. The amount of adsorbed metal ions (Cd, Pb and Cu) by *Phanerochaete chryosporium* was very high at the beginning with more than 60% of the

Table 6.2 Various physical and chemical methods used in pretreatment of biomass

Biomass type	Pretreatment method	Metal studied	Effect on biosorption[a]	Reference
R. arrhizus	0.2 M $Ca(NO_3)_2$ at 100°C for 24 h	Cd, Pb, Cu, Zn	2	Yin et al. (1999)
A. niger	0.5 N NaOH	Cu, Pb	2	Dursun (2006)
N. crassa	NaOH	Pb, Cu	2	Kiran et al. (2005)
	Acetic acid	Pb, Cu	2	
	Dimethyl sulfoxide	Pb, Cu	2	
	detergent	Pb, Cu	2	
A. flavus	Dried at 60°C overnight	Pb, Cu	2	Akar and Tunali (2006)
	0.5 N NaOH, 15 min boiled	Pb, Cu	2	
	10% acetic acid, 15 min boiled	Pb, Cu	2	
	50% dimethyl sulfoxide, 15 min boiled	Pb, Cu	2	
	2.5 g of commercial laundry detergent in 50 mL of water, 15 min boiled	Pb, Cu	2	
A. niger	NaOH	Cu	2	Mukhopadhyay et al. (2007)
A. niger	Acid pretreated	Cr(VI)	2	Mungasavalli et al. (2007)
	Alkali pretreated	Cr(VI)	2	
	Acetone pretreated	Cr(VI)	2	
	Formaldehyde pretreated	Cr(VI)	2	
	Cetyl trimethyl ammonium bromide	Cr(VI)	2	
	Polyethylemine	Cr(VI)	2	
	3-(2-amino ethyl amino) propyl trimethoxy silane	Cr(VI)	2	
N. crassa	Heat inactivated	Cr(VI)	2	Tunali et al. (2005)
	Sodium hydroxide	Cr(VI)	2	
	Acetic acid	Cr(VI)	2	
Botrytis cinerea	NaOH	Pb	1	Akar et al. (2005)
	Dimethyl sulfoxide	Pb	1	
	Acetic acid	Pb	1	
R. oryzae	0.1 M NaOH	Cu	2	Bhainsa and D'Souza (2008)
P. chrysogenum	Alkaline pretreatment	Cr(III), Ni, Zn	2	Tan and Cheng (2003)
A. niger	Autoclaved	Pb, Cd, Cu, Ni	1,1,1,1	
	0.5 N NaOH,15 min boiled	Pb, Cd, Cu, Ni	2,2,2,1	
	50% dimethyl sulfoxide, 15 min boiled	Pb, Cd, Cu, Ni	2,2,2,1	
	2.5 g of laundry detergent in 500 mL water, 15 min boiled	Pb, Cd, Cu, Ni	2,2,2,1	

Table 6.2 (continued)

Biomass type	Pretreatment method	Metal studied	Effect on biosorption[a]	Reference
A. niger (continued)	15% formaldehyde, 15 min boiled	Pb, Cd, Cu, Ni	2,2,2,1	
	10% acetic acid, 15 min boiled	Pb, Cd, Cu, Ni	2,1,2,1	
	10% o-phosphoric acid, 15 min boiled	Pb, Cd, Cu, Ni	1,1,2,1	
	2 N ammonium persulfate, 15 min boiled	Pb, Cd, Cu, Ni	2,2,2,1	
	10% hydrogen peroxide, 15 min boiled	Pb, Cd, Cu, Ni	1,1,1,1	
	Ether, 15 min boiled under reflux	Pb, Cd, Cu, Ni	2,2,2,1	
	Ethanol, 15 min boiled under reflux	Pb, Cd, Cu, Ni	2,2,2,1	
M. rouxii	NaOH	Pb, Cd, Ni, Zn	2,2,2,2	Yan and Viraraghavan (2000)
	Na_2CO_3	Pb, Cd, Ni, Zn	2,2,2,2	
	$NaHCO_3$	Pb, Cd, Ni, Zn	2,2,2,2	
	HCl	Pb, Cd, Ni, Zn	1,1,1,1	
	H_2SO_4	Pb, Cd, Ni, Zn	1,1,1,1	
	$C_2H_4O_2$	Pb, Cd, Ni, Zn	1,1,1,1	
	$CaCl_2$	Pb, Cd, Ni, Zn	1,1,1,1	
	NaCl	Pb, Cd, Ni, Zn	1,1,1,1	

[a] 1, Decrease in metal uptake in comparison to live cells; 2, increase in metal uptake as compared to living cell

metal ions adsorbed within the initial 2 h and equilibrium was reached at 6 h (Say et al. 2001). The biosorption capacity was observed to be related to ionic radii of the metal ions. The biosorption was higher for metals with a larger ionic radius, the exceptions being chromium and the alkali metal ions. The alkali metal ions were not biosorbed, as they lack the ability to form complexes with the various ligand groups present on the fungal surface. The biosorption of metal ions such as Sr, Mn, Zn, Cd, Cu and Pb has been observed to be proportional to their covalent index (X_m^2r), where X_m is electronegativity and r is the ionic radius.

Biosorption of metal ions has been observed to be pH-dependent. Optimum removal is usually achieved in the pH range of 4–5, and biosorption is substantially reduced at a pH of 2.5. Zhou (1999) observed that adsorption of Zn by *R. arrhizus* increased as the pH rose from 4.0 to 5.5. In the case of dead *M. rouxii* biomass, low pH resulted in a decrease in the biosorption capacity. At pH 3.0 or less, the inhibition of biosorption of metal ions took place. At pH 4.0 or higher, biosorption of metal ions increased sharply (Yan and Viraraghavan 2003). Say et al. (2001) observed an increase in metal ion adsorption by *P. chrysosporium* with increasing pH from 2.0 to 6.0. Similarly, adsorption of metal ions by *Penicillium purpurogenum* showed maximum capacity at a pH of 5.0. The pH variation changes the speciation and availability of the metallic elements in solution and also the chemical state of the functional groups responsible for biosorption. For Cu, Cd, and Zn, the biosorption capacity of *S. cerevisiae* at pH 4.5 was found to be far higher than that at pH 2.5 and pH 3.5. At pH 4.5, the most important active molecular group is phosphate, carboxyl and sulfate and electrostatic attraction to negatively charged functional group may be the biosorption mechanism (Wang and Chen 2006). The optimum pH for removal of a particular metal ion depends not only on the fungal strain being used but also on the extent of pretreatment the biomass receives.

Biosorption of metal on certain fungal strains has been found to be both selective and in some cases competitive. Say et al. (2001) observed that the competitive biosorption capacities of *P. chryosporium* containing 100 mg/L each of Cd, Pb and Cu metal ions (7.8, 16.91 and 7.57 mg/g, respectively) were lower than non-competitive conditions (13.24, 45.24 and 10.72, respectively). Pb accumulation by *S. cerevisiae* was seriously inhibited by Hg (Suh and Kim 2000) and Cr(VI) biosorption significantly reduced in the presence of lower concentration of Fe(III) (Goyal et al. 2003). However, Fe(III) did not affect the uptake of Cr(VI) by *A. niger* (Wang and Chen 2009). Cocations have certain effects on lead and zinc biosorption by *P. chrysosporium* in binary and multi-metal systems. The uptake of lead is preferred in a multi-metal solution by the biomass. The representative sorption order of *P. chrysosporium* is Pb > Cu > Zn > Cd > Ni > Co (Singh 2006).

The presence of anions also affects the biosorption of metal ions. Kapoor and Viraraghavan (1997a) reported that the biosorption capacity of *A. niger* decreased in the presence of ethylenediamine tetracetate (EDTA), sulfate, chloride, phosphate, carbonate, glutamate, citrate and pyrophostate. Zhou (1999) observed that uptake of Zn by *R. arrhizus* was reduced in the presence of three anions, the effect being most significant with EDTA, followed by sulfate and chloride. The inhibition was found to increase with an increase in anion/metal molar ratio. The stability constants of metal-anion complexes can be high and if these constants are greater than stability constants of metal biosorption sites on the cell surface, biosorption can be expected to be considerably reduced. Therefore, anion complexation can decrease biosorption of metal ions.

Temperature also affects biosorption of metal ions. Adsorption reactions are generally exothermic and the extent of adsorption increases with decreasing temperature. The maximum biosorption capacity for Ni and Pb by *S. cerevisiae* was obtained at 25°C and found to decrease as the temperature was increased to 40°C (Goyal et al. 2003). However, biosorption capacity for Cr(VI) by *S. cerevisiae* was found to increase with increasing temperature indicating that in this case it was endothermic.

6.4 Use of Immobilized Fungal Biomass in Biosorption

The biomass in contact with water becomes soft and can have low mechanical strength. The small size and low density of free cells cause difficulties in column applications tending to plug the bed, resulting in large pressure drops. Therefore, industrial applications of fungal biosorption prefer immobilized or palletized biomass. Support matrices suitable for biomass immobilization include alginate, polyacrylamide, polyvinyl alcohol, polysulfone, silica gel, cellulose and glutaraldehyde (Wang and Chen 2009). The immobilization of the biomass in solid structures would create a biosorbent material with the right size, mechanical strength, rigidity and porosity necessary for use in practical processes. *P. chrysosporium* was immobilized into Ca-alginate beads via entrapment for the removal of Hg and Cd ions from aqueous solution and the alginate-fungus beads could be regenerated using 10 mM HCl, up to 97% recovery (Kacar et al. 2002). *Trametes versicolor* mycelia were immobilized in carboxymethylcellulose to form beads via entrapment and the maximum biosorption capacity for immobilized live *T. versicolor* was 1.51 mmol Cu, 0.85 mmol Pb and 1.33 mmol Zn per g of dry biosorbent, respectively (Bayramoglu et al. 2003). *M. rouxii* biomass was immobilized in a polysulfone matrix and was packed in a column to remove metal ions such as Pb, Cd, Ni and Zn from aqueous solutions. For single component metal solutions, the metal removal capacities of the beads for Pb, Cd, Ni and Zn were 4.06, 3.76, 0.36 and 1.36 mg/g, respectively. For a multi-component metal solution containing Cd, Ni and Zn, the capacities were 0.36, 0.31 and 0.4 mg/g for Cd, Ni and Zn, respectively (Yan and Viraraghavan 2001). *A. niger* was immobilized into beads in a polysulfone matrix to remove metal ions such as Cd, Cu, Pb, Ni. The metal removal capacities of the beads were 3.60, 2.89, 10.05, 1.08 mg/g for Cd, Cu, Pb, Ni, respectively (Kapoor and Viraraghavan 1998b). Use of polysulfone seems to be a good choice as an immobilizing agent because it is an amorphous, rigid, heat resistant and chemically stable thermoplastic material. Selection of good and cheap support materials for biosorbent immobilization, improvement of reuse methods, and enhancement of properties of immobilized biomaterials such as porous ratio, mechanical intensity and chemical stability are important factors for application (Wang and Chen 2009).

6.5 Regeneration of Fungal Biomass and Elution of Biosorbed Metals

The application of fungi and yeast as biosorbents is dependent not only on the biosorptive capacity, but also on the ease with which biomass can be regenerated and reused. Various elutants have been screened for recovery of Pb, Cd, Ni and Zn from pretreated *M. rouxii.* Nitric acid proved to be an effective elutant than $CaCl_2$ and NaCl with more than 90% elution for all the four metal ions (Yan and Viraraghavan 2003). *M. rouxii* biomass regenerated with NaOH regained its initial metal removal capacity. The caustic regeneration decreases protonation and substitutes sodium ions on functional groups. Zhang et al. (1998) reported that more than 80% of Pb could be desorbed from non-living *R. nigricans* with the use of mineral acids such as HCl and HNO_3. The use of mineral acids as an elutant has been widely studied (Kapoor and Viraraghavan 1997b). Elution of Cu, Pb and Zn biosorbed on *T. versicolor* and Cd on *T. versicolor* entrapped in Ca-alginate beads with 10 mM HCl desorbed more than 95% of the metal ions (Bayramoglu et al. 2003; Arlca et al. 2001). Concerns over the damage to biosorbent structure from using mineral acids as elutants prompted investigating less aggressive elutants. The 0.1 M calcium chloride and magnesium sulfate solutions were able to elute cadmium and nickel biosorbed by *A. niger* (Kapoor and Viraraghavan 1997b).

6.6 Biosorption Mechanisms

Biosorption of metal by biomass has been attributed to many different mechanisms which include metabolism-dependent transport, ion-exchange or complexation, adsorption of simple ionic species and hydrolysis products of metal ions. The metabolic-dependent uptake involves mechanisms which may be a part a of the cell metabolic cycle. Biosorption of oxy-anions (molybdate, vandate) was explained by electrostatic attraction to the positively charged functional groups. Anions such as Cr(VI) were found to be biosorbed on *S. cerevisiae, R. nigricans* at low solution pH as the anions could be sorbed on the protonated active sites of the biosorbent which is primarily electrostatic in nature (Wang and Chen 2006). Functional groups such as phosphate, carboxyl, amine, amide and sulfahydryl groups can complex metal ions (Kapoor and Viraraghavan 1995). Uptake of metals has been found to bind on the carboxyl and phosphate groups along with hydroxyl groups and depends on the ratio of phosphate to carboxyl groups in the biomass (Kapoor and Viraraghavan 1995). Kapoor and Viraraghavan (1997a) suggest that carboxylate and amine groups are important in metal ion biosorption on *A. niger* biomass. Sarret et al. (1998) showed that *Penicillium chrysogenum* cell wall possessed strong Zn- and Pb- complexing properties associated predominantly with phosphoryl groups. Amide groups on the surface of *Neurospora crassa* contributed substantially in the mechanism of biosorption of Pb and Cu (Kiran et al. 2005). Ion exchange was believed to be the principal mechanism rather than complexation.

Chitin and chitosan present in fungal cells can also sequester metal ions. Zn was adsorbed more actively on the cell wall of *R. arrhizus*, particularly on its chitin/chitosan components. Hefnawy and Razak (1998) reported that the amount of protein, total sugars and chitin in the cell-wall of *Fusarium oxysporum* increased in the presence of copper in the growth medium. Research has indicated that heavy metal ions are exchanged with calcium, hydrogen, magnesium and potassium ions. Biosorption of lead and cadmium displaced Ca^{2+}, Mg^{2+} and K^{+} ions present on the *A. niger* biomass surface, indicating that biosorption took place as a result of an ion-exchange process (Kapoor and Viraraghavan 1997a). Intracellular cations, K^{+},

Mg^{2+}, Na^{+} and Ca^{2+} of the cells were observed to be released during the process of Zn^{2+} biosorption by *Saccharomyces cerevisiae* (Chen and Wang 2007).

6.7 Application to Practice

Earlier reviews on biosorption (Gadd 2009; Singh 2006; Volesky 2003; Kapoor and Viraraghavan 1995) have shown that a lot of scientific information is already available on the use of a number of biosorbents for metal removal; however no detailed economic and market analyses are available. A number of factors govern the application of an adsorbent to be used in practice. The important ones include: (a) the effectiveness in removing pollutants; (b) the availability of the adsorbent; (c) the cost of the adsorbent; (d) the regeneration of the adsorbent; and (e) the ease with which the adsorbent can be used. Immobilization of biosorbents is a key aspect for the purpose of biosorption application. It is important to decrease the cost of immobilization and consequently distribution, regeneration and reuse of biosorbents (Tsezos 2001). Another important consideration is the availability of the biosorbents in large amounts for industrial application. Economic considerations, such as isolation of microorganisms, screening and harvesting on a large scale, including transportation to a processing area is of practical importance (Lesmana et al. 2009). Large scale fermentation might be required to supply large amounts of microorganisms for industry. The fate of exhausted biosorbent should also be considered (Vijayaraghavan and Yun 2008). Precipitation and electrowinning procedures recover metals from concentrated solution; however, the final disposal of the material should be addressed.

There is still reluctance in opting for non-viable immobilized biomass-based systems compared to the use of biological reactors with living cells, although advantages of the former systems are well-established. There have been many attempts in the past at commercializing immobilized biomass biosorbents such as AlgaSORB™, AMT-Bioclaim, B.V. Sorbex's biosorbents and Bio-Fix, but none have made a successful commercial entry in the market (Wang and Chen 2009; Tsezos 2001; Wase and Forster 1997). Peat is considered as the most successful biosorbent in use either in natural state or in a modified form; however it is not regarded as the best biomaterial for commercialization because it is a finite resource and it is also not available everywhere in the world (Wase and Forester 1997).

6.8 Future Research Needs

The physico-chemical nature of the fungal biosorbents needs to be studied and ways to tailor the nature of fungal cells for increased biosorption capacity of heavy metals needs to be developed. Techniques for regeneration and their effects on the biosorbent quality should also be evaluated. The rates of biosorption for various metals, and multi-component sorption, should be studied to serve as a basis for the design of biosorption treatment systems. Further testing in real wastewater should be conducted and finally preliminary design and economic studies should follow.

The challenge with biosorption as with many in the biotechnology industry is to move this process to an industrial scale. It is relatively less difficult to demonstrate it in a laboratory; it is a little more challenging to demonstrate it at a pilot scale, but to really scale it up to a large scale would call for a significant financial and technological effort. This mismatch between scientific progress in biosorption research (biosciences) and stagnation in industrial biotechnology innovation needs to be corrected through translational research and technology transfer with a push for commercialization of research. Universities can play an active role in this process through more formalized approach to technology transfer and protection of intellectual property (West and Nightingale 2009).

Acknowledgements The authors would like to thank the Natural Sciences and Engineering Research Council of Canada for partial funding through a grant to the first author.

References

Ahluwalia SS, Goyal D (2007) Microbial and plant derived biomass for removal of heavy metals from wastewater. Biores Technol 98:2243–2257

Akar T, Tunali S, Kiran I (2005) *Botrytis cinerea* as a new fungal biosorbent for removal of Pb(II) from aqueous solutions. Biochem Eng J 25:227–235

Akar T, Tunali S (2006) Biosorption characteristics of *Aspergillus flavus* biomass for removal of Pb(II) and Cu(II) ions from an aqueous solution. Biores Technol 97:1780–1787

Aksu Z, Balibek E (2007) Chromium(VI) biosorption by dried *Rhizopus arrhizus*: effect of salt (NaCl) concentration on equilibrium and kinetic parameters. J Hazard Mater 145: 210–220

Al-Saraj M, Abdel-Latif MS, El-Nahal I, Baraka R (1999) Bioaccumulation of some hazardous metals by sol–gel entrapped microorganisms. J Non-Cryst Solids 248:137–140

Arlca MY, Kaçar Y, Genç O (2001) Entrapment of white-rot fungus *Trametes versicolor* in Ca-alginate beads: preparation and biosorption kinetic analysis for cadmium removal from an aqueous solution. Biores Technol 80:121–129

Bayramoglu G, Bekta S, Arlca MY (2003) Biosorption of heavy metal ions on immobilized white-rot fungus *Trametes versicolor*. J Hazard Mater 101:285–300

Bayramoglu G, Celik G, Yalcın E, Yılmaz M, Arıca MY (2005) Modification of surface properties of *Lentinus sajor-caju* mycelia by physical and chemical methods: evaluation of their Cr^{6+} removal efficiencies from aqueous medium. J Hazard Mater B 119:219–29

Bhainsa KC, D'Souza SF (2008) Removal of copper ions by the filamentous fungus, *Rhizopus oryzae* from aqueous solution. Biores Technol 99:3829–3835

Ceribasi H, Yetis U (2001) Biosorption of Ni(II) and Pb(II) by *Phanerochaete chrysosporium* from a binary metal system—Kinetics. Water SA 27:15–20

Chen C, Wang J-L (2007) Characteristics of Zn^{2+} biosorption by *Saccharomyces cerevisiae*. Biomed Environ Sci 20:478–482

Deng SB, Ting YP (2005a) Characterization of PEI-modified biomass and biosorption of Cu(II), Pb(II) and Ni(II). Water Res 39:2167–2177

Deng SB, Ting YP (2005b) Fungal biomass with grafted poly(acrylic acid) for enhancement of Cu(II) and Cd(II) biosorption. Langmuir 21:5940–5948

Donmez G, Aksu Z (1999) The effect of copper(II) ions on the growth and bioaccumulation properties of some yeasts. Process Biochem 35:135–142

Dursun AY (2006) A comparative study on determination of the equilibrium, kinetic and thermodynamic parameters of biosorption of copper(II) and lead(II) ions onto pretreated *Aspergillus niger*. Biochem Eng J 28:187–195

Dursun AY, Uslu G, Cuci Y, Aksu Z (2003a) Bioaccumulation of copper(II), lead(II) and chromium(VI) by growing *Aspergillus niger*. Process Biochem 38:1647–1651

Dursun AY, Uslu G, Tepe O, Cuci Y, Ekiz HI (2003b) A comparative investigation on the bioaccumulation of heavy metal ions by growing *Rhizopus arrhizus* and *Aspergillus niger*. Biochem Eng J 15:87–92

Fan T, LiuY, Feng B, Zeng G, Yang C, Zhou M, Zhou H, Tan Z (2008) XinWang biosorption of cadmium(II), zinc(II) and lead(II) by *Penicillium simplicissimum*: isotherms, kinetics and thermodynamics. J Hazard Mater 160:655 661

Gadd GM (2009) Biosorption: critical review of scientific literature, environmental importance and significance of pollution treatment. J Chem Technol Biotechnol 84:13–28

Goyal N, Jain SC, Banerjee UC (2003) Comparative studies on the microbial adsorption of heavy metals. Adv Environ Res 7:311–319

Gulati R, Saxena RK, Gupta R (2002) Fermentation waste of *Aspergillus terreus*: a potential copper biosorbent. World J Microbiol Biotechnol 18:397–401

Hefnawy MA, Razak AA (1998) Alteration of cell-wall compostion of *Fusarium oxysporum* by copper stress. Folia Microbiol 43:453–458

Kacar Y, Arpa C, Tan S, Denizli A, Genc O, Arica MY (2002) Biosorption of Hg(II) and Cd(II) from aqueous solutions: comparison of biosorptive capacity of alginate and immobilized live and heat inactivated *Phanerochaete chrysosporium*. Process Biochem 37:601–610

Kapoor A, Viraraghavan T (1995) Fungal biosorption—an alternative treatment option for heavy metal bearing wastewaters: a review. Biores Technol 53:195–206

Kapoor A, Viraraghavan T (1997a) Heavy metal biosorption sites in *Aspergillus niger.* Biores Technol 61:221–227

Kapoor A, Viraraghavan T (1997b) Fungi as biosorbents. In: Wase J, Forster C (eds) Biosorbents for metal ions. Taylor and Francis, London, pp 67–86

Kapoor A, Viraraghavan T (1998a) Biosorption of heavy metals on *Aspergillus niger:* effect of pretreatment. Biores Technol 63:109–113

Kapoor A, Viraraghavan T (1998b) Removal of heavy metals from aqueous solutions using immobilized fungal biomass in continuous mode. Water Res 32:1968–1977

Kapoor A, Viraraghavan T, Cullimore DR (1999) Removal of heavy metals using the fungus *Aspergillus niger*. Biores Technol 70:95–104

Khambhaty Y, Mody K, Basha S, Jha B (2009) Kinetics, equilibrium and thermodynamic studies on biosorption of hexavalent chromium by dead fungal biomass of marine *Aspergillus niger.* Chem Eng J 145:489–495

Kiran I, Akar T, Tunali S (2005) Biosorption of Pb(II) and Cu(II) from aqueous solutions by pretreated biomass of *Neurospora crassa*. Process Biochem 40:3550–3558

Kogej A, Pavko A (2001) Comparison of *Rhizopus nigricans* in a pelleted growth form with some other types of waste microbial biomass as biosorbents for metal ions. World J Microbiol Biotechnol 17:677–685

Kumar R, Bishnoi NR, Bishnoi K (2008) Biosorption of chromium(VI) from aqueous solution and electroplating wastewater using fungal biomass. Chem Eng J 135:202–208

Lesmana SO, Febrianaa N, Soetaredjoa FE, Sunarsob J, Ismadjia S (2009) Studies on potential applications of biomass for the separation of heavy metals from water and wastewater. Biochem Eng J 44:19–41

Mapolelo M, Torto N (2004) Trace enrichment of metal ions in aquatic environments by *Saccharomyces cerevisiae*. Talanta 64:39–47

Mukhopadhyay M, Noronha SB, Suraishkumar GK (2007) Kinetic modeling for the biosorption of copper by pretreated *Aspergillus niger* biomass. Process Biochem 98:1781–1787

Mullen MD, Wolf DC, Beveridge TJ, Bailey GW (1992) Sorption of heavy metals by the soil fungi *Aspergillus niger* and *Mucor rouxi*. Soil Biol Biochem 24:129–135

Mungasavalli DM, Viraraghavan T, Jin Y-C (2007) Biosorption of chromium from aqueous solutions by pretreated *Aspergillus niger*: Batch and column studies. Colloids Surf A-Physicochem Eng Asp 301:214–223

Ozer A, Ozer D (2003) Comparative study of the biosorption of Pb(II), Ni(II) and Cr(VI) ions onto *S. cerevisiae*: determination of biosorption heats. J Hazard Mater 100:219–229

Park JK, Lee JW, Jung JY (2003) Cadmium uptake capacity of two strains of *Saccharomyces cerevisiae* cells. Enzyme Microb Technol 33:371–378

Sarret G, Manceau A, Spadini L, Roux JC, Hazemann JL, Soldo Y, EybertBerard L, Menthonnex JJ (1998) Structural determination of Zn and Pb binding sites in *Penicillium chrysogenum* cell walls by EXAFS spectroscopy. Environ Sci Technol 32:1648–1655

Say R, Denizil A, Arica MY (2001) Biosorption of cadmium(II), lead(II) and copper(II) with the filamentous fungus *Phanerochaete chrysosporium*. Biores Technol 76:67–70

Say R, Yilmaz N, Denizli A (2003a) Biosorption of cadmium, lead, mercury, and arsenic ions by the fungus *Penicillium purpurogenum*. Sep Sci Technol 38:2039–2053

Say R, Yimaz N, Denizli A (2003b) Removal of heavy metal ions using the fungus *Penicillium canescens*. Adsorpt Sci Technol 21:643–650

Say R, Yilmaz N, Denizli A (2004) Removal of chromium(VI) ions from synthetic solutions by the fungus *Penicillium purpurogenum*. Eng Life Sci 4:276–280

Singh H (2006) Mycoremediation. Wiley, New York

Skowronski T, Pirszel J, Pawlik-Skowronska B (2001) Heavy metal removal by the waste biomass of *Penicillium chrysogenum*. Water Qual Res J Can 36:793–803

Spanelova M, Machovic V, Brezina M (2003) Characterization and sorption properties of *Aspergillus niger* waste biomass. Cent Eur J Chem 1:192–200

Su HJ, Zhao Y, Li J, Tan TW (2006) Biosorption of Ni^{2+} by the surface molecular imprinting adsorbent. Process Biochem 41:1422–1426

Suh JH, Kim DS (2000) Effects of Hg^{2+} and cell conditions on Pb^{2+} accumulation by *Saccharomyces cerevisae*. Bioprocess Eng 23:327–329

Tan TW, Cheng P (2003) Biosorption of metal ions with *Penicillium chrysogenum*. Appl Biochem Biotechnol 104:119–128

Tan TW, Hu B, Su HJ (2004) Adsorption of Ni^{2+} on amine-modified mycelium of *Penicillium chrysogenum*. Enzyme Microb Technol 35:508–513

Tewari N, Vasudevan P, Guha BK (2005) Study on biosorption of Cr(VI) by *Mucor mhiemalis*. Biochem Eng J 23:185–192

Tsezos M (2001) Biosorption of metals. The experience accumulated and the outlook for technology development. Hydrometallurgy 59:241–243

Tunali S, Kiran I, Akar T (2005) Chromium(VI) biosorption characteristics of *Neurospora crassa* fungal biomass. Miner Eng 18:681–689

Vijayaraghavan K, Yun YS (2008) Bacterial biosorbents and biosorption. Biotechnol Adv 26:266–291

Volesky B (2003) Sorption and biosorption. BV Sorbex, Montreal

Wang J, Chen C (2006) Biosorption of heavy metals by *Saccharomyces cerevisiae*: a review. Biotechnol Adv 24:427–451

Wang J, Chen C (2009) Biosorbents for heavy metals removal and their future. Biotechnol Adv 27:195–226

Wase J, Forster C (1997) Biosorbents for metal ions. Taylor and Francis, London

Weber WJ Jr (1972) Adsorption. In: Weber WJ Jr (ed) Physicochemical processes for water quality control. Wiley, New York, pp 199–260

West W, Nightingale P (2009) Organizing for innovation: towards successful translational research. Trends Biotechnol 27:558–561

Yan G, Viraraghavan T (2000) Effect of pretreatment on the bioadsorption of heavy metals on *Mucor rouxii*. Water SA 26:119–123

Yan G, Viraraghavan T (2001) Heavy metal removal in a biosorption column by immobilized *M. rouxii* biomass. Biores Technol 78:243–249

Yan G, Viraraghavan T (2003) Heavy-metal removal from aqueous solution by fungus *Mucor rouxii*. Water Res 37:4486–4496

Yin P, Yu Q, Jin B, Ling Z (1999) Biosorption removal of cadmium from aqueous solution by using pretreated fungal biomass cultured from starch wastewater. Water Res 33:1960–1963

Zhang L, Zhao L, Yu Y, Chen C (1998) Removal of lead from aqueous solution by non-living *Rhizopus nigricans*. Water Res 32:1437–1444

Zhou JL (1999) Zn biosorption by *Rhizopus arrhizus* and other fungi. Appl Microbiol Biotechnol 51:686–693

Chapter 7
Algal Biosorption and Biosorbents

Felisa González, Esther Romera, Antonio Ballester, María Luisa Blázquez, Jesús Ángel Muñoz and Camino García-Balboa

Abstract Algae belong to the kingdom Protista which contains all the Eucaryotes organisms that cannot be classified within other eucaryotic kingdoms: Fungi, Animalia or Plantae. They are autotrophic organisms that carry out an oxygenic photosynthesis. Maybe the most well-known use of algae since ancient times is in food, especially in the Asian coast. In addition, the phycocolloid industry uses algae as raw material in the manufacture of a wide variety of additive products in the cosmetic, pharmaceutical and food industries. Lately algae have been proposed for the treatment of wastewaters due to their high heavy metal sorption capacity. Although, traditionally they have been used in less extent than other biomass, algae have important advantages such as: high efficiency metal removal, non-toxic chemical sludge and low cost. The main kinds of algae (green, red and brown) have constituents (cellulose, carrageenan and alginate, respectively) that provide binding sites such as: hydroxyl, carboxyl, amino and sulfhydryl, which are responsible for the selectivity of these biomass for heavy metals. In this way, *Fucus spiralis*, a brown alga very common in the Galician coast, has been proved very selective in the sorption of copper versus other heavy metals. Like for other types of biomass, one way to improve its biosorbent capacity is by pre-treatment with different reagents.

Keywords Algae types • Binding sites • Sorption uptake • Biosorbent pretreatment

F. González (✉)
Departamento de Ciencia de los Materiales e Ingeniería Metalúrgica, Facultad de Ciencias Químicas, Universidad Complutense, 28040 Madrid, Spain
e-mail: fgonzalezg@quim.ucm.es

P. Kotrba et al. (eds.), *Microbial Biosorption of Metals*,
DOI 10.1007/978-94-007-0443-5_7,

7.1 Introduction

7.1.1 A Few Words About Taxonomy

Algae are a diverse group of organisms that cause mayhem even when professional systematists and biologists try to define them in relation to other organism. Through the years, algal classification has been very unstable. Placements at the kingdom and lower levels are currently still dynamic. It is interesting to trace the changes in their position in the hierarchy of life. Algae, like bacteria and fungi, were often assigned to the plant kingdom. Properties for including the algae into the plant kingdom were their ability to make their own food by photosynthesis, their structural similarity to land plants and the fact that the larger forms were observed to be sedentary.

Unicellular creatures that move and ingest food—protozoa—were called animals. Microbes such as *Euglena* that move but are photosynthetic were claimed by both botanists and zoologists, and showed up in the taxonomies of both the plant and animal kingdoms. Clearly, the 2-Kingdom System was riddled with conceptual and operational problems. Technological advances, such refined microscopy further proved that the system was oversimplified and therefore not widely acceptable.

In 1969, Robert H. Whittaker suggested a scheme with additional kingdoms as an alternative to the 2-Kingdom System (Animalia and Plantae). The 5-Kingdom System proposed considered Monera, Protista, Plants, Animals and Fungi. Most algae are traditionally considered as a plant subkingdom within the 5-kingdom classification. However, their phylogenetic relationship and evolutionary lineages are still not reflected adequately by the 5-kingdom system.

Major revisions were made to the 5-Kingdom System to form the 8-Kingdom System. The 8-Kingdom System was the most recently proposed and also it is the most controversial. Its construction stresses that the term "algae" wrongly tried to combine organisms with such diversity and differential evolutionary histories into a single group. Under the 8-Kingdom System, algal members are dispersed in vastly different kingdoms that are distinct from each other. Based on advanced genetic and ultrastructural studies, this system is generally supported by a handful of biologists who advocate the need to either elevate various algal members to a higher taxonomic status or to qualify the more distinctively. This has led to the reorganization of many groups in such a way that their diagnostic characters suggest their evolutionary relationship to other groups accurately.

7.1.2 Algae Habitat and Uses

Algae are a very diverse group of all photosynthetic organisms that are not plants. Algae are important in marine, freshwater, and some terrestrial ecosystems. Seaweeds are large marine algae. Algae may be unicellular, colonial, or multicellular. Some algae, like the diatoms, are microscopically small. Other algae, like kelp, are as big as trees. Some algae, the phytoplankton, drift in the water. Other algae, the epiphytic or benthic algae, grow attached to rocks, docks, plants, and other solid objects (Raven et al. 1999; Lembi and Waaland 1988).

Probably the most well-known use of algae since ancient times is the food industry, especially in the Asian coast (Korea is the world's largest consumer). The species more consumed are: *Undaria pinnatifida, Himantalia elongata, Ascophyllum nodosum, Gelidium sesquipedale, Porphydadioica* and *Palmaria palmata.* Algae in Eastern cultures have also been used in fertilizers, glass manufacturing or iodine extraction.

In the Western world, algae were initially used as food by human populations of low standard of living. Paradoxically, it was in these countries where algae acquired a major relevance in the modern industry. In the middle of the twentieth century, the phycocolloid industry was developed. That industry is characterized by the use of algae as raw material for the manufacture of a wide variety of compounds used as additives in the cosmetic, pharmaceutical and food industries. More recently, especially in the last decade, there has been an increasing interest for algae as biosorbent material due to its high sorption capacity and its almost unlimited availability in oceans.

However, in the field of biosorption, it has not been until recently that biomass of different origin have been used, such as: waste materials of food and agricultural industry (Nilanjana 2008). Traditionally, bacteria and fungi have been the biomass more studied. Unlikely, algae have been used in less proportion.

According to our own statistical study, of the 2057 bibliographic records found on biosorption until 2004 only 315 were related to algae that is 15.31% of the total (Romera et al. 2006). In addition, despite the great variety of algae in nature, the number of species studied for its sorption properties and the works providing equilibrium parameters are rather scarce. Considering the latter aspect, Table 7.1 shows the range of values collected from literature of the maximum sorption uptake (q_{max}) and constant b, related with the affinity of the biomass (green, red or brown algae) for the metal. Both parameters were derived from the Langmuir model applied to experimental results obtained with dead biomass without any treatment. Such table provides information on the more frequently algae used and the metal commonly studied. These data seems consistent with the excellent biosorbent capacity of algae and makes them a potentially attractive raw material.

An evidence of the interest for using algae is the increasing number of research groups working on biosorption of metals with algae (Apiratikul and Pavasant 2008; Grimm et al. 2008; Akhtar et al. 2008; Deng et al. 2007; Han et al. 2007; Vilar et al. 2007; Al-Rub et al. 2006; Aksu and Dönmez 2006; Gupta et al. 2006; Luo et al. 2006; Martins et al. 2006; Vijayaraghavan et al. 2006; Aksu and Acikel 2000; Figueira et al. 2000a). These authors have investigated different aspects of the process such as the use of free or immobilized biomass in batch or column experiments, respectively; the factors affecting the process (pH, metal initial concentration, metal speciation, biomass concentration, temperature, presence of other cations and anions, pretreatment of biomass); mathematical modelling; mechanistics; thermodynamic and kinetic aspects; and the possibility of biomass regeneration using

Table 7.1 Sorption parameters of biomass without a previous treatment in monometallic systems. (According to Romera et al. 2006)

Alga	Metal	q_{max} (m mol/g)	B (l/m mol)
Ascophyllum nodosum (B)	Cd^{2+}	0.34–1.91	6.29–31.24
	Ni^{2+}	1.35–2.32	3.28–9.16
	Pb^{2+}	1.31–2.31	21.75–42.06
Chaetomorha linum (G)	Cd^{2+}	0.48	1.43
Chlorella miniata (G)	Cu^{2+}	0.366	26.586
	Ni^{2+}	0.21–1.02	2.93–16–63
Chlorella vulgaris (G)	Cd^{2+}	0.30	56
	Pb^{2+}	0.47	38
	Zn^{2+}	0.37	6
	Cu^{2+}	0.25–0.76	4.44–13–34
Cladophora glomerata (G)	Pb^{2+}	0.355	447.53
Chondrus crispus (R)	Ni^{2+}	0.443	12.920
	Pb^{2+}	0.941	3.315
Codium fragile (G)	Cd^{2+}	0.082,7	1.124
Codium taylori (G)	Ni^{2+}	0.099	29.06
	Pb^{2+}	1.815	5.38
Corallina officinalis (R)	Cd^{2+}	0.264,2	21.356
Fucus vesiculosus (B)	Cd^{2+}	0.649	2.360
	Ni^{2+}	0.392	16.140
	Pb^{2+}	1.10–2.90	32.11–62.77
Galaxaura marginata (R)	Ni^{2+}	0.187	4.638
	Pb^{2+}	0.121	9.110
Gracilaria corticata (R)	Pb^{2+}	0.201,7	480.68
Gracilaria changüí (R)	Cd^{2+}	0.23	9.65
Gracilaria corticata (R)	Pb^{2+}	0.260,6	327.36
Gracilaria edulis (R)	Cd^{2+}	0.24	4.82
Gracilaria salicornia (R)	Cd^{2+}	0.16	9.04
Padina sp. (B)	Cd^{2+}	0.53	5.37
Padina gymnospora (B)	Ni^{2+}	0.170	3.933
	Pb^{2+}	0.314	108.980
Padina tetrastomatica (B)	Pb^{2+}	1.049	149.177
	Cd^{2+}	0.53	4.65
Polysiphonia violacea	Pb^{2+}	0.492,3	2,734.908
Porphira columbina (R)	Cd^{2+}	0.404,8	4.496
Sargassum sp. (B)	Cd^{2+}	1.40	6.52
	Cu^{2+}	1.08	0.236,4
Sargassum baccularia (B)	Cd^{2+}	0.74	4.67
Sargassum fluitans (B)	Ni^{2+}	0.409	17.260
	Pb^{2+}	1.594	44.340
Sargassum hystrix (B)	Pb^{2+}	1.375,5	89.09
Sargassum natans (B)	Cd^{2+}	1.174	23.490
	Ni^{2+}	0.409	6.980
	Pb^{2+}	1.15–1.22	54.07–116.03
Sargassum siliquosum (B)	Cd^{2+}	0.73	6.60
Sargassum vulgare (B)	Ni^{2+}	0.085	30.880
	Pb^{2+}	1.100	19.270

B brown, *G* green and *R* red

Table 7.2 Advantages and disadvantages of using algal biomass for heavy metals removal from wastewater. (According to Brinza et al. 2007)

Advantages	Disadvantages
• No need to synthesise the product • Can be applied in wastewater with higher metal concentrations than for membrane processes • High efficiency of metals removal • High uptake capacity • Biomass can be regenerated • Biomass can be re-used in many adsorption/desorption cycles • Selectivity for heavy metals ions • Macroalgal biomass does not need to be inmobilized • No toxic chemical sludge generated • Few chemicals needed for desorption and regeneration of biosorbent • Can be used in continuous and discontinuous regimes • If dead biomass used, no oxygen or nutrient supply required • Suitable for aerobic and anaerobic effluent treatment units • Can be used all year round • Low cost	• If dead biomass used, energy needed for drying • Microalgae need to be inmobilized • Microalgae have limited applicability in batch systems

different extractant reagents. In the last decade, several noteworthy revisions on the use of algae have been published (Brinza et al. 2007; Mehta and Gaur 2005; Davis et al. 2003). Further interesting information on both the structure of algae cell walls, directly related to sorption mechanisms, and process yields can be found in three recent works (Lesmana et al. 2009; Wang and Chen 2009; Nilanjana et al. 2008).

Table 7.2 summarizes the main advantages and disadvantages of the use of algal biomass for metal uptake from wastewaters (Brinza et al. 2007). On that basis, algae can be considered as an alternative to conventional adsorbent materials used in the treatment of effluents contaminated with heavy metals. Nevertheless, further research studies are necessary on aspects such as: the selectivity of algal species, the design of new immobilization matrices for microalgae, interferences with other compounds in the effluents and modelling and simulation of processes.

7.2 Types of Algae

7.2.1 The Biodiversity of Algae: Classification

Algae classification may consider several aspects such as the habitat or the colour—which depends on the living site and the receiving radiation. Other criteria such as the nature of chlorophyll(s), the cell wall chemistry, flagellation, form in which food or assimilatory products of photosynthesis are stored, cell morphology, habitat, reproductive structures, life story patterns, etc, are other characteristics that could have been used for the classification of algae. Other characteristics used to classify algae have been the nature of chlorophyll(s) present, the carbon reserve polymers produced, the cell wall structure and the type of motility (Wang and Chen 2009). All algae contain chlorophyll a. Some, however, also contain other chlorophylls that differ in minor ways from chlorophyll a. The presence of these additional chlorophylls is characteristic of particular algal groups. The nature of the reserve polymer synthesized as a result of photosynthesis is another differential aspect between the different groups.

Table 7.3 presents a classification of algae (Wang and Chen 2009) based on the type of pigments, cell wall and stored food materials. The potential of the different species as biosorbents of metal removal was taken into consideration.

The main characteristics of the major five groups are as follows:

1. **Green Algae.** They are the algae most closely related to plants. They have the same pigments (chlorophyll a and b and carotenoids), the same chemicals in their cell walls (cellulose), and the same storage product (starch) as plants. Green algae may be unicellular or form filaments, nets, sheets, spheres, or complex mosslike structures. There are both freshwater and marine species. Some species of green algae live on snow, or in symbiotic associations as lichens, or with sponges or other aquatic animals. Edible green algae include *Chlorella* and the so called "sea lettuce". There are at least seventeen thousand species of green algae.
2. **Diatoms.** They are often regarded as the most beautiful of the algae. Each diatom has a cell wall made of glass that is very finely etched with a species-specific pattern of dots and lines. The patterns on the diatom cell walls are so precise that they were used for years to test the optics of new microscopes. Diatoms are also the most abundant algae in the open ocean and responsible for about one-quarter of all the oxygen gas produced on the earth each year. Diatom populations often bloom in lakes in the spring, providing a major food for zooplankton, forming the base of the aquatic food chain. There are over one hundred thousand species of diatoms.
3. **Red Algae.** They are almost exclusively marine and include many edible and economically important species, including nori and laver. These algae are also the source of carrageenan and agar, which are used as food thickeners and stabilizers. Red algae are mostly large, complex seaweeds. There are four thousand to six thousand species.
4. **Brown Algae.** They algae are most exclusively marine and include the largest and most complex seaweeds. Kelp, for example, may be more than 60 m high, and forms dense underwater forests off the California coast. Other important brown algae include the rockweeds and *Sargassum*, for which the Sargasso Sea is named. There are about fifteen hundred species of brown algae.
5. **Dinoflagellates.** They are unicellular algae with armor made of cellulose and flagella that cause them to spin as they swim. Dinoflagellates are found in both

Table 7.3 A classification of major groups of algae and their primary characteristics. (According to Wang and Chen 2009)

Group	Common name	Morphology	Pigments	Typical representative	Carbon reserve materials	Cell walls	Major habitats	Kingdom
Chrysophyta	Yellow-green and golden-brown algae; diatoms	Unicellular	Chlorophylls a and c	*Navacul* sp.	Lipids	Many have two overlapping components made of silica	Freshwater, marine, soil	Protista (single cell or colonia; eukaryotic)
Euglenophyta	Euglenoids	Unicellular, Photosynthetic euglenoid flagellates	Chlorophylls a and b	*Euglena* sp.	Pramylon (β-1,2-glucan)	No wall present	Freshwater, a few marine	Protista
Pyrrhophyta	Dinoflagellates							Protista
Charophyta	Stoneworts							Protista
Chlorophyta	Green algae	Unicellular to leafy	Chlorophylls a and b	*Chlamydomonas* sp.	Starch (α-1,4-glucan)	Cellulose	Freshwater, soil, a few marine	Protista
Phaeophyta	Brown algae	Filamentous leafy, occasionally massive and plantlike	Chlorophylls a and c, xanthophylls	*Laminaria* sp.	Laninarin (β-1,3-glucan), mannitol	Cellulose	Marine	Plantae (multicellular: eukaryotic)
Rhodophyta	Red algae	Unicellular, filamentous to leafy	Chlorophylls a and d, phycocyanim, phycoerythrin	*Polysiphonia* sp.	Floridean starch (α-1,4- and α-1,6-glucan), fluoridoside (glycerol-galactoside)	Cellulose	Marine	Plantae

freshwater and marine ecosystems. Some species of dinoflagellates emit an eerie blue when disturbed, called bioluminescence. Other dinoflagellates are toxic and responsible for red tides and outbreaks of shellfish poisoning. There are two thousand to four thousand species of dinoflagellates.

7.2.2 Algal Metal Binding Sites

In biosorption various algae have been used and investigated as biosorbents for metal removal. Metal biosorption mainly depends on the components of the biomass cell, especially the cell surface and the spatial structure of the cell wall. Peptidoglycan, teichoic acids and lipoteichoic acids are all important chemical components of bacterial surface structures. Various polysaccharides, including cellulose, chitin, alginate, glycan, etc. present in fungi or algae cell walls, have been proved to play a very important role in metal binding for certain kinds of biomass. Some functional groups have been found to bind metal ions, especially carboxyl group. There are evidences that confirm that functional groups containing O-, N-, S-, or P-, participate directly in binding certain metals (Wang and Chen 2009). The carboxyl, hydroxyl, sulphate and amino groups in algal cell wall polysaccharides act as binding sites for metals (Lesmana et al. 2009). In addition, the acid treatment of algae can dissolve polysaccharide compounds in the outer layer of the cell wall to a certain extent, thus creating additional binding sites (usually amino groups).

Table 7.4 summarizes the most representative functional groups and classes of organic compounds in biomass potentially involved in biosorption processes (Wang and Chen 2009). The symbol R is shorthand for residue, and its placement in the formula indicates that is attachment at that site varies from one compound to another, according to Talaro and Talaro (2002).

Some active sites involved in metal uptake can be determined by using techniques of titration, infrared and Raman spectroscopy (XPS), electron microscopy (scanning and/or transmission), nuclear magnetic resonance (NMR), X-ray diffraction analysis (XRD), X-ray absorption fine structure spectroscopy (XAFS), etc. (Wang and Chen 2009). The most important of these groups have been summarized by Volesky (2007) and include: carbonyl (ketone), carboxyl, sulfhydryl (thiol), sul-

Table 7.4 Functional groups and classes of organic compounds in biomass. (According to Talaro and Talaro 2002)

Functional group	Class of compounds
Hydroxyl	Alcohols, carbohydrates
Carboxyl	Fatty acids, proteins, organic acids
Amino	Proteins, nucleic acids
Ester	Lipids
Sulfhydryl	Cysteine (amino acid), proteins
Carbonyl, terminal end	Aldehydes, polysaccharides
Carbonil, internal	Ketones, polysaccharides
Phosphate	DNA, RNA, ATP

fonate, thioeter, amine, secondary amine, amide, imine, imidazole, phosphonate, phosphodiester (Volesky 2007).

Although Algae are a large and diverse group of simple plant-like organisms, ranging from unicellular to multicellular forms, which can be seen in bodies of water and terrestrial environments, a classification considering three main types is widely extended. This classification emphasizes the cellular composition and principal sorption active sites.

Some specific examples of metal biosoprtion with different algae are given in the following paragraphs. It has been preferred to maintain the classification widely extended considering four main groups: marine brown macroalgae, marine red macroalgae, fresh water green microalgae and marine green macroalgae (Lesmana et al. 2009). The particular binding site for each species has also been specified.

Marine Green Macroalgae Diverse marine green macroalgae—such *Caulerpa lentillifera* or *Cladophora fascicularis*—are described as biosorbents for divalent cations such as Cu(II), Pb(II) and Zn(II) (Pasavant et al. 2006; Deng et al. 2006). The possible functional groups involved in metal sorption properties of *Caulerpa lentillifera* were carboxylic acids, amines, amides, sulfonyl and sulfonate. In the case of *Cladophora fascicularis*, amino, phosphate and carboxyl were the main groups involved in biosorption processes.

Fresh Water Green Microalgae Several types of green microalgae have also been employed for the removal of heavy metals from solutions (Lesmana 2009). Divalent metals such as Hg, Cd, Pb, Ni, Cu, Zn have been tested with *Chlamydomonas reinhardtii, Chlorella miniata, Chlorella vulgaris, Cladonia rangiformis* and *Fucus spiralis.* Trivalent (Fe, Cr) and hexavalent cations (Cr) have also been removed using *Chlorella vulgaris, Spirulina platensis* and *Chlorella miniata* and *vulgaris*, respectively. The evaluation of Ni(II) and Cu(II) adsorption onto *Sphaeroplea* algae and its acid treated form was performed by Rao et al. (2005).

Chlamydomonas reinhardtii is a motile single cell green algae whose cell wall structure is composed of hydroxyproline-rich glycoproteins. The use of this alga as a biosorbent for heavy metals has been attempted by Tuzun et al. (2005). In their study, they found that the functional groups responsible for biosorption onto *Chlamydomonas reinhardtii* cells were amino, carboxylic acid, hydroxyl and carbonyl groups.

Chlorella is a genus of single cell green algae belonging to the phylum *Chlorophyta. Chlorella vulgaris* is of particular interest as a biosorbent. This planktonic unicellular green alga has been widely exploited as a food supplement. Interesting sorption properties exhibited by this alga originate from its porous cell wall, allowing the free passage of molecules and ions in aqueous solution. Moreover, its cell wall's constituents provide an array of ligands with different functional groups capable of binding various heavy metals (Lesmana et al. 2009).

Marine Red Macroalgae The red marine macroalgae *Gelidium sesquipedale* has been employed to study the complex mechanisms involved in biosorption processes (Vilar et al. 2008).

Marine Brown Macroalgae The cell walls of brown algae generally contain three components (Lesmana et al. 2009): (1) cellulose, the structural support; (2) alginic acid, a polymer of mannuronic and guluronic acids and the corresponding salts of sodium, potassium, magnesium and calcium; and (3) sulphated polysaccharides. Thus, carboxyl and sulphate are the predominant active groups in this alga. An example of the biotechnological application of this alga is related to *Laminaria japonica.* A modified version was optimally used for biosoption of lead (Pb) due to the increasing amount of carboxylic acid groups exposed after the modification process (Luo et al. 2006).

7.3 Comparative Study on Metal Sorption Uptakes and Affinities by Different Types of Algae

One of the mathematical models more cited in the literature is the Langmuir model that provides a quantification of the biosorption phenomenon, such is the case for algae used as biosorbents. The maximum sorption uptake and the equilibrium constant or affinity degree of the biomass for the adsorbate can be derived from that model. The equilibrium constant represents a fundamental tool in order to define the chemical equilibria in this type of systems. From the knowledge of those constants it is possible to predict the behaviour of different biomass with a given metal. Therefore, this could be a useful tool in order to make the right choice between different biosorbents for each occasion.

On the other hand, the great difference in the chemical composition of the cell walls of the different types of algae is also responsible for its different biosorbent behaviour. Our studies in this field have corroborated such idea after studying the sorption uptake of a series of algae with five toxic metals (Cd, Cu, Ni, Pb and Zn) commonly found in industrial effluents (Romera et al. 2007). Table 7.5 collects the parameters of Langmuir, q_{max} and b, obtained with two green (*Codium vermilara* and *Spirogyra insignis*), two red (*Asparagopsis armata* and *Chondrus crispus*) and two brown algae (*Ascophyllum nodosum* and *Fucus spiralis*). Two important conclusions were drawn from that study: (1) there is, in general, a similar decreasing order of affinity of the biomass for the five metals tested, independently of the alga used; which is in agreement with the proposal of Chong and Volesky (1996) that the binding of metals to active sites on the cell walls of the biomass is related to some intrinsic properties of the own metal ion such as: ionic radii and electronegativity. (2) The brown algae present the highest sorption uptake followed by red algae, especially those with carragenates in their cell walls (*Chondrus crispus*), and the worse sorption uptake correspond to green algae. The presence of alginates in cell walls of brown algae, like carragenates in red algae, is responsible for the metal binding on the biomass. Although that is the case for most of the studies reported in the literature, the values of q_{max} for brown algae are higher than for other kinds of algae (Romera et al. 2006).

Table 7.5 Values of Langmuir constants

Algae	Cadmium		Nickel		Zinc		Copper		Lead	
	b (l/m mol)	q_{max} (m mol/g)	b (l/m mol)	q_{max} (m mol/g)	b (l/m mol)	q_{max} (m mol/g)	b (l/m mol)	q_{max} (m mol/g)	b (l/m mol)	q_{max} (m mol/g)
C.odium vermilara	11.15	0.19	5.34	0.22	1.80	0.36	8.92	0.27	23.45	0.30
Spirogyra insignis	13.59	0.20	2.57	0.30	2.58	0.32	5.51	0.30	117.87	0.25
Asparadopsis. armata	10.61	0.29	7.23	0.29	4.92	0.33	8.37	0.33	9.10	0.31
Chondrus crispus	6.37	0.67	2.83	0.63	4.63	0.70	2.47	0.64	2.08	0.98
Fucus spiralis	12.67	1.02	7.92	0.85	6.94	0.81	10.87	1.12	26.72	0.98
Ascophyllum nodosum	17.34	0.78	7.90	0.74	14.38	0.64	10.39	0.93	19.15	0.86

Despite that, there is a lack of experimental data on multimetallic systems. In our studies, we have only employed the brown alga *Fucus spiralis* (Romera et al. 2008a) in an attempt to determine the degree of interference and/or competence between different metals for the same active sites of the biomass. We have studied both bimetallic and trimetallic systems and experimental data could be adjusted to the multimetallic type-Langmuir model given by the following expression (Eq. 7.1):

$$q(M_n) = \frac{q_{\max} b_n C_e(M_n)}{1 + b_1 C_e(M_1) + b_2 C_e(M_2) + \cdots + b_n C_e(M_n)} \tag{7.1}$$

Table 7.6 collects the values of q_{max}, b_1 and b_2 for ten bimetallic systems obtained from the combination of Cd, Cu, Ni, Pb and Zn. The first relevant fact is that the maximum sorption uptake of the biomass remained practically constant around 1 m mol/g. That is an indication that the number of active sites present in the alga is fixed and independent of the solution tested. Nevertheless, the highest values of q_{max} obtained corresponded to the systems performed with the metals showing a higher affinity for the biomass and the lowest to the metals with a worse recovery in monometallic systems.

A similar conclusion was reached in trimetallic systems. Thus, the ternary-type Langmuir equation defined for the Cd-Ni-Cu system the following parameters: $q_{max} = 0.93$, b(Cd) = 6.39 l/m mol, b(Ni) = 1.82 l/m mol and b(Cu) = 17.89 l/m mol, i.e. the maximum sorption uptake was achieved around 1 m mol/g and the affinity order obtained from the values of b was the same that in the mono- and bimetallic systems. Since the biosorption process depended on both the type of biomass and the metal adsorbed, the alga had a limited number of binding sites with a different preference for the metals tested. In fact, *Fucus spiralis* shown specificity for copper.

The treatment of sorption data using the MATLAB software is a useful tool to obtain a graphic representation of the sorption isotherms and allows an analysis on the competition between the different metals for the same sorption sites. For instance, in the trimetallic system shown in Fig. 7.1, at higher copper concentration the sorption uptake of both Cd and Ni decreased significantly (Fig. 7.1a and b); however, none of these two metals affected the sorption of Cu (Fig. 7.1c and d).

Table 7.6 Values of Langmuir constants in ten bimetallic systems

	q_{max} (m mol/g)	b_1 (l/m mol)	b_2 (l/m mol)	R_{12}	R_{22}
Cd-Ni	0.75	21.79	5.32	0.99	0.99
Cd-Zn	0.80	21.50	7.57	0.99	0.95
Cd-Cu	0.96	7.28	17.61	0.97	0.98
Cd-Pb	1.09	3.03	44.25	0.81	0.86
Ni-Zn	0.64	9.89	12.61	0.94	0.99
Ni-Cu	0.92	1.98	18.66	0.85	0.98
Ni-Pb	0.97	1.03	93.46	0.69	0.70
Zn-Cu	0.95	1.98	15.70	0.71	0.99
Zn-Pb	1.04	1.36	50.00	0.70	0.84
Cu-Pb	1.07	5.48	44.44	0.92	0.89

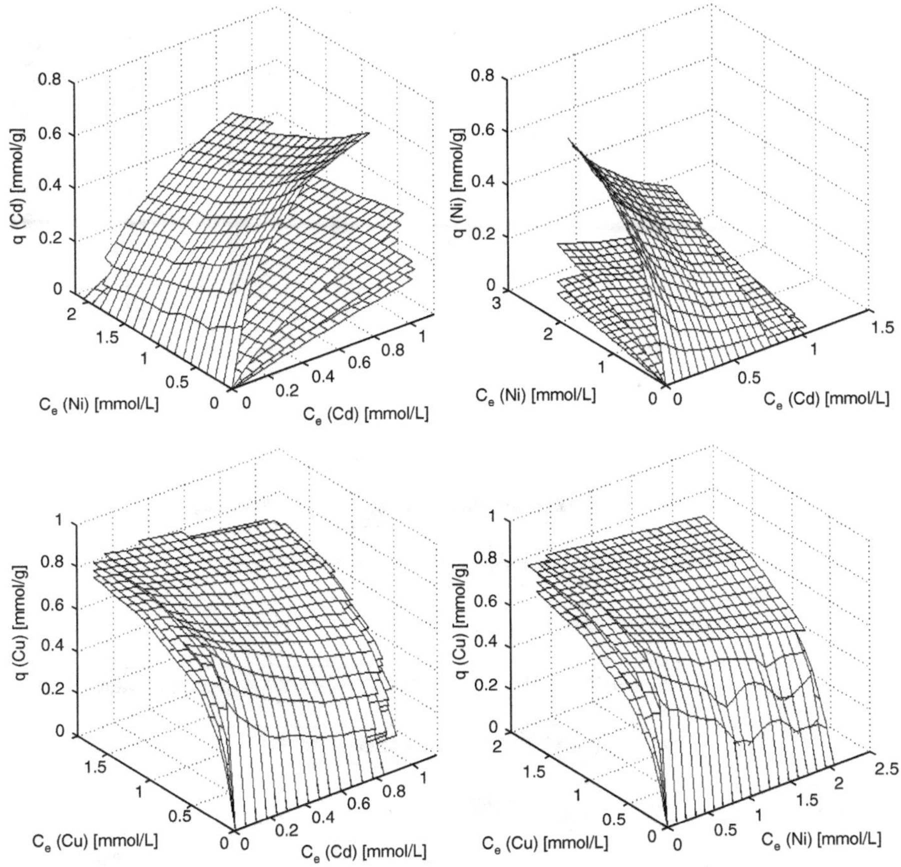

Fig. 7.1 Sorption isotherms in the trimetallic system Cu-Cd-Ni. (According to Romera et al. 2008a)

Therefore, copper was adsorbed on *Fucus spiralis* in the same proportion independently of the presence of Cd and Ni and of its concentration.

All these results indicate that biosorption can be a very attractive process for the decontamination of waters polluted with heavy metals and suggest the idea that biomass, in this case algae, can be specific for a given metal. The specificity of *Fucus spiralis* was corroborated in multimetallic systems and, as shown in Fig. 7.2, such alga has a preference uptake for copper against the rest of metals tested.

In addition, we have also tested different binary and ternary combinations of algae in order to check changes of behaviour with respect to single-algal systems. For that purpose, a green (*Codium vermilara*), a red (*Chondrus crispus*) and a brown alga (*Ascophyllum nodosum*) were selected (Romera et al. 2008b). We found that metal uptake was improved with any kind of mixture compared to the green alga but not for the brown alga (Fig. 7.3). Again, these results confirmed that brown algae presented better biosorbent characteristics than red or green algae. Furthermore,

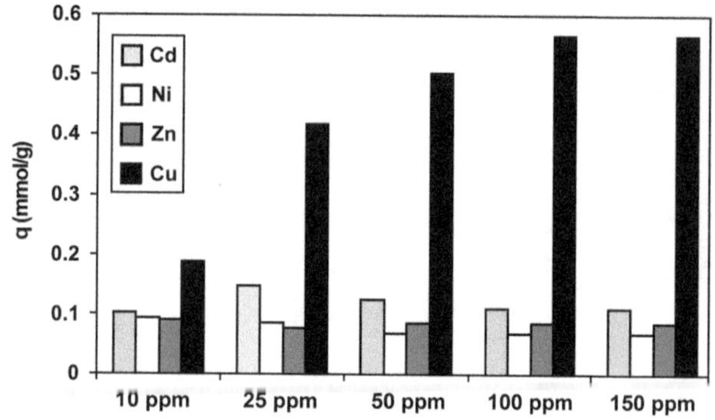

Fig. 7.2 Sorption uptake values using metal concentrations in Cd-Cu-Ni-Zn system. (According to Romera et al. 2008a)

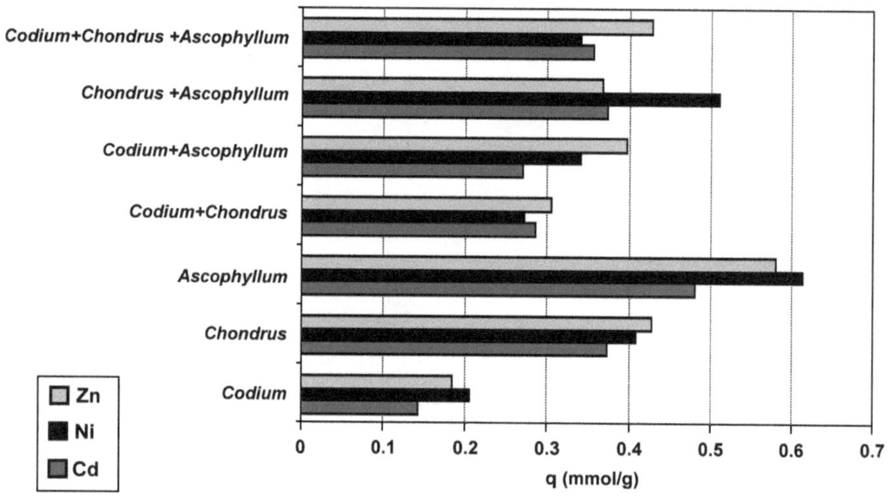

Fig. 7.3 Sorption of several metals with different algae (single and complex biomass)

metals with a worse single-algal sorption in multimetallic systems behaved better when biomass (Fig. 7.4). These results suggest that the mixture of algae can be an interesting choice for the treatment of real solutions that contain several metals since algae usually are not solely found in nature.

7.4 Pretreatment of Algal Biomass

From the scientific point of view, physico-chemical modifications of biomass, particularly algae, could lead to better sorption uptakes, better metal yields or higher mechanical stability. The literature describes a wide variety of reagents that could be used for the pre-treatment of biomass: acids, alkalis, salts, organic compounds (acetone, methanol, ethanol, formaldehyde, glutaraldeyde, etc), enzimes (lipase, lisozine, alkaline phosphatase), metal chelates (EDTA) and distilled water (Nad-

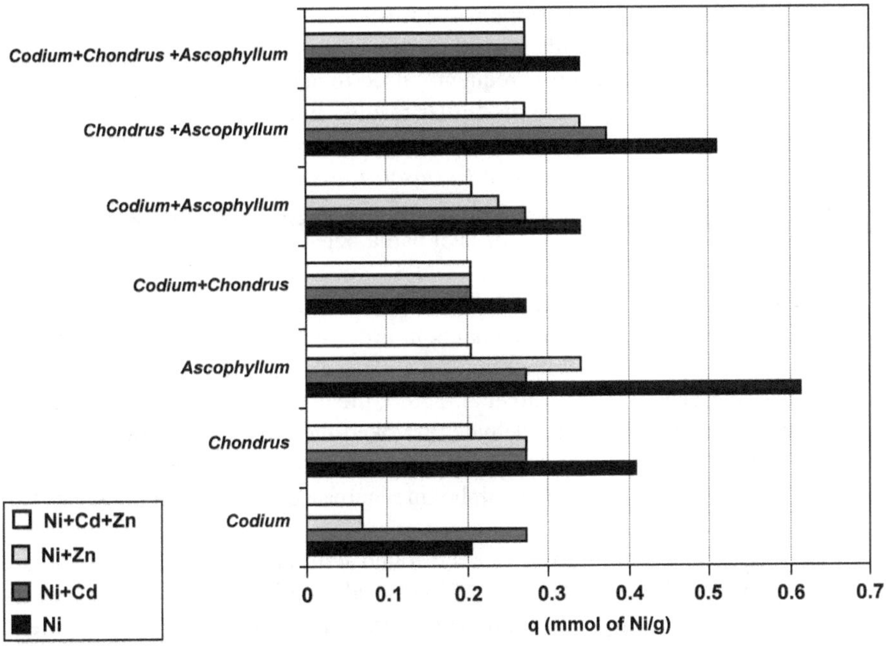

Fig. 7.4 Specific sorption uptake of nickel in monometallic, bimetallic and trimetallic systems for different algal biomass. (According to Romera et al. 2008b)

dafi and Saeedi 2009; Southichak et al. 2009; Yinghui et al. 2009; Mehta and Gaur 2005). In general, the action of these reagents can be summarized as follows: (1) the biomass cleanup which facilitates the access of metal cations to the active centres; (2) the chemical modification of specific compounds of the biomass (proteins, esters or polysaccharides) that promotes the formation of new active centres; (3) the improvement of the binding efficiency between active centres and metals facilitating its ion exchange with some cations (protons, sodium or calcium) or the formation of electrostatic bonds that modifies the total surface charge by increasing the number of active groups with negative charge (carboxylic or nitrogenated) and finally, (4) the increase of the structural stability of the biomass, principally when using continuous flow reactors, by provoking the cross-linking of the biomass through the formation of covalent bridges between their functional groups.

However, chemical modification of the biomass not always produces the desire effects. Some methods, unlike others, were able to improve the metal uptake in comparison with the untreated biomass. Our studies with the brown alga *Fucus vesiculosus* using HCl, $CaCl_2$, formaldehyde, Na_2CO_3 and NaOH as pretreatment reagents (Rincon et al. 2005) showed that only the pretreatment of the biomass with $CaCl_2$ improved the sorption yield of Cu, Pb and Ni by 38, 20 and 16%, respectively. On the other hand, HCl and formaldehyde had no effect on the sorption uptake of the biomass and, conversely, the alkaline reagents provoked the metal precipitation by increasing pH.

Several studies have also shown the positive effect of pretreating algal biomass with $CaCl_2$ (Southichak 2008; Feng and Aldrich 2004; Kaewsarn and Yu 2001).

According to them, calcium is responsible for the biosorption process through an ion exchange mechanism. With respect to the treatment of the biomass with dilute acid, HCl has been the most frequently used to achieve the protonation of the functional groups, although with dissimilar results. In some cases, such treatment did not affect the sorption uptake of the alga, i.e. *Fucus vesiculosus* (Rincon et al. 2005); however, with the brown alga *Cystoseira myrica* its protonation with HCl provoked a significant sorption increase of Cu^{2+} (Naddafi and Saeedi 2009) whereas with *Scenedesmus obiquus* the use of HCl had a negative effect on the sorption of U (Zhang et al. 1997).

Sometimes, the aim of a chemical modification of the biomass is to achieve cross-linking of the polyssacharide chains in order to improve structural stability. However, according to the literature this not always improve metal sorption uptake. For instance, in a recent study (Lio et al. 2009), the pretreatment of the brown alga *Laminaria japonica* with glutaraldehyde provoked the cross-linking but destroyed some of the binding sites which decreased metal uptake compared with the raw biomass; on the contrary, dimethyl sulphoxide increased the sorption uptake of Cd, Zn, Cu and Ni.

Similar unequal effects have also been found with another pretreating reagent frequently used as NaOH: a decrease of the metal sorption uptake due to the solubilization of different alginate forms (Figueira et al. 2000b) or dissimilar metal uptakes on *Chlorella vulgaris* treated with NaOH 0.1 mM which significantly improved sorption of Cu but slightly decreased sorption of Ni (Mehta and Gaur 2001).

Therefore, there is not a generalized criterion that can be used whether on the biomass or the right type of treatment to be employed. Then, the viability of the pretreatment process should be considered for each specific case.

7.5 Comparison with Other Biosorbents and Future Perspectives

Taking into account that, in general, biological materials present good metal uptakes the next step should be to make a choice among the large number of different types of biomass potentially available for biosorption. A considerable number of bacteria, fungi, algae and yeasts, and different wastes and by-products of the agriculture and food industry have been investigated for their biosorbent metal properties (Das et al. 2008). Indeed, the choice of the biosorbent should consider both efficiency and economy.

The efficiency of the process will depend on the biomass chemical composition which varies significantly for different species within the same genus or order (Gadd 2009). In this way, in our statistical study of 2004 the average values of q_{max} published for five metals and different types of biomass (bacteria, fungi and algae) was already an indication that algae had better metal sorption uptakes than the rest of biomass (Fig. 7.5). Nevertheless, at that moment, the amount of data available for algae was lower than for bacteria and fungi. In addition, in recent years there has

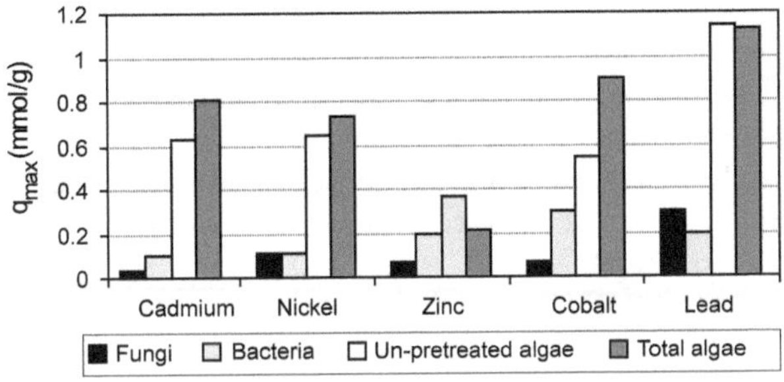

Fig. 7.5 Average values of q_{max} obtained with different types of biomass

been an increasing interest in the use of new sorbent and very efficient biomaterials, mainly agriculture and industrial wastes which were not included in the mentioned comparative study.

At present, the various bibliographic revisions on the sorption uptake by different biomass do not allow to assert which one is more effective due to the many determining factors involved (Wang and Chen 2009; Das et al. 2008; Gadd 2009). Table 7.7 shows sorption data of non-algal biomass frequently used in investigations classified according to the corresponding type: bacteria, fungi or plant wastes.

Table 7.7 Biosorption uptakes by biomass from various sources

	q_{max} (m mol/g)					References
	Cu	Cd	Ni	Pb	Zn	
Bacteria						
Bacillus cereus	0.79			0.18		Pan et al. (2007)
Bacillus sp.	0.26			0.45		Wang and Chen (2009)
Geobacillus toebii	0.76	0.26	0.36		0.21	Ozdemir et al. (2009)
Pseudomonas aeruginosa	0.36	0.38		0.38		Wang and Chen (2009)
Pseudomonas putida	1.53	0.07		0.27	0.27	Wang and Chen (2009)
Streptomyces rimosus		0.58		0.15	1.22	Wang and Chen (2009)
Fungi						
Aspergillus niger	0.08			0.15		Ahluwalia and Goyal (2007)
Mucor rouxii		0.06	0.09	0.08	0.08	Ahluwalia and Goyal (2007)
Saccharomyces cerevisiae	0.16				0.61	Ahluwalia and Goyal (2007)
Rhizopus arrhizus	0.15	0.24	0.31	0.27	0.21	Ahluwalia and Goyal (2007)
Penicillium chrysogenum	0.14	0.10	1.41	0.56	0.10	Wang and Chen (2009)
Plant or agricultural wastes						
Lemon peel		0.47		0.87		Lesmana et al. (2009)
Orange peel		0.33				Lesmana et al. (2009)
Olive pomace	0.48	0.10				Lesmana et al. (2009)
Rice husk	0.22			0.07	0.21	Lesmana et al. (2009)
Saw dust	0.22	0.65	0.18			Ngah and Hanafiah (2008)
Sugar cane bagasse	1.79	1.68		0.95		Ngah and Hanafiah (2008)

The comparison of these data with those for some algae (Tables 7.1 and 7.5) suggest that algae, in general, and brown algae in particular, are promising biosorbents for heavy metals. Furthermore, algae are natural products collected in large amounts, especially the marine and brown ones. In addition, most algae are innocuous and easily welcomed by the public and environmental agencies for its biotechnological use since they do not generate toxic or non-removable by-products (Cuizano and Navarro 2008).

References

Ahluwalia SS, Goyal D (2007) Microbial and plant derived biomass for removal of heavy metals from wastewater. Biores Technol 98:2243–2257

Akhtar N, Iqbal M, Zafar SI, Iqbal J (2008) Biosorption characteristics of uinicellular green alga *Chlorella sorokiniana* immobilized in loofa sponge for removal of Cr(III). J Environ Sci 20:231–239

Aksu Z, Acikel U (2000) Modeling of a single-staged bioseparation process for simultaneous removal of iron(III) and chromium(VI) by using *Chlorella vulgaris*. Biochem Eng J 4:229–238

Aksu Z, Dönmez G (2006) Binary biosorption of cadmium (II) and nickel(II) onto dried *Chlorella vulgaris:* co-ion effect on mono-component isotherm parameters. Process Biochem 41:860–868

Al-Rub FAA, El-Naas MH, Ashour I, Al-Marzouqi (2006) Biosorption of copper by *Chlorella vulgaris* from, single, binary and ternary metal aqueous solutions. Process Biochem 41:457–464

Apiratikul R, Pavasant P (2008) Batch and column studies on biosorption of heavy metals by *Caulerpa lentillifera*. Biores Technol 99:2766–2777

Brinza L, Dring MJ, Gavrilescu M (2007) Marine micro algal species as biosorbents for heavy metals. Environ Eng Manag J 6:237–251

Chong KH, Volesky B (1996) Metal biosorption equilibra in a ternary system. Biotechnol Bioeng 4:629–638

Cuizano NA, Navarro AE (2008) Biosorción de metales pesados por algas marinas: posible solución a la contaminación a bajas concentraciones. An Quim 104:120–125

Das J, Vimala R, Karthika P (2008) Biosorption of heavy metals—an overview. Indian J Biotechnol 7:159–169

Davis TA, Volesky B, Mucci A (2003) A review of the biochemistry of heavy metal biosorption by brown algae. Water Res 37:4311–4330

Deng L, Su Y, Su H, Wang X, Zhu X (2006) Biosorption of copper(II) and lead(II) from aqueous solutions by nonliving green algae *Cladophora fascicularis:* equilibrium, kinetics and environmental effects. Adsorption 2:267–277

Deng L, Zhu X, Wang X, Su Y, Su H (2007) Biosorption of copper(II) from aqueous solutions by green alga *Cladophora fascicularis*. Biodegradation 18:393–402

Feng D, Aldrich C (2004) Adsorption of heavy metals by biomaterials derived from marine alga *Ecklonia maxima*. Hydrometallurgy 73:1–10

Figueira MM, Volesky B, Azarian K, Ciminelli ST (2000a). Biosorption column performance with a metal mixture. Environ Sci Technol 34:4320–4326

Figueira MM, Volesky B, Ciminelli VST, Roddick FA (2000b) Biosorption of metals in brown seaweed biomass. Water Res 34:196–204

Gadd GM (2009) Biosorption: critical review of scientific rationale, environmental importance and significance for pollution treatment. J Chem Technol Biotechnol 84:13–28

Grimm A, Znzi R, Björnbom E, Cukierman AL (2008) Comparison of different types of biomasses for copper biosorption. Biores Technol 99:2559–2565

Gupta VK, Rastogi A, Saini VK, Jain N (2006) Biosorption of copper(II) from aqueous solutions by *Spirogyra* species. J Colloid Interface Sci 296:59–63

Han X, Wong YS, Wong MH, Tam NFY (2007) Biosorption and bioreduction of Cr(VI) by a microalgal isolate, *Chlorella miniata.* J Hazard Mater 146:65–72

Kaewsarn P, Yu Q (2001) Cadmium (II) removal from aqueous solutions by pre-treated of marine alga *Padina* sp. Environ Pollut 112:209–213

Lembi CA, Waaland JR (1988) Algae and human affairs. Cambridge University Press, Cambridge

Lesmana SO, Febriana N, Soetaredjo FE, Sunarso J, Ismadji S (2009) Studies on potential applications of biomass for the separation of heavy metals from water ad wastewater. Biochem Eng J 44:19–41

Lio Y, Cao Q, Luo F, Chen J (2009) Biosorption of Cd^{2+}, Cu^{2+}, Ni^{2+} and Zn^{2+} ions from aqueous solutions by pretreated biomass of brown algae. J Hazard Mater 163:931–938

Luo F, Liu Y, Li X, Xuan Z, Ma J (2006) Biosorption of lead ion by chemically-modified biomass of marine brown algae *Laminaria japonica.* Chemosphere 64:1122–1127

Martins BL, Cruz CCV, Luna AS (2006) Sorption and desorption of Pb^{2+} ions by deal *Sargassum sp* biomass. Biochem Eng 27:310–314

Mehta SK, Gaur JP (2001) Characterization and optimization of Ni and Cu sorption from aqueous solution by *Chlorella vulgaris.* Ecol Eng 18:1–13

Mehta SK, Gaur JP (2005) Use of algae for removing heavy metal ions from wastewater: Progress and prospects. Crit Rev Biotechnol 25:113–152

Naddafi K, Saeedi R (2009) Biosorption of copper(II) from aqueous solutions by brown macroalga *Cistoseira myrica.* Environ Eng Sci 26:1009–1015

Ngah WSW, Hanafiah MAKM (2008) Removal of heavy metal ions from wastewater by chemically modified plant wastes as adsorbents: A review. Biores Technol 99:3935–3948

Nilanjana D, Vimala R, Kartjika P (2008) Biosorption of heavy metals - An overview. Chemical Biomed Eng 7:159–169

Ozdemir S, Kilinc E, Poli A, Nicolaus B, Guven K (2009) Biosorption of Cd, Cu, Ni, Man and Zn from aqueous solutions by thermophilic bacteria *Ceobacillus toebii* subsp. *decanicus* and *Geobacillus thermoleovorans* subsp. *stromboliensis:* equilibrium, kinetic and thermodynamic studies. Chem Eng J 152:195–206

Pan JH, Liu RX, Tang HX (2007) Surface reaction of *Bacillus cereus* biomass and its biosorption for deal and copper ions. J Environ Sci China 19:403–408

Pasavant P, Apiratikul R, Sungkhum V, Suthiparinyanont P, Wattanachira S, Marhaba TF (2006) Biosorption of Cu^{2+}, Cd^{2+}, Pb^{2+} and Zn^{2+} using dried marine green macroalga *Caulerpa lentillifera.* Biores Technol 97:2321–2329

Rao PS, Kalyani S, Reddy KVNS, Krishnaiah A (2005) Comparison of nickel(II), and copper(II) ions from aqueous solution by *Sphaeroplea* algae and acid treated *Sphaeroplea* algae. Sci Technol 40:3149–3165

Raven PH, Evert FE, Eichorn SE (1999) Biology of plants, 6th edn. W.H. Freeman and Company, New York

Rincon J, Gonzalez F, Ballester A, Blázquez ML, Muñoz JA (2005) Biosorption of heavy metals by chemically activated algae *Fucus vesiculosus.* J Chem Technol Biotechnol 80: 1403–1407

Romera E, González F, Ballester A, Blázquez ML, Muñoz JA (2006) Biosorption with algae: A statistical review. Crit Rev Biotechnol 26:223–235

Romera E, González F, Ballester A, Blázquez ML, Muñoz JA (2007) Comparative study of biosorption of heavy metals using diffierent types of algae. Biores Technol 98:3344–3353

Romera E, González F, Ballester A, Blázquez ML, Muñoz JA (2008a). Biosorption of heavy metals by *Fucus spiralis.* Biores Technol 99:4684–4693

Romera E, González F, Ballester A, Blázquez ML, Muñoz JA (2008b) Biosorption of Cd, Ni and Zn with mixtures of different types of algae. Environ Eng Sci 25:999–1007

Southichak B, Nakano K, Nomura M, Chiba N, Nishimura O (2008) Marine macroalgae *Sargassum horneiri* as biosorbent for heavy metal removal: roles of calcium in ion exchange mechanism. Water Sci Technol 58:697–704

Southichak B, Nakano K, Nomura M, Chiba N, Nishimura O (2009) Differences in adsorption mechanisms of heavy metal by two different plant biomasses: reed and brown seaweed. Water Sci Technol 59:339–346

Talaro KP, Talaro A (2002) Foundations in microbiology, 4th edn. McGraw-Hill Collegue, Blacklick

Tuzun I, Bayramoglu G, Yalcin E, Basaran G, Celik G, Arica MY (2005) Equilibrium and kinetic studies on biosorption of Hg(II), Cd(II), and Pb(II) ions onto microalgae *Chlamydomonas reinhardtii.* J Environ Manag 77:85–92

Vilar VJP, Botelho CMS, Boaventura RAR (2007) Modeling equilibrium and kinetics of metal uptake by algal biomass in continuous stirred and packed bed adsorbers. Adsorption 13:587–601

Vilar VJP, Botelho CMS, Boaventura RAR (2008) Effect of Cu(II), Cd(II) and Zn(II) on Pb(II) biosorption by algae *Gelidium*-derived materials. J Hazard Mater 154:711–720

Vijayaraghavan K, Padmesh VN, Palanivelu K, Velan M (2006) Biosorption of nickel(II) ions onto *Sargassum wightii:* application of two-parameter and three-parameter isotherm models. J Hazrd Mater B 133:304–308

Volesky B (2007) Biosorption and me. Water Res 41:4017–4029

Wang J, Chen C (2009) Biosorbents for heavy metals removal and their future. Biotechnology Adv 27:195–226

Yinghui L, Cao Q, Luo F, Chen J (2009) Biosorption of Cd^{2+}, Cu^{2+}, Ni^{2+} and Zn^{2+} ions from aqueous solutions by pretreated biomass of brown algae. J Hazard Mater 163:931–938

Zhang X, Luo S, Yang Q, Zhang H, Li J (1997) Accumulation of uranium at low concentration by green alga *Scenedesmus obliquus.* J Phycol 9:65–71

Chapter 8
Removal of Rare Earth Elements and Precious Metal Species by Biosorption

Yves Andrès and Claire Gérente

Abstract In Rare Earth Elements (REE) or Precious Metal Species (PMS) removal many types of biological phenomena can take place, such as biosorption, bioaccumulation, resistance/detoxification mechanisms, and direct or indirect utilization in the microbial metabolism. The high demand for the REE or PMS implies demand on increased production of ores containing REE or PMS (i.e. mining) and recycling of solutions to recover the elements contained in waste. But the use of REE and PMS in many anthropogenic applications and devices has led to an increased of public and environment exposure. The aim of this chapter is to compare under different operating conditions, the biosorption capacities of various microbial species and natural by-products for rare earth elements, to investigate the involved sorption mechanisms and to evaluate the potential industrial use of this process to metal ion removal.

Keywords Biosorbent • Batch reactor • Dynamic reactor • Pilot scale biosorption

8.1 Introduction

The need for an effective and economical process to remove heavy metal ions and/or radionuclides from large volumes of diluted industrial wastewater streams has stimulated increasing interest in the metal binding capacities of various microorganisms (Wang and Chen 2009; Mack et al. 2007; Andrès et al. 2003) and biopolymers (Guibal 2004). Biosorption is known as a potential purification process for sequestering metallic cations from diluted aqueous solutions and the applicability of biosorption in continuous metal recovery processes has received an increasing attention from researchers (Wang and Chen 2009; Kratochvil and Volesky 1998). Atkinson et al. (1998) have described the numerous factors that must be considered for application of biosorption technology. These authors have shown the need of

Y. Andrès (✉)
Ecole des Mines de Nantes, GEPEA UMR CNRS 6144, BP 20722, 44307 Nantes Cedex 3, France
e-mail: yves.andres@emn.fr

P. Kotrba et al. (eds.), *Microbial Biosorption of Metals*,
DOI 10.1007/978-94-007-0443-5_8, © Springer Science+Business Media B.V. 2011

multiplying investigations to answer various and not yet solved questions prior to system design at larger scale. The purpose of the present chapter is to review evidence supporting suitability and prospects of research in microbial biosorption.

8.2 Rare Earth Elements and Precious Metal Species

8.2.1 Rare Earth Elements

Rare earth elements (REE) include 15 elements in the Periodic Table: lanthanum (La), cerium (Ce), praseodymium (Pr), neodymium (Nd), promethium (Pm), samarium (Sm), europium (Eu), gadolinium (Gd), terbium (Tb), dysprosium (Dy), holmium (Ho), erbium (Er), thulium (Tm), ytterbium (Yb), lutetium (Lu), also known as lanthanides, plus yttrium (Y) and scandium (Sc). The lanthanide series consists of the fourteen elements following lanthanum in the periodic table and is formed by the successive addition of a 4f electron to the electronic configuration of lanthanum. Because the 4f shell is an inner shell, the chemical properties of the lanthanide ions are in general very similar.

Rare Earth Elements (REE) -based materials are gaining increasing importance, both in terms of research activity, and in terms of commercial products. They are often used in industry for the production of glass additives, fluorescent materials, catalysts, ceramics, lighters, supra-conductors, magnets or condensers. For example, the electroluminescent properties of various lanthanide complexes have made them popular as emitters in electroluminescent devices. Lanthanide-based reagents have also become highly interesting in organic synthesis, as well as asymmetric catalysis. Lanthanides are an important component of plastic fibre optic lasers, ionic conducting oxides (a key component of fuel cells), metallorganic chemical vapour deposition of oxides used in microelectronics, and even liquid crystals and surfactants (Yantasee et al. 2009). They are even more widely used in agriculture, forestry and aqua-culture in which they are found in micro-element fertilizers or animal food. Moreover, discoveries that yield the quality of the agricultural products can be significantly improved by applying REE micro-fertilizers have led to a large-scale usage of REEs in agricultural fields in China since 1970s. More and more REEs are getting into the environment as a result of their usage (Wei et al. 2005). Thus, some authors (Diatloff et al. 1995) have shown that REE lanthanum and cerium could have a negative effect on the root elongation of corn and mungbean. Early reports demonstrated that REE can be accumulated in different parts of plants and the concentrations in roots, leaves, and stems may increase remarkably when REE-containing fertilizers are applied. This may lead to the transfer of REE through the food chain to the human body (Liang et al. 2008). The median lethal dose (LD50) for the rat is 10 mg of cerium per kg of body weight. A published review, on REE toxicity, has reported that cerium could be a potent antiseptic drug for Gram-negative bacteria and fungi (Hirano and Suzuki 1996). On the other hand, some natural plants collected from a rare earth ore area located in China contain significantly high levels of REE from 675 to 3358 μg g^{-1} (Wang et al. 1997) without any apparent toxicity. Although REEs are less toxic than other heavy metals such as copper and cadmium, the long-term hazardous effect of REEs on human health remains serious. For example, it was reported that the mean intelligence quotient and the memory of children in REE polluted areas were significantly lower than those in the control areas. REEs have already been classified as main environmental pollutants in China since 1990s (Wei et al. 2005).

Many studies have shown the absorption, accumulation, and distribution of cerium in mammalian tissues. For example, cerium residues were found in mice, with the highest concentrations in eyes, bone, testes, brain, heart and fat, and the tissues accumulation enhanced with higher doses or increased feed times (Huang et al. 2010). Controversies exist about any biological effects of REE. Toxicity studies have shown that cerium can cause liver damage in rodents. Elevated alanine aminotransferase (ALT) and aspartate aminotransferase (AST) levels are the most specific markers of liver cell damage in mammals. Administration of $CeCl_3$ caused an increase in the activities of ALT and AST (Kobayashi et al. 2005). In another study, orally administered cerium generated reactive oxygen species (ROS) and increased the antioxidants metallothionein (MT) and glutathione (GSH) in the mouse liver (Kawagoe et al. 2005). However, some experts consider the low concentrations of REE in these studies could not have been responsible for any toxicity; furthermore they suggest that REE may have significant antioxidative potential. Schubert et al. (2006) reported that cerium and yttrium nanoparticles acted as direct antioxidants limiting the amount of ROS required to kill cells. These data suggest that this group of nanoparticles could be used to modulate oxidative stress in biological systems.

In Germany, the annual emission of gadolinium by hospitals is estimated at 484 kg (Kümmerer and Helmers 2000), and up to 3,400 kg yr^{-1} are emitted from automobile catalysts. Moreover, some lanthanide ions are produced in nuclear fission and could be dispersed in the environment like ^{140}La or $^{141}Ce/^{143}Ce$ in the case of the Tchernobyl accident in 1986 (Gattavecchia et al. 1989). Finally, europium and gadolinium ions could be used as model elements in the extraction chemistry of the high alpha radioactive isotopes of Americium and Curium found in nuclear waste.

8.2.2 Precious Metals

Precious metals discussed here involve the Platinum-group elements (PGE; i.e. Ir, Os, Pd, Pt, Rh, Ru) and gold. PGE are concentrated in the Earth's core and mantle and have low natural abundances in the continental crust. Average concentrations in the upper continental crust range from 0.02 ng g^{-1} (ppb) for Ir to 0.5 ng g^{-1} for Pt and Pd and represent <0.01% of the Earth's PGE concentration. PGE are used in a wide range of applications like autocatalyst, chemical, electrical, electronic, jewellery, and glass. Emissions might occur during PGE production, manufacture of PGE-containing products, and use and disposal of these products. This increasing use results in elevated environmental concentrations of these normally rare metals. Automobile exhaust catalysts, which use Pd, Pt, and Rh as active components, are the main source of PGE emitted into urban and roadside environments, and they contribute to a global increase in PGE concentrations. The average emission rate is 0.1–0.8 μg km^{-1} mainly in the elemental form of (0)-oxidation state. The nano-crystalline Pt particles are attached to μm-sized aluminium oxide particles (Merget and Rosner 2001; Rauch and Morisson 2008). From recent occupational studies conducted in catalytic converter production, a conservative no-effect level (NOEL) of 1.5 ng m^{-3} can be derived for the sensitizing effect of halogenated Pt salts. In a reasonable worst case approach, it is assumed that such compounds comprise between 1% and 0.1% of the total Pt emissions. Applying a safety factor of 10 to account for inter individual variability, a guidance value of 15–150 ng m^{-3} is derived for catalyst-borne (Rauch and Morisson 2008). Platinum species contained in road dusts can be soluble; consequently, the metal enters the waters, sediments, soil and finally, the food chain. The effect of chronic occupational exposure to Pt compounds is well-documented and certain Pt species are known to exhibit allergenic potential. However, the toxicity of biologically available anthropogenic Pt is not clear. There is an evidence of spread and bioaccumulation of these elements in the environment (Ravindra et al. 2004).

Palladium exists in three oxidation states: Pd^0 (metallic), Pd^{2+}, and Pd^{4+}. It is found ubiquitously in the earth's crust ($\leq 1 \pm 10$ μg/kg) and in sea water (<1 ng/kg) and is among the scarcest of the metallic elements. The fraction of Pd within PGE is approximately 20%. It seems that Pd is more mobile in the environment than Pt and Rh. A major source of health concern is the sensitization risk of Pd as very low doses are sufficient to cause allergic reactions in susceptible individuals. Persons with known nickel allergy may be especially susceptible. Workers occupationally exposed to Pd include miners, dental technicians and chemical workers. The latter are exposed mainly to Pd salts several of which may cause primary skin and eye irritations. The general population may come into contact with palladium mainly through mucosal contact with dental restorations and jewellery containing palladium and possibly via emissions from Pd catalysts. In general, in dental patients who are sensitive to Pd, restorations using Pd-containing materials should not be used although Pd has been used without allergic effects in some of these individuals (Kielhorn et al. 2002).

8.3 Biosorption Capacities

Table 8.1 summarises results some of the major published REE biosorption studies. The sorption capacities range from 2 to 10,473 μmol g^{-1}, depending on the micro-organism or biosorbent type as well as on the experimental conditions. Biosorption rate can be related to the nature and the composition of the microbial cell wall layers, to the rare earth cation chemistry, and to the medium conditions. Generally,

Table 8.1 Comparison of biosorption capacities of various biosorbents for several rare earth metal ions. (According to Andrès et al. 2003)

	Element (trivalent)	Biosorption µmol g^{-1} (dry weight)	Experimental Medium conditions	References
Pseudomonas aeruginosa (CIP A 22)	La	397	pH 5.0	Texier et al. (1999)
	Eu	290	pH 5.0	Texier et al. (1999)
	Yb	326	pH 5.0	Texier et al. (1999)
	Gd	322	pH 5.0	Andrès et al. (2000)
Pseudomonas putida (CCUG 28920)	Eu	330	pH 6.4, 0.01 M KCl	Ledin et al. (1997)
Pseudomonas aeruginosa	La	144	pH 4.0, 10 mM $Ca(NO_3)_2$	Mullen et al. (1986)
Pseudomonas aeruginosa (MTCC-1223)	La	1,000	pH 5.0 acetate buffer	Philip et al. (2000)
	Pr	940	pH 5.0 acetate buffer	Philip et al. (2000)
	Nd	1,100	pH 5.0 acetate buffer	Philip et al. (2000)
	Eu	830	pH 5.0 acetate buffer	Philip et al. (2000)
	Dy	1,000	pH 5.0 acetate buffer	Philip et al. (2000)
Bacillus cereus	La	33	pH 4.0, 10 mM $Ca(NO_3)_2$	Mullen et al. (1986)
Bacillus subtilis	La	114	pH 4.0, 10 mM $Ca(NO_3)_2$	Mullen et al. (1986)
Bacillus subtilis (CIP 52.65)	Gd	350	pH 5.0	Andrès et al. (2000)
Myxococcus xanthus	La	990	pH 4.5	Merroun et al. (2003)
Myxococcus xanthus	La	650	pH 4.5	Ben Omar et al. (1997)
Mycobacterium smegmatis (CIP 73.26)	Yb	103	pH 1.5	Andrès et al. (1993)
	La	57	pH 1.5	Texier et al. (1997)
	Eu	101	pH 1.5	Texier et al. (1997)
	Gd	110	pH 5.0	Andrès et al. (2000)
Cupravidus metallidurans CH34 (*Ralstonia matelidurans* CH34)	Gd	147	pH 5.0	Andrès et al. (2000)
Shewanella putrefaciens (CCUG-22948)	Pm	2	pH 4	Pedersen and Albisson (1991)
Escherichia coli	La	70	pH 4.0, 10 mM $Ca(NO_3)_2$	Mullen et al. (1986)
Brewer's yeast waste	La	650	pH 4.5 dry at 100°C, sieved 50–60 mesh	Li Shoujian et al. (1996)
Saccharomyces cerevisiae	Gd	5.1	pH 5.0	Andrès et al. (2000)
Saccharomyces cerevisiae Brewery strain Rh from VLB	La	560	pH 4.5	Ben Omar et al. (1997)
Bakers' yeast	Nd	2169	pH 1.5	Palmieri et al. (2000)

Table 8.1 (continued)

	Element (trivalent)	Biosorption µmol g^{-1} (dry weight)	Experimental Medium conditions	References
Saccharomyces cerevisiae	Sc	134	pH 0.6	Karavaiko et al. (1996)
Candida valida	Sc	100	pH 0.6	Karavaiko et al. (1996)
Penicillium sp.	Nd	1,234	pH 1.5	Palmieri et al. (2000)
Rhizopus arrhizus	La	350	pH 3.5–4.0	Tobin et al. (1984)
Rhizopus arrhizus	Sc	366	pH 0.6	Karavaiko et al. (1996)
Aspergillus niger	Sc	51	pH 0.6	Karavaiko et al. (1996)
Aspergillus terreus	Sc	153	pH 0.6	Karavaiko et al. (1996)
Monoraphidium sp.	Nd	10,473	pH 1.5	Palmieri et al. (2000)
Sargassum polycystum	La	800–900	pH 3–4.5	Diniz and Volesky (2005)
	Eu	800–900	pH 3–4.5	Diniz and Volesky (2005)
	Yb	700–900	pH 3–4.5	Diniz and Volesky (2005)
Sargassum fluitans	La	730	pH 5	Palmieri et al. (2002)
Fucus serratus	Ce	956	pH 5.5	Ahmady-Asbchin et al. (2009)
Crab shell particles (HCl 0.1 M pre-treat)	La	1,000	pH 5	Vijayaraghavan et al. (2009)
Platanus orientalis leaf powder	La	206	pH 4	Sert et al. (2008)
	Ce	216	pH 4	Sert et al. (2008)
Alfalfa (native)	Er	253	pH 5	Gardea-Torresdey et al. (2004)
	Ho	258	pH 5	Gardea-Torresdey et al. (2004)
	Gd	213	pH 5	Parsons et al. (2005)
	Nd	237	pH 5	Parsons et al. (2005)
Alfalfa (hydrolysed)	Er	334	pH 5	Gardea-Torresdey et al. (2004)
	Ho	402	pH 5	Gardea-Torresdey et al. (2004)
	Gd	524	pH 5	Parsons et al. (2005)
	Nd	288	pH 5	Parsons et al. (2005)
Carboxyl resin	Er	385	pH 5	Gardea-Torresdey et al. (2004)
	Ho	482	pH 5	Gardea-Torresdey et al. (2004)
	Gd	596	pH 5	Parsons et al. (2005)
	Nd	297	pH 5	Parsons et al. (2005)
Activated carbon	Nd	423	pH 1.5	Palmieri et al. (2000)

the experiments take place at pH close to 5, below the pH of precipitation of REE hydroxide species.

Some experiments have been carried out under acidic conditions (pH 1.5 and 0.6), thereby mimicking the acid nature of effluents (Palmieri 2000; Texier et al. 1997; Karavaiko et al. 1996; Andrès et al. 1993). As described in Sect. 8.3, pH has an important role in the global surface charge of the cell wall. In the range of pH 3–6, the sorption capacities for lanthanum ions by *Pseudomonas aeruginosa* are slightly affected with an optimal fixation at pH 5 (Texier et al. 1999). Among the results shown in Table 8.1, a few experiments have been carried out with imposed ionic strength (Philip et al. 2000; Texier et al. 1997; Andrès et al. 1993; Mullen et al. 1986). Furthermore, a pre-treatment (e.g., heat, acid or caustic washing, or organic solvent) of the biomass before utilization has been proposed and could lead to an increase in the biosorption capacities (Wang and Chen 2006). The activated carbon and carboxyl resins present in table are use as reference and show that biomass could be use as an alternative sorbent to remove REE. The use of multi-component system allowed proposing a sequence of preferential adsorption: Eu(III) > Yb(III) > La(III) for *P. aeruginosa* (Texier et al. 2002) and Eu(III) > La(III) > Yb(III) for *Sargassum polycystum* (Diniz and Volesky 2005).

The precious metal biosorption is well documented in the literature (see for example Cui and Zhang 2008; Mack et al. 2007; Gerente et al. 2007). For gold, as Au(III), ranged the biosorption capacities from 0.026 mmol g^{-1} for *Saccharomyces cerevisiae* at pH 5 to 2.1 mmol g^{-1} for *Sargassum natans* at pH 2 (Cui and Zhang 2008). For Pd(II) range the biosorption capacity values from 1.2 mmol g^{-1} for *Desulfovibrio vulgaris* to 2.44 mmol g^{-1} for cross-linked chitosan at pH 2 (Mack et al. 2007). Finally, the Pt(IV) biosorption capacities are comprise between 0.17 mmol g^{-1} for *D. vulgaris* and 1.6 mmol g^{-1} for cross-linked chitosan (Cui and Zhang 2008; Gerente et al. 2007).

8.4 Biosorbent Characteristics and Metal Binding

For better understanding of the effect the biosorbent nature and structure has on biosorption, many studies were devoted to characterisation of the microbial surfaces and their functional groups. The first observations have indicated that biosorption of heavy metals and REE on bacterial surface was accompanied of a release of calcium and/or magnesium ions, corresponding to ion exchange mechanism (Diniz and Volesky 2005; Philip et al. 2000; Andrès et al. 1993). Further studies employing potentiometric titrations have shown that metal ions fixation takes place on acidic moieties. This method can be used to measure the global pK_a, and be extrapolated to give the intrinsic pK_a. For example, with *P. aeruginosa* are the pK_a values 2.8 and 6.1, which led to conclusion that the cell wall layers present two kinds of acidic function as the REE-binding functionalities (Texier et al. 1999). The acid/base properties of the cell wall of *Bacillus subtilis* have been characterized and it has been shown three distinct types of surface organic functional groups with pK_a of 4.8, 6.9 and 9.4 (Fein et al. 1997). These various values are generally attributed to carboxyl, phosphate functions and hydroxyl moieties. Concerning Gram-negative bacteria, such functional groups are present mainly in the lipopolysaccharide of the outer layer and in the peptidoglycan, and within Gram-positive wall in the techoic acid. Some electron microscopy studies have indicated that the lanthanum biosorption at the surface of *P. aeruginosa* involved crystalline precipitation (Mullen et al. 1986). Microprecipitation of lanthanum carbonate ($La_2(CO_3)_3$) was described with biosorbents made of crab shell (Vijayaraghavan et al. 2009). The major cell wall component involved in metal biosorption is the alginic acid. In this case three kinds of acidic functional groups with three intrinsic pK_a were determined at 3.5, 8.2 and 9.6 corresponding respectively to the carboxylic (alginic acid) and phosphate moieties, amine functions and phenol groups (Ahmady-Asbchin et al. 2009).

Time-resolved laser-induced fluorescence spectroscopy (TRLIFS) observations have shown that europium is bound at bacterial cells within two distinct chemical environments. The comparison of lifetimes of standard compounds (Eu-oxalate and $Eu(PO_4)$) suggests that this cation interacts with carboxylic and/or phosphate functions in the cell wall of *P. aeruginosa* (Texier et al. 2000a). Markai et al. (2003) have studied the interaction between Eu(III) and *B. subtilis*. The data on the Eu(III)/*B. subtilis* system at pH 5 were satisfactorily described by one site at which Eu(III) was bound through one carboxylic function of the bacteria. With increasing pH, another site should be considered, involving a phosphate-bound environment as confirmed by TRLIFS.

In alfalfa biomass, Parsons et al. (2002) used EXAFS spectra to show that Eu(III) nitrate was bound via a nitrogen or oxygen ligand with bond lengths ranging from 2.44 to 2.49 Å. However, chemical esterification experiments demonstrated that oxygen may be the primary ligand for coordination with the approximate coordination number of 9. Later, Gardea-Torresdey et al. (2004) and Parsons et al. (2005) respectively showed that the carboxyl groups on the alfalfa biomass play the most important role in Er(III), Ho(III), Gd(III) and Nd(III) binding. In addition, the amino groups has also an high effect on the metal ions binding, and especially for Nd(III). Studies using XAS (X-ray Absorption Spectroscopic) analysis showed that Gd(III) and Nd(III) ions were bound to the alfalfa biomass via oxygen (or nitrogen ligands), which were coordinated to carbon atoms. As far as lanthanides were concerned, the complexes within the biomass included some coordinated water molecules (Parsons et al. 2005).

More recently, Ngwenya et al. (2009) has used XAS and EXAFS (Extended X-ray Absorption Fine Structure) measurements to identify lanthanide sorption sites on the bacterial surface. For this purpose they studied the adsorption of selected representatives for light (La and Nd), middle (Sm and Gd) and heavy (Er and Yb) lanthanides. The surface complexation modelling revealed the weak lanthanide adsorption on the phosphate sites, whereas the adsorption of middle and heavy lanthanides could be modelled equally well as mediated by carboxyl as phosphate sites. The existence of such mixed mode coordination was confirmed by EXAFS analysis, which was also consistent with adsorption to phosphate sites at low pH, with secondary involvement of carboxyl sites at high adsorption density (high pH). Furthermore, spectroscopic analysis suggested that the coordination to phosphate sites was monodentate at the metal/biomass ratios used. Based on the best-fitting pK_a site, Ngwenya et al. (2009) conclude that the phosphate sites are located on N-acetylglucosamine phosphate, the most likely polymer on gram-negative cells, with non-protonated phosphate sites around neutral pH.

Guibal et al. (1999) studied the recovery of Pt(IV) from dilute solutions by chitosan cross-linked with glutaraldehyde. At acidic pH, chloride and nitrate anions significantly decreased the uptake of the platinum chloride anions. This suggests competition for protonated binding sites, implying a simple electrostatic binding mechanism. Godlewska-Żyłkiewicz (2003) observed the highest Pd(II) and Pt(IV) biosorption on *Chlorella vulgaris* and *S. cerevisiae* at pH 1.6–2.2, suggesting an ion-exchange mechanism of biosorption. In acidic solutions Pt(II) and Pd(IV) occur mostly as anionic complex, while functional groups of the cell wall are protonated. The functional groups such as protonated amine, imidazole or thiol groups were suggested as possible ligands interacting with the precious metal anions via the N or S atoms.

Chassary et al. (2005) have show that cross-linked chitosan and derivative sorbents have a marked preference for Pd(II) compared to Pt(IV), in single and two-metal solutions, although both metals compete for the same sorption sites. X-ray photoelectron (XPS) and FTIR spectroscopy were used by Pethkar et al. (2001) to determine the mechanism involved in the adsorption of gold ions to two strains of the fungus, *Cladosporium cladosporioides*. These methods confirmed that that no chemical change to the biosorbent took place after metal loading, suggesting that the acidic conditions merely favoured electrostatic interaction between gold anions ($AuCl_4^-$) and protonated biomass. Fourier-transform infra-red analysis suggested the involvement of carboxyl and hydroxyl groups in binding and later reduction of Au(III) to Au(0) and Ag(I) to Ag (0) in the case of metal binding to calcium alginate beads (Torres et al. 2005). Lin et al. (2005) characterized the biosorption of Au(III) to brewery waste *S. cerevisiae* using numerous spectroscopic techniques to find with. X-ray diffraction that Au(III) ions bound to the cell wall were reduced to elemental Au(0).

8.5 Sorption Modelling in Batch Reactor

Results of bisorption experiments with REE and precious metal species may best fit the Freundlich, the Langmuir or the Brunauer-Emmett-Teller (BET) isotherms. The Langmuir approach assumes a single-layer adsorption whereas the BET reflects multi-layer adsorption. Each layer of this type of model can be reduced to a Langmuir one (de Rome and Gadd 1987). The metal uptake from aqueous solutions can be fitted to their classical isotherm equations given below.

The Freundlich equation is:

$$Qe = K_f {C_e}^{1/n} \tag{8.1}$$

where K_f is the Freundlich constant ($l^{1/n}$ $\mu mol^{1-1/n}$ g^{-1}), $1/n$ the Freundlich parameter, Qe the adsorption capacity at equilibrium (µmol g^{-1}) and Ce the solution concentration at equilibrium (µmol l^{-1}).

The Langmuir equation presents the following form:

$$Qe = \frac{Q\text{max} \cdot b \cdot Ce}{1 + b \cdot Ce} \tag{8.2}$$

where Qmax is the maximum adsorption capacity (µmol g^{-1}) and b (l µmol^{-1}) is an equilibrium constant relating to the interaction energy with the surface. The Langmuir equation is widely used to calculate the maximum biosorption capacities in many published works.

Unlike the precious models, the BET model assumes a multi-layer adsorption process in which one layer does not necessarily need to be completely filled before another is commenced. Moreover, each adsorption layer of the BET model can be reduced to Langmuir behaviour with homogeneous surface energy.

The BET equation has the following form:

$$Qe = \frac{Q\text{max} \cdot Ce}{(Cs - Ce) \cdot \left[1 + (b - 1) \cdot \left(\frac{Ce}{Cs}\right)\right]} \tag{8.3}$$

where Cs is the solute saturation concentration (µmol l^{-1}) and Qmax and b, the Langmuir parameters. The use of the BET model leads various authors to describe a multi-layer adsorption of rare earth elements by different bacterial biomass (Andrès et al. 1993, 2000; Texier et al. 1999).

The Toth isotherm, derived from potential theory, has proven to be useful in describing sorption in heterogeneous systems such as phenolic compounds on carbon. It assumes an asymmetrical quasi-Gaussian energy distribution with a widened left-hand side, i.e. most sites have sorption energy less than the maximum adsorption energy. The Toth equation has the following form:

$$Qe = \frac{Q\text{max} \cdot b_T \cdot Ce}{\left[1 + (b_T Ce)^{\frac{1}{n_T}}\right]^{n_T}} \tag{8.4}$$

where b_T the Toth model constant (L mg^{-1}) and n_T is the Toth model exponent. Vijayaraghavan et al. (2009) conclude that the successful application of the Toth model to La(III) sorption data on crab shell particles is due to the surface heterogeneity of the biosorbent.

The Scatchard model has been used (Andrès et al. 2000; Philip et al. 2000) to evaluate the affinity constants (K) of the binding sites for the Gd(III) and the binding capacities. The linear form used is the following,

$$\frac{Qe}{Ce} = K[B\text{max} - Qe] \tag{8.5}$$

where K is the Scatchard affinity constant (l µmol^{-1}) and Bmax (µmol g^{-1}) the maximal ion binding value. If the Scatchard plot (Qe/Ce versus Qe) results in a hyperbolic shape, this modelling leads to the identification of at least two types of binding sites, with high or low binding affinity. If the shape is linear, this approach leads to the identification of only one kind of binding site (Dahlquist 1978).

For example, the pK values for binding of Gd(III) calculated based on Scatchard plot were within the range of 5.2 (*P. aeruginosa* walls) and 3.6 (cell walls of *B. subtilis*) (Andrès et al. 2000). These values are sufficient to give an exchange between gadolinium fixed on sand and resting cells of *Cupravidus metallidurans* (renamed from *Ralstonia metallidurans*), strain CH34 (Andrès et al. 2000). Another interesting finding was that the medium composition for the biomass production has an influence on the biosorption capacity. For example *C. metallidurans* CH34 cultivated on synthetic medium showed biosorption capacity for gadolinium decreased by 73% compared to control cells from complex medium. Interestingly, the biosorption data obtained with biomass raised on minimal medium follow the Langmuir isotherm, whereas data obtained with cells from complex medium fit best the BET two-layer model. Moreover, the Scatchard plots present an hyperbolic shape corresponding to at least two kinds of binding site for the complex medium and a linear shape representative of a single binding site for the synthetic medium. These observations could be correlated with the variation in the composition of the macromolecular compounds or in their quantity at the microbial surface and with the growth conditions. It has been shown (Daughney et al. 2001) that the amount of the functional groups present at the cell surface could change according with the physiological state and the growth phase of the bacteria. The deprotonation constant value of these carboxyl or phosphate functional groups also varies as these are influenced by consequent changes in the global electronegativity of the cell wall structure, depending on nutrition and the culture growth phase.

8.6 Effects of Competing Ions

Multi-elemental studies with *P. aeruginosa* biomass show that La(III), Eu(III) and Yb(III) are bound to the wall sites of the same characteristics (Texier et al. 1999). This work has also pointed out a preferential adsorption for europium ions. The influence of cations (Al^{3+}, Ca^{2+}, Na^{+}, K^{+}) and anions (NO_3^{-}, SO_4^{2-}, Cl^{-}) on biosorption performance has also been studied and Al^{3+} was shown to be the most potent inhibitive ion for the fixation of La(III), Eu(III) and Yb(III) by *P. aeruginosa* biomass. The presence of glutamate, sulphate, phosphate, carbonate, chloride and ethylenediaminetetraacetate (EDTA) in solution, affects the biosorption of lanthanum by a *Rhizopus arrhizus* biomass (Tobin et al. 1987). Furthermore, the biosorption of La(III) by *Mycobacterium smegmatis* at pH 1.0 is substantially impaired by the presence of thorium or uranyl ions (Andrès et al. 1993). Scandium(III) biosorption by various microorganisms could be decreased by 56 to 94% in the presence of Al^{3+}, Fe^{3+}, Ti^{3+} (Karavaiko et al. 1996). Tsezos et al. (1996) have studied the effect of competing ions on the biosorption of metals in relation to their Pearson's classification (soft, hard and borderline species). Significant ionic competition effect was observed for metals belonging to the same class of the hard and soft classification.

For example, in case of the "hard-hard" pair represented by U(IV) and Y(III) was biosorption of Y(III) strongly depressed by uranyl.

Dziwulska et al. (2004) pointed out that at very low pH, strong competition between chlorides and Pt(IV) species inhibits biosorption onto immobilized *Chlorella vulgaris* biosorbent. They also showed that among all metals (Zn^{2+}, Ni^{2+}, Co^{2+}, Cu^{2+}, Mn^{2+}, Fe^{3+} and Na^{+}) tested for their impact on buiosorption of Pd(II) and Pt(IV), only Fe^{3+} and Zn^{2+} interfered with the binding of Pt and Pd species, probably due to their ability to form chloride complexes in excess chloride ions at acidic pH.

8.7 Adsorption in Dynamic Reactors

The applicability of biosorption in continuous metal recovery processes has received increasing attention because of the potential industrial applications (Tsezos 1986). In fact, metal sorption on cell surfaces can occur with non-living micro-organisms and other biological materials. By-products of industrial fermentations, activated sludges, or specific micro-organisms produced using inexpensive growth media. can be obtained in large quantities and converted to low-cost biosorbents. The numerous factors which must be taken into consideration for the applications of biosorption technologies has been described (Atkinson et al. 1998). The breakthrough curves performed in a dynamic process could be successfully modelled by using classical models or a statistical approach like neural networks (Texier et al. 2002).

Promising are results from equilibrium batch studies presented for for REE ions in Table 8.1, which prompted many researchers to continue investigations in flow-through sorption column systems. Column experiments are a simple and reproducible method, commonly used to assess the metal removal performances of the biosorbent material in a continuous system. The dynamic studies were performed in a laboratory-scale fixed bed reactor by using immobilized microbial cells as a biosorbents. An advantage of this process, using a fixed bed of immobilized micro-organisms, is to avoid the separation of the two phases: the biomass and the effluent. A successful method emerging in laboratories and that could be useful in commercial applications is the encapsulation of cells in a polymer gel-matrix (Cassidy et al. 1996) or by immobilization onto activated alumina (Philip et al. 2000). The immobilisation procedure converts the biomass to a spherical shape for use like conventional adsorbents (ion exchange resins or activated carbon materials). The sizes range from 0.5 to 1.5 mm with a good external porosity, and chemical and physical resistance is generally achieved for the sorbent particles. The materials with lattice structures, such as polyacrylamide gels, polyurethane polymers or calcium alginate, used to encapsulate the microorganisms has also been described useful for immobilization of biosorbent materials (Texier et al. 2002). In the case of selective biosorption, calcium alginate, commonly used for cell immobilisation, contains carboxyl groups which could interfere with the metal ions. Polyacrylamide gels are being extensively used as immobilisation materials at a laboratory scale. Their major advantage is a simple and rapid experimental protocol for encapsulation. Moreover, the small pore size and the mechanical stability of these biosorbent beads may be useful for applications in bioreactors because of the low leakage of cells from the matrix. It has been shown that the selective adsorption properties of a mycobacterial biomass were not affected by the presence of this synthetic polymer (Andrès et al. 1995). In this work the authors observed an initial stage of non-selective adsorption of La(III), Th(IV) and U(IV). In the second stage, cation competition occurred with continuing fixation of Th(IV) and displacement first of La(III) and then of U(IV) ions. The uranyl release was followed by a concentration peak in the eluent corresponding to the sum of the displaced ions and those introduced from the feeding solution. A similar phenomenon has been described for the selective adsorption of europium ions from a solution containing La(III), Eu(III) and Yb(III) ions, by immobilized cells of *P. aeruginosa* (Texier et al. 2000b). The affinity of biomass for these ions followed the order Eu(III) > Yb(III) > La(III).

Several studies also inspected the performance of biosorbents after the recycling step. Use of EDTA for desorption of lanthanide ions from mycobacterial biosobent allowed to concentrate the lanthanide over its inlet concentration by factor of 6 (Andrès et al. 1995). However, with three adsorption/desorption cycles, the La(III) biosorption capacity between the first and the third cycle decreased by 50%. In the case of *P. aeruginosa* cells immobilized onto activated alumina the efficiency of La(III) adsorption was 80% in the second cycle (Philip et al. 2000). Desorption of La(III) from yeast biomass has been performed using HCl, Na_2CO_3, $CaCl_2$ and EDTA at 0.1 M, with 88.3, 68.7, 11.4 and 99.4% recovery, respectively (Li Shoujian et al. 1996). Alternative desorption agents include sodium carbonate or bicarbonate (Ben Omar et al. 1997) or citrate at pH 4.0 (Philip et al. 2000).

More recently Diniz et al. (2008) have studied the column biosorption of La(III) and Eu(III) using protonated dry broken *S. polycystum* biomass. They showed that the ion exchange sorption mechanism was the predominant mechanism involved in biosorption of these REE. Column runs with a single metal feed further showed the intra-particle mass transfer coefficients for La(III) and Eu(III) of $6.0 \cdot 10^{-4}$ and $3.7 \cdot 10^{-4}$ min^{-1}, respectively. A modelling of a column binary system with proton as the common ion allowed to to predict reasonably well the behaviour of a ternary system containing protons. Finally, a series of consecutive sorption/desorption runs demonstrated that the metal could be recovered by using 0.1 N HCl for desorption with no apparent loss in the metal uptake capacity of the biosorbent.

The sorption of Pd(II) and Pt(IV) by fixed-bed columns containing cross-linked chitosan also confirmed results obtained from batch tests (Chassary et al. 2005). While Pd breakthrough curves were not influenced by the presence of Pt(IV), biosorption of Pt(IV) was strongly impaired in the presence of Pd(IV) as a competing ion. Both metals were simultaneously sorbed in a first step of the sorption process, while in the later stages, Pb(II) displaced Pt(IV) from the sorbent and, consequently, the Pt(IV) outlet concentration significantly exceeded the inlet concentration. Desorption of Pd(II) and Pt(IV) from loaded chitosan using thiourea is sufficiently efficient to allow reuse of the sorbent for at least 3 sorption-desorption cycles (Chassary et al. 2005).

8.8 Pilot Scale Studies

Industrial fermentation by-products composed of *Streptomyces* sp. mycelium was used for the biosorption of uranium and iron ions from low acidic solution in a 40 m^3 batch reactor (Glombitza et al. 1995). AlgaSORB™ contains algal biomass immobilized in a silica matrix which is used in batch or column systems. Columns are slurry-packed with immobilized algal particles. Selective metal recovery can be achieved by treatment with appropriate reagents after which the regenerated biomass retains approximately 90% of the original metal uptake capacity even after 18 months of regular use. AlgaSORB™ has been successfully used for the removal of Ag, Al, Au, Co, Cu, Cr, Hg, Ni, Pb, Pd, Pt, U and Zn from contaminated effluents and process streams (White et al. 1995).

The AMT-Bioclaim process (Advanced Mineral Technologies, Colorado USA) is composed of *B. subtilis* biomass obtained from industrial fermentation. This biosorbent was used to treat solutions containing lead, copper, zinc, cadmium and silver (Brierley et al. 1986). A mixed powder of cyanobacteria (*Spirulium*) yeast and plant biomass (*Lemnu*, *Sphagnum* sp.) immobilized with xanthan was commercially proposed under Bio-Fix trademark for zinc removal with the possibility of 120 adsorption/elution cycles (Brierley 1993). Some pilot scale studies were conducted using a fluidized bed reactor for the remediation of 150 l of water polluted with Zn^{2+} (Fourest et al. 1994). Table 8.2 shows some American patents of biosorption activities for metallic ions recovery. Many of them concern the biomass immobilisation technique.

Table 8.2 Example of American patents concerning biosorption processes

Patents number	Date	Inventor	Subject
4769223	09/1988	Volesky et al.	Biosorbent for gold
4898827	02/1990	Brierley et al.	Metal recovery
5055402	10/1991	Greene et al.	Removal of metal ions with immobilized metal ion-binding microorganisms
5279745	01/1994	Jeffers et al.	Polymer beads containing an immobilized extractant for sorbing metals from solution
5538645	07/1996	Yannai et al.	Process for the removal of species containing metallic ions from effluents
5602071	02/1997	Summers, Jr. et al.	Bead for removing dissolved metal contaminants
5648313	07/1997	Pohl	Method for production of adsorption material
5789204	08/1998	Kogtev et al.	Biosorbent for heavy metals prepared from biomass
5976847	11/1999	Hermann	Hydrophilic urethane binder immobilizing organisms having active sites for binding noxious materials
6013511	01/2000	Diels et al.	Precipitating metals or degrading xenobiotic organic compounds with membrane immobilized microorganisms
6027543	02/2000	Yoshizaki et al.	Method for removing a heavy metal from sludge
7326344	05/2008	Cotoras Tadic et al.	Process for the removal of metals by biosorption from mining or industrial effluents

8.9 Conclusion

The industrial use of low-cost biosorbents, made of microorganisms of brown algae, has achieved increasing attention for environmental protection and remediation of metal ions as well as REE. Number of studies was devoted to optimisation of the biosorption conditions, characterization of metal-binding sites and the studies on the biosorption capacities of immobilized cells and their results provide decent arguments for implementation of biosorption in the industrial removal of heavy metal from solutions. Due to the increased use of REE and precious metals in many electronic devices, the metal recovery will be a challenge for the future.

References

Ahmady-Asbchin S, Andrès Y, Gérente C, Le Cloirec P (2009) Natural seaweed waste as sorbent for heavy metal removal from solution. Environ Technol 30:755–762

Andrès Y, MacCordick HJ, Hubert JC (1993) Adsorption of several actinide (Th, U) and lanthanide (La, Eu, Yb) ions by *Mycobacterium smegmatis*. Appl Microbiol Biotechnol 39:413–417

Andrès Y, MacCordick HJ, Hubert JC (1995) Selective biosorption of thorium ions by an immobilized mycobacterial biomass. Appl Microbiol Biotechnol 44:271–276

Andrès Y, Thouand G, Boualam M, Mergeay M (2000) Factors influencing the biosorption of gadolinium by micro-organisms and sand. Appl Microbiol Biotechnol 54:262–267

Andrès Y, Texier AC, Le Cloire P (2003) Rare earth elements removal by microbial biosorption: a review. Environ Technol 24:1367–1375

Atkinson BW, Bux F, Kasan HC (1998) Considerations for application of biosorption technology to remediate metal-contaminated industrial effluents. Water SA 24:129–135

Ben Omar N, Lardi Merroun M, Arias Penalver JM, Gonzalez Munoz T (1997) Comparative heavy metal biosorption study of brewery yeast and *Myxococcus xanthus* biomass. Chemosphere 35:2277–2283

Brierley CL (1993) Environmental biotechnology applications in mining. In: Adelaide A (ed) Biomine'93-conference proceedings. Australian Mineral Foundation, Linden Park, SA, Australia, pp 62–71

Brierley JA, Brierley CL, Goyak GM (1986) AMT-Bioclaim: a new wastewater treatment and metal-recovery technology. In: Lawrence RW, Branion RMR, Ebner GG (eds) Fundamental and applied biohydrometallurgy. Elsevier, Amsterdam, p 291

Cassidy MB, Lee H, Trevors JT (1996) Environmental applications of immobilized microbial cells: a review. J Ind Microbiol 16:79–101

Chassary P, Vincent T, Sanchez Marcano J, Macaskie LE, Guibal E (2005) Palladium and platinum recovery from bicomponent mixtures using chitosan derivatives. Hydrometallurgy 76:131–147

Cui J, Zhang L (2008) Metallurgical recovery of metals from electronic waste: a review. J Hazard Mater 158:228–256

Dahlquist FW (1978) Scatchard and Hill plots. Methods Enzymol 48:270–299

Daughney CJ, Fowle DA, Fortin D (2001) The effect of growth phase on proton and metal adsorption by *Bacillus subtilis*. Geochim Cosmochim Acta 65:1025–1035

de Rome L, Gadd G (1987) Copper adsorption by *Rhizopus arrhizus*, *Cladosporium resinae* and *Penicillium italicum*. Appl Microbiol Biotechnol 26:84–90

Diatloff E, Smith FW, Asher CJ (1995) Rare earth elements and plant growth: 1. Effects of lanthanum and cerium on root elongation of corn and mungbean. J Plant Nutr 18:1963–1976

Diniz V, Volesky B (2005) Biosorption of La, Eu and Yb using *Sargassum* biomass. Water Res 39:239–247

Diniz V, Weber ME, Volesky B, Naja G (2008) Column biosorption of lanthanum and europium by *Sargassum*. Water Res 42:363–371

Dziwulska U, Bajguz A, Godlewska-Żyłkiewicz B (2004) The use of algae *Chlorella vulgaris* immobilized on Cellex-T support for separation/preconcentration of trace amounts of platinum and palladium before GFAAS determination. Anal Lett 37:2189–2203

Fein JB, Daughney CJ, Yee N, Davis T (1997) A chemical equilibrium model for metal adsorption onto bacterial surfaces. Geochim Cosmoschim Acta 61:3319–3328

Fourest E, Canal C, Roux, JC (1994) Improvement of heavy metal biosorption by mycelial dead biomasses (*Rhizopus arrhizus*, *Mucor miehei* and *Penicillium chrysogenum*): pH control and cationic activation. FEMS Microbiol Rev 14:325–332

Gardea-Torresdey JL, Tiemann KJ, Peralta-Videa JR, Parsons JG, Delgado M (2004) Binding of erbium(III) and holmium(III) to native and chemically modified alfalfa biomass: a spectroscopic investigation. Microchem J 76:65–76

Gattavecchia E, Ghini S, Tonelli D (1989) Fallout from Tchernobyl in Bologna and its environs: radioactivity in airborne, rain water and soil. J Radioanal Nucl Chem 133:407–419

Gerente C, Lee VKC, Le Cloirec P, McKay G (2007) Application of chitosan for the removal of metals from wastewaters by adsorption—mechanisms and models review. Crit Rev Environ Sci Technol 37:41–127

Glombitza F; Eckardt L, Hummel A (1995) Fundamentals of the application of biosorption to the separation of uranium from mining drainage waters. In: Bacterial interactions with heavy metals in the environment. VITO/MT/95023, Belgium, pp 36–37

Godlewska-Żyłkiewicz B (2003) Biosorption of platinum and palladium for their separation/preconcentration prior to graphite furnace atomic absorption spectrometric determination. Spectrochim Acta Part B Atom Spectrosc 58:1531–1540

Guibal E (2004) Interactions of metal ions with chitosan-based sorbents: a review. Sep Purif Technol 38:43–74

Guibal E, Larkin A, Vincent T, Tobin JM (1999) Chitosan sorbents for platinum sorption from dilute solutions. Ind Eng Chem Res 38:4011–4022

Hirano S, Suzuki KT (1996) Exposure, metabolism, and toxicity of rare earths and related compounds. Environ Health Perspect 104:85–95

Huang SF, Li ZY, Wang XQ, Wang QX, Hu FF (2010) Cerium caused life span shortening and oxidative stress resistance in *Drosophila melanogaster*. Ecotoxicol Environ Saf 73:89–93

Karavaiko GI, Kareva AS, Avakian ZA, Zakharova VI, Korenevsky AA (1996) Biosorption of scandium and yttrium from solutions. Biotechnol Lett 18:1291–1296

Kawagoe M, Hirasawa F, Wang SC, Liu Y, Ueno Y, Sugiyama Y (2005) Orally administrated rare earth element cerium induces metallothionein synthesis and increases glutathione in the mouse liver. Life Sci 77:922–937

Kielhorn J, Melber C, Keller D, Mangelsdorf I (2002) Palladium—a review of exposure and effects to human health. Int J Hyg Environ Health 205:417–432

Kobayashi K, Shida R, Hasegawa T, Satoh M, Seko Y, Tohyama C, Kuroda J, Shibata N, Imura N, Himeno S (2005) Induction of hepatic metallo-thionein by trivalent cerium: role of interleukin-6. Biol Pharm Bull 28:1859–1863

Kratochvil D, Volesky B (1998) Advances in the biosorption of heavy metals. Trends Biotechnol 16:291–300

Kümmerer K, Helmers E (2000) Hospital effluents as a source of gadolinium in the aquatic environment. Environ Sci Technol 34:573–577

Ledin M, Pedersen K, Allard B (1997) Effects of pH and ionic strength on the adsorption of Cs, Sr, Eu, Zn, Cd, and Hg by *Pseudomonas putida*. Water Air Soil Pollut 93:367–381

Li Shoujian, He Jiang, Yi Changhou, Li Yankun (1996) The study of using waste brewer's east as a new sorbent for accumulation of lanthanides. J Sichuan University (Nat Sci Ed) 33:568–574

Liang T, Ding S, Song W, Chong Z, Zhang C, Li H (2008) A review of fractionations of rare earth elements in plants. J Rare Earths 26:7–15

Lin Z, Wu J, Xue R, Yang Y (2005) Spectroscopic characterization of Au^{3+} biosorption by waste biomass of *Saccharomyces cerevisiae*. Spectrochim Acta A 61:761–765

Mack C, Wilhelmi B, Duncan JR, Burgess JE (2007) Biosorption of precious metals. Biotechnol Adv 25:264–71
Markai S, Andrès Y, Montavon G, Grambow B (2003) Study of the interaction between europium (III) and *Bacillus subtilis*: fixation sites, biosorption modeling and reversibility. J Coll Interface Sci 262:351–361
Merget R, Rosner G (2001) Evaluation of the health risk of platinum group metals emitted from automotive catalytic converters. Sci Total Env 270:165–173
Merroun ML, Ben Chekroun K, Arias JM, Gonzalez-Munoz MT (2003) Lanthanum fixation by *Myxococcus xanthus*: cellular location and extracellular polysaccharide observation. Chemosphere 52:113–120
Mullen MD, Wolf DC, Ferris FG, Beveridge TJ, Flemming CA, Bailley GW (1986) Bacterial sorption of heavy metals. Appl Environ Microbiol 55:3143–3149
Ngwenya BT, Mosselmans JFW, Magennis M, Atkinson KD, Tourney J, Olive V, Ellam RM (2009) Macroscopic and spectroscopic analysis of lanthanide adsorption to bacterial cells. Geochim Cosmochim Acta 73:3134–3147
Palmieri MC, O Garcia Jr, Melnikov P (2000) Neodymium biosorption from acidic solutions in batch system. Process Biochem 36:441–444
Palmieri MC, Volesky B, O Garcia Jr (2002) Biosorption of lanthanum using *Sargassum fluitans* in batch system. Hydrometallurgy 67:31–36
Parsons JG, Gardea-Torresdey JL, Tiemann KJ, Gonzalez JH, Peralta-Videa JR, Gomez E, Herrera I (2002) Absorption and emission spectroscopic investigation of the phytoextraction of europium(III) nitrate from aqueous solutions by alfalfa biomass. Microchem J 71:175–183
Parsons JG, Peralta-Videa JR, Tiemann KJ, Saupe GB, Gardea-Torresdey JL (2005) Use of chemical modification and spectroscopic techniques to determine the binding and coordination of gadolinium(III) and neodymium(III) ions by alfalfa biomass. Talanta 67:34–45
Pedersen K, Albinsson Y (1991) Effect of cell number, pH and lanthanide concentration on the sorption of promethium by *Shewanella putrefaciens*. Radiochim Acta 54:91–95
Pethkar AV, Kulkarni SK, Paknikar KM (2001) Comparative studies on metal biosorption by two strains of *Cladosporium cladosporioides*. Bioresour Technol 80:211–215
Philip L, Iyengar L, Venkobachar C (2000) Biosorption of U, La, Pr, Nd, Eu, and Dy by *Pseudomonas aeruginosa*. J Ind Microbiol Biotechnol 25:1–7
Rauch S, Morrison GM (2008) Environmental relevance of the platinum-group elements. Elements 4:259–263
Ravindra K, Bencs L, Van Grieken R (2004) Platinum group elements in the environment and their health risk. Sci Total Environ 318:1–43
Schubert D, Dargusch R, Raitano J, Chan SW (2006) Cerium and yttrium oxide nanoparticles are neuroprotective. Biochem Biophys Res Commun 342:86–91
Sert Ş, Kütahyali C, İnan S, Talip Z, Çetinkaya B, Eral M (2008) Biosorption of lanthanum and cerium from aqueous solutions by *Platanus orientalis* leaf powder. Hydrometallurgy 90:13–18
Texier AC, Andrès Y, Le Cloirec P (1997) Selective Biosorption of Lanthanide (La, Eu) ions by *Mycobacterium Smegmatis*. Environ Technol 18:835–841
Texier AC, Andrès Y, Le Cloirec P (1999) Selective Biosorption of Lanthanide (La, Eu, Yb) ions by *Pseudomonas aeruginosa*. Environ Sci Technol 33:489–495
Texier AC, Andrès Y, Illemassene M, Le Cloirec P (2000a) Characterization of lanthanide ions binding sites in the cell wall of *Pseudomonas aeruginosa*. Environ Sci Technol 34:610–615
Texier AC, Andrès Y, Le Cloirec P (2000b) Selective biosorption of lanthanide (La, Eu, Yb) ions by an immobilized bacterial biomass. Wat Sci Technol 42:91–94
Texier AC, Andrès Y Faur-Brasquet C, Le Cloirec P (2002) Fixed-bed study for lanthanide (La, Eu, Yb) ions removal from aqueous solutions by immobilized *Pseudomonas aeruginosa*: experimental data and modelization. Chemosphere 47:333–342
Tobin JM, Cooper DG, Neufeld RJ (1984) Uptake of metal ions by *Rhizopus arrhizus* biomass. Appl Environ Microbiol 47:821–824
Tobin JM, Cooper DG, Neufeld RJ (1987) Influence of anions on metal adsorption by *Rhizopus arrhizus* biomass. Biotechnol Bioeng 30:882–886

Torres E, Mata YN, Blázquez ML, Muñoz JA, González F, Ballester A (2005) Gold and silver uptake and nanoprecipitation on calcium alginate beads. Langmuir 21:7951–7958

Tsezos M (1986) Adsorption by microbial biomass as a process for removal of ions from process or waste solutions. In: Eccles H, Hunt S (eds) Immobilization of ions by biosorption. Ellis Horwood, Chister, pp 201–208

Tsezos M, Remoudaki E, Angelatou V (1996) A study of the effects of competing ions on the biosorption of metals. Int Biodeter Biodeg 38:19–29

Vijayaraghavan K, Mahadevan A, Joshi UM, Balasubramanian R (2009) An examination of the uptake of lanthanum from aqueous solution by crab shell particles. Chem Eng J 152:116–121

Wang J, Chen C (2006) Biosorption of heavy metals by Saccharomyces cerevisiae: a review. Biotechnol Adv 24:427–451

Wang J, Chen C (2009) Biosorbents for heavy metals removal and their future. Biotechnol Adv 27:195–226

Wang YG, Sun JX, Chen HM and Guo FQ (1997) Determination of the contents and distribution characteristics of REE in natural plants by NAA. J Radioanal Nuc Chem 219:99–103

Wei Z, Hong F, Yin M, Li H, Hu F, Zhao G, Wong JW (2005) Subcellular and molecular localization of rare earth elements and structural characterization of yttrium bound chlorophyll a in naturally grown fern *Dicranopteris dichotoma*. Microchem J 80:1–8

White C, Wilkinson SC, Gadd GM (1995) The role of microorganisms in biosorption of toxic metals and radionuclides. Int Biodeterior Biodegrad 35:17–40

Yantaseea W, Fryxell GE, Addleman RS, Wiacek RJ, Koonsiripaiboon V, Pattamakomsan K, Sukwarotwat V, Xub J, Raymond KN (2009) Selective removal of lanthanides from natural waters, acidic streams and dialysate. J Hazard Mater 168:1233–1238

Chapter 9
Biosorption and Metal Removal Through Living Cells

Pavel Kotrba, Martina Mackova, Jan Fišer and Tomas Macek

Abstract Unicellular and higher organisms have a variety of properties that can affect chemical speciation, mobility and toxicity of metals and radionuclides. Apart from their importance in the environmental cycling of elements, the passive and active interactions of microbes, fungi, algae and plants with metals and radionuclides could have beneficial consequences in context of bioremediation of polluted water, soils and sediments. The capacity of living cells to remove metals and radionuclides from waste waters is well documented. Besides biosorption, the bioremediation using living organisms may exploit their bioaccumulation capacity or metabolic pathways. Useful microbial properties exploited to immobilize or volatize metals and radionuclides involve mainly production of phosphates, carbonates of sulphides that precipitate soluble toxic species and reductive transformations to insoluble ionic or metallic forms, including production of catalytically active nanoparticles directly from waste water. Various bioprocesses, bioreactor setups or in situ bioremediation approaches employing isolated microbes or microbial consortia were proposed and brought to pilot scale. One of the approaches achieving compliance using mixed-function consortia at low cost is construction of artificial wetlands, the systems that rely on collective action of physical, chemical and biological processes. Besides the microbial activities promoting in wetlands formation of insoluble metal sediments, the heavy metal removal performance of wetland is largely contributed by plants, namely through metal biosorption, root uptake and precipitation induced by changes in redox potential within the rhizosphere. The possibility of altering the microbes and plants to improve their bioremediation potential by genetic engineering is under study in many laboratories.

Keywords Heavy metal • Bioprecipitation • Biotransformation • Activated sludge • Artificial wetland • Metal-accumulating plants • Phytoremediation

---

P. Kotrba (✉)
Department of Biochemistry and Microbiology, Institute of Chemical Technology Prague, Technicka 3, 166 28 Prague, Czech Republic
e-mail: pavel.kotrba@vscht.cz

P. Kotrba et al. (eds.), *Microbial Biosorption of Metals*,
DOI 10.1007/978-94-007-0443-5_9,

9.1 Introduction

Heavy metals have been abundant on the Earth since the beginning of life dated back to 3.8 billion years ago. The chemistry involving heavy metals seemingly played a role in synthesis of simple and more complex organic molecules also during prebiotic era. For example, the theory of a chemoautotrophic origin of life in a volcanic iron–sulphur world postulates that "life" at this early stage consisted of twodimensional monomolecular layer of more and more complex organic molecules anionically bounded to positively charged surfaces (e.g., pyrite) at the interface of hot water (Wächtershäuser 2006). Within the inorganic surfaces $Fe^{2+/3+}$, Co^{2+}, Ni^{2+} and other transition metal centers with sulphido, carbonyl and other ligands were catalytically active and promoted the growth of the organic superstructure through carbon fixation, driven by the reducing potential of the volcanic exhalations. Certain polyvalent metal ions, namely Zn^{2+}, $Cu^{+/2+}$, Ni^{2+}, Co^{2+}, $Fe^{2+/3+}$, Mn^{2+}, Mg^{2+}, Ca^{2+} are essential components of numerous metalloproteins and are directly involved in essential biochemistry of modern life. Many other metal ions (e.g., Cd^{2+}, Pb^{2+}, Hg^{2+}, Al^{3+}, Sn^{2+}) have no known essential metabolic function but they can still be accumulated by cells and cause cell damage. Similarly, an excess of essential metal ions may potentially disturb established biological equilibrium.

Coexistence of metalliferrous environment and life since its evolution established many ways by which bacteria, fungi, algae and plants may interact with exogenous and indigenous metals (Fig. 9.1). These involve namely mechanisms of metal uptake (passive by biosorption and active by intracellular bioaccumulation), maintanance of homeostasis and active detoxification, lithotrophic lifestyle employing metal ions as electron sources, active or passive mobilization of metals

---

Fig. 9.1 Mechanisms involved in transformations and detoxification of metal species. Operation of particular mechanism or collective action of several mechanisms is possible depending on the organism and physico-chemical properties of cellular environment. Extracellular immobilization of metal species can occur by biosorption to cell walls or excreted polymers; precipitation of insoluble metal species (biogenic minerals); entrapment of biomineral colloids and particulates in excreted polymers; and binding of metal ions to biomineral bodies by adsorption, chemisorption or absorption. Precipitation can result from release of inorganic or organic metabolites involving phosphate, CO_2, oxalate and increased extracellular pH (bioprecipitation of biosorbed metals) and/or production of sulphide (amorphous metal sulphide precipitation); metabolic reduction of metal ion to elemental metal and autotrophic reduction of metals to insoluble forms (reductive transformations); and metabolism induced increase in redox potential of cellular environment. Intracellular detoxification can occur by metal ion efflux by specific (usually metal-inducible) transporters; sequestration of metal by intracellular ligands; transport of metal complexes into metabolically-inactive vacuoles in eukaryotes; translocation to specialized organs serving as the final metal sink in multicelular organisms; enzymatic reductive transformation resulting in intracellular elemental metal or insoluble metal species and/or volatile Hg^0; and enzymatic methylation resulting in volatile species. Solubilization and mobilization of metals contained in minerals and bound to matrices (e.g., clays, sediments, humic and fulvic acid) can occur by chemolithotrophic utilization of mineral components or of elemental S^0 accompanied by production H_2SO_4 (autotrophic leaching); excretion of metal ligands and complex-forming chemicals (heterotrophic leaching); and biodeterioration of minerals by lowering extracellular pH and by metal/proton competition for metal ion-binding sites at soil or sediment matrices

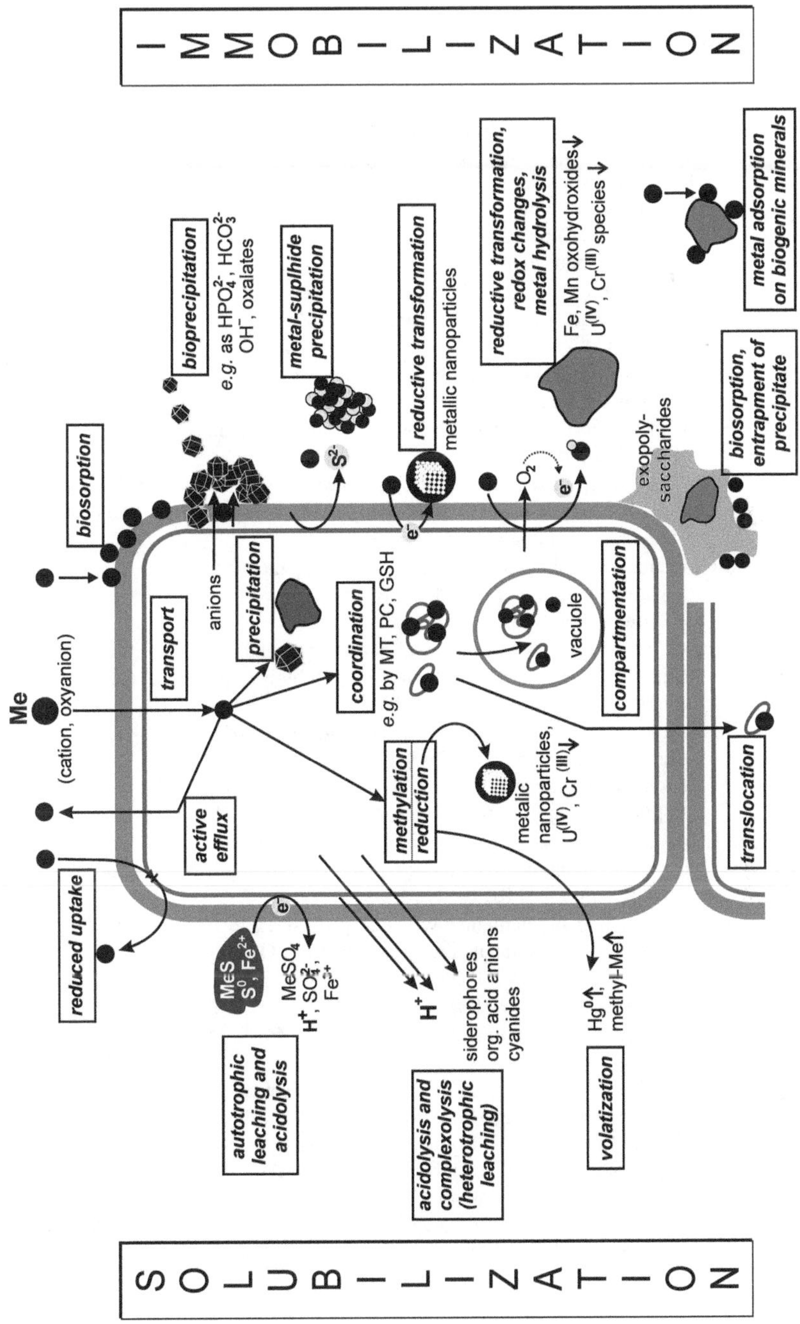
IMMOBILIZATION
SOLUBILIZATION
Me
(cation, oxyanion)
biosorption
transport
active efflux
reduced uptake
anions
precipitation
coordination
e.g. by MT, PC, GSH
compartmentation
vacuole
translocation
methylation
reduction
metalic nanoparticles, U(IV), Cr (III)
bioprecipitation
e.g. as HPO4 2-, HCO3 2-, OH-, oxalates
metal-suplhide precipitation
S2-
reductive transformation
metallic nanoparticles
e-
reductive transformation, redox changes, metal hydrolysis
Fe, Mn oxohydroxides
U(IV), Cr(III) species
O2
exopoly-saccharides
biosorption, entrapment of precipitate
metal adsorption on biogenic minerals
autotrophic leaching and acidolysis
MeS S0, Fe2+
MeSO4
H+, SO4 2-, Fe3+
acidolysis and complexolysis (heterotrophic leaching)
H+
siderophores
org. acid anions
cyanides
volatization
Hg0, methyl-Me

Table 9.1 Some approaches employing living (micro)organisms for bioremediation of heavy metals

Method	Principle and application	Reference
Biofilm and immobilized cells	Mixed consortia on moving bed sand filter, removal of Cu, Zn, Ni and Co by biosorption and bioprecpitation (MERESAFIN system)	Diels et al. (2003)
	Rotating biological contactors (discs) with cyanide-utilizing *Pseudomonas* sp., removal of Cu^{2+} by biosorption (at Homestake Mine, Lead, SD, USA)	Whitlock (1990)
	Packed bed reactor with Hg-reducing consortium, removal of Hg^{2+} by its reduction to Hg^0 followed by separation on activated carbon filter	von Canstein et al. (2002b)
	Consortia from industrial sludge in column filled with marble support, chromium by reduction of Cr^{VI} to precipitating Cr^{III}	Dermou et al. (2005)
	C. metallidurans CH34 immobilized on polysulfone membrane, removal of Cd^{2+}, Zn^{2+}, Cu^{2+}, Pb^{2+} and Y^{3+}, Ni^{2+}, $Co^{2+}$and Pd^{3+} by biosorption and bioprecipitation (BICMER system)	Diels et al. (1993a, b)
	SRB on polypropylene hollow fibers in bioreactor for removal of Cd^{2+}, Cu^{2+}, Zn^{2+}, Ni^{2+} and Co^{2+} from AMD by sulphate precipitation	Tabak and Govind (2003)
Sludge reactor and separate reactor systems	SRB in sludge-blanket reactor, removal of heavy metals by sulphate precipitation (at Budelko BV, Budel-Droplein and Parques BV, Balk, The Netherlands)	Barnes (1993)
	C. metallidurans CH34 in stirred tank with contaminated water and soil, removal of metals (Cd^{2+}) by bioprecipitation (BMSR system)	Diels et al. (2001)
	Four stage system with SRB and aerobic stages for AMD treatment, removal of Cd^{2+}, Cu^{2+}, Pb^{2+} and Zn^{2+} by sulphate precipitation and Mn^{4+} and Fe^{3+} by reduction/sorption/precipitation (Surething Mine, Elliston, MT, USA)	USEPA (2005a)
"In situ" bioreactor (biostimulation)	Dissimilatory *Geobacter* sp. in anoxic groundwater compartment to remove U^{VI} by reduction to precipitating U^{IV} (Old Rifle UMTRA, Rifle, CO., USA)	Anderson et al. (2003)
	SRB in mineshaft supplemented with cheap substrate to remove metals by sulphate precipitation (Lilly-Orphan Boy Mine, Elliston, MT, USA)	USEPA (2005b)
Artificial wetland	Meander system for removal of Pb^{2+}, Cu^{2+}, Zn^{2+}, Mn^{2+}, Ni^{2+}, $Fe^{2+/3+}$ and Cd^{2+} from mine water (Lead Belt Mines, MO, USA)	Gadd (1992)
	Ponds with SRB and floating cattalis in ARUM system (Acid Reduction Using Microbiology) for removal of Fe^{3+}, Ni^{2+}, Zn^{2+} and Al^{3+} from AMD (Copper Cliff, Sudbury, Canada; Minas Gerais, Nova Lima, Brazil)	Kalin et al. (1993); Kalin et al. (2003)
	Inspired by natural wetland employs mainly Zn-accumulator sedge *Carex aquatilis* to remove Zn^{2+} from mine water (United Keno Hill Mine, Elsa, Yukon Territory, Canada)	Sobolewski (1996)

from minerals or formation of biominerals by inorganic or organic precipitation of insoluble metal species (Clemens et al. 2002; Ruml and Kotrba 2003; Malik 2004; Clemens 2006; Gadd 2007, 2010). These mechanisms did not evolve in a response to anthropogenic contamination of the environment in the last few hundreds of years, but, as shown in the following paragraphs, may be effective today when used to remedy polluted waters, soils and sediments (see also Table 9.1 for selected examples).

9.2 Bacterial Immobilization of Metals in Solutions

9.2.1 Immobilization of Metals by Bioprecipitation

Bioprecipitation (biomineralization) occurs when initial biosorption of metal ion creates the supersaturated environment at the cell wall and insoluble metal compound precipitates (and crystallize) as the metal ions combine with anionic species produced by the cell metabolism. This process is generally non-specific and depends on the insolubility of the particular metal compound. The considerable benefits of the process are (1) a high metal-to-biomass ratio, (2) the highly crystalline material easy to separate and (3) possessing low organic matter content that might be advantageous for recuperation of the metal.

The exceptional metal and radionuclide (Cd^{2+}, Pb^{2+}, Cu^{2+}, UO_2^{2+}, PuO_2^{2+} and AmO_2^{2+}) accumulation by growth-decoupled resting cells of *Citrobacter* sp. is due to bioprecipitation of corresponding (hydrogen)phosphate species, which is triggered by periplasmic acid-type phosphatase that releases phosphate from an appropriate substrate (Jeong et al. 1997, 1998). This enzyme is stable, metal and CN^- resistant, and operates in a wide range of pH from 5 up to 9 and at temperatures from 2 to 45°C. In the particular case of bioprecipitation of polycrystalline $HUO_2PO_4 \cdot 4H_2O$, which is the most thoroughly studied aspect of the *Citrobacter* sp. mediated metal remediation, was formation of the nucleation site identified as the rate-limiting step (Macaskie et al. 2000). Tethered HUO_2PO_4 crystallites localize within the periplasm and around the cell wall of *Citrobacter* sp. associated with lipopolysaccharide (LPS), which is seemingly the site of initial deposition and formation of micronuclei of uranium precipitate. In addition to immobilization of uranium salt, the removal of Ni^{2+} is possible via intercalative ion exchange, forming crystals of $Ni(UO_2PO_4)_2{\cdot}7H_2O$ (Finlay et al. 1999; Bonthrone et al. 1996). The cells that were used either as immobilized in the polyacrylamide gels or rather as a biofilm on an inert support can reach loadings of 4 g of Pb^{2+}, 7 g of Cd^{2+}, and 9 g of UO_2^{2+} per 1 g of cell dry weight (Finlay et al. 1999; Macaskie et al. 1992; Macaskie and Dean 1984, 1985).

The use of a laboratory-scale sequencing batch reactor performing Enhanced Biological Phosphorus Removal allowed Renninger et al. (2001) to enrich a mixed microbial consortium with ability to cycle phosphate. This consortium was able to precipitate more than than 98% of the uranyl from a 1.5 mM uranyl nitrate, reaching accumulation levels of more than 0.5 g of U per g of cell dry weight. The genetic engineering approach was employed to promote accumulation of the polyphosphate in *Pseudomonas aeruginosa* overexpressing polyphosphate kinase gene (Renninger et al. 2004). Degradation of this polyphosphate resulted in release of phosphate from cells and concomitant formation of $HUO_2PO_4 \cdot 4H_2O$ precipitate, resulting in the removal of nearly 80% of the uranyl ion from 1 mM uranyl nitrate solution. More recently, Kazy et al. (2009) characterized the mechanism of uranium and thorium accumulation in *Pseudomonas* strain MTCC 3087 as combined ion-exchange-complexation-microprecipitation. An interesting feature of MTCC 3087 strain is high capacity for apparently metabolism-independent accumulation of uranium, with loads of 540 mg g^{-1} of dry biomass dry weight (D'Souza et al. 2006; Sar and D'Souza 2001), and thorium, with loads of 430 mg g^{-1} of dry biomass dry weight (Sar and D'Souza 2002). The accumulated radionuclides precipitated as their corresponding phosphate species both at cell wall and in cytoplasm, with over 50% of accumulated radionuclide localizing intracellularly.

Cupravidus metallidurans CH34 (formerly classified as *Alcaligenes eutrophus*, *Ralstonia eutropha* and/or *Ralstonia metallidurans* CH34) is chemilithoptrophic bacterium, originally isolated from decantation tank in a zinc factory. This bacterium employs specific metal efflux transporters providing resistance towards numerous heavy metal ions, including Cd^{2+}, Co^{2+}, Cu^{2+}, Hg^{2+}, Ni^{2+}, Pb^{2+}, Tl^{+}, Zn^{2+}, and CrO_4^{2-} (Mergeay et al. 2003). Due to effective efflux system, the high concentrations of metals are accumulated at the outside of the cells and are thus bound to cell surface structures via biosorption. At the same time, the bacterial metabolism increases the pH, produces CO_2 that is converted into CO_3^{2-} and HCO_3^{-}. Subsequently the metal carbonates, bicarbonates and hydroxides precipitate at the cell surface, serving as a nucleation sites for the growth of the crystalline structures of diameter above 10 μm (Diels et al. 1993a, b). The potential of such physiology for remediation of contaminated water was soon recognized. BICMER (Bacteria Immobilized Composite MEmbrane Reactor) concept was developed, which employs bacterium immobilized in Zirfon® membrane. The reactor is processed in the arrangement of the flat sheet reactor or of the (continuous) tubular membrane reactor (Diels et al. 1993a, b, 1995). It reduced Cd^{2+}, Zn^{2+}, Cu^{2+}, Pb^{2+} and Y^{3+}, to less than 50 μg l^{-1} and other metals (Ni^{2+}, Co^{2+}, Pd^{3+}) to less than 100 μg l^{-1}. Another viable process termed MERESAFIN (MEtal REmoval by SAnd Filter INoculation) used active moving-bed sand filter with biofilm consisting of mixed consortia of *C. metallidurans* CH34, and metalloresistant *Pseudomonas mendocina* AS302, *Arthrobacter* sp. BP7/26, *Methylobacillus* sp. MB127 (Diels et al. 2003; Pümpel et al. 2001). The bioreactor set-up allowed periodic harvest of biomass and subsequent separation and recovery of metals (Ni, Zn, Cu, Co, Fe, Al, Ag and Cr) via standard pyrometallurgical processing technologies.

Hydrogen sulphide is produced by anaerobic sulphate-reducing bacteria (SRB), such as *Desulphovibrio* and/or *Desulphotomaculum* sp. (Gadd 2000) Precipitation of metal sulphide can also be promoted by aerobic or aerotolerant bacteria, such as members of *Desulfobacteriaceae* or *Klebsiella planticola* (Sharma et al. 2000).

Most metal sulphides, such as CdS, ZnS, CuS, FeS, PbS or Ag_2S, are readily formed in the solution and their solubility is extremely low. The bacteria-promoted metal sulphide precipitation does not completely fulfill the stated definition of bioprecipitation as the metal binding to bacteria and subsequent crystal does not (necessarily) occur. Instead, amorphous precipitate flocs are formed in the solution. White and Gadd (1998, 2000) suggested that mixed consortia, rather than pure cultures of SRB, are to be investigated in an effort to increase the bioremediation potential of these microorganisms as demonstrated with biofilm reactors designed to remove Cu and Cd. One of the first examples of commercialized use of selected (but unidentified) consortium of sulfate reducing bacteria was the sludge-blanket bioreactor at Budelco BV. Use of SRB appear to be the most promising for the treatment of acid mine drainage (AMD) (Pol et al. 2001) and several reactor designs have been reported (Martins et al. 2010; Costa and Duarte 2005; Costa et al. 2008; Gonçalves et al. 2007; Kaksonen and Puhakka 2007; Huisman et al. 2006; Tabak and Govind 2003; Tabak et al. 2003; Veeken and Rulkens 2003; Pol et al. 2001). The major limitation of bioremediation of AMD using SRB is sensitivity of SRB to low pH. Designed reactor systems thus often use SRB maintained in separate bioreactor and biogenic H_2S is then fed into tanks containing metal polluted wastewater. The passive systems are based on promoting microbial activity in groundwater aquifers, permeable reactive barriers and wetland systems, generally achieved by injection of nutrients (Kaksonen and Puhakka 2007).

9.2.2 Immobilization of Metals by Reductive Transformation

Many microorganisms have been reported to catalyze changes in the metal redox state that result in insoluble metal species sustaining the bacterial growth in the metal-contaminated environment. One of the most thoroughly studied heavy metal resistance mechanisms is the mercury resistance mediated by machinery comprising transport of Hg^{2+} into cytoplasm and subsequent enzymatic reduction of Hg^{2+} leading to elemental, volatile Hg^0 that escapes from the cell with the aeration (Barkay et al. 2003). This ability is widely spread among both Gram-negative and Gram-positive bacteria. Essentially all the species use similar mechanism, despite the differences in the localization (chromosomal vs. extrachromosomal), organization and transcriptional control of the responsible *mer* operon. The crucial components are the specific mercury transporters, mercuric reductase and organomercurial lyase (Barkay et al. 2003; Silver and Phung 2005). The exploitation of *mer*-mediated biotransformation of Hg^{2+} and organomercurials for wastewater treatment was proposed more than 25 years ago (Hansen et al. 1984). Since then, several approaches employing naturally occurring or genetically engineered bacteria have been described. For example, the highly radiation-resistant strain *Deinococcus radiodurans* transformed with *mer* genes was constructed, which reduced Hg^{2+} in the presence of 6,000 rad h^{-1} (Brim et al. 2000). Immobilized wild-type Hg^{2+}-reducing *Aeromonas hydrophyla* as well as the recombinant *P. putida* strains expressing *merA* gene encoding mercuric reductase were shown at model scale to remove 95–99% of input Hg^{2+} and thus lower the metal concentration in the effluent to less than 50 μg l^{-1} (Horn et al. 1994; Brunke et al. 1993). Moreover, the leakage of volatile metal from the culture was diminished as virtually all of released Hg^0 formed droplets of about 1–5 μm diameter outside the cells and associated with the matrix material (ceramics or alginate). However, when tested under varying environmental conditions, these strains proved inferior to naturally isolated Hg^{2+}-reducing strains (von Canstein et al. 2002a).

The industrial use for the treatment of chlor-alkali process wastewater found an approach employing packed bed bioreactor inoculated with an artificially formulated consortium of mercury-resistant bacteria, mostly of *Pseudomonas* sp. (Wagner-Döbler et al. 2000a). The planktonic phase was followed by formation of a thick biofillm of cells embedded in exopolysaccharides associated with pumice granules of packed bed (von Canstein et al. 2002b). The bioreactor was fed with a neutralized wastewater containing Hg^{2+} at up to 10 mg l^{-1} and supplemented with low concentrations of nutrients. Under these conditions was the bioreactor operating stably for up to 240 days with 99% efficiency in mercury removal. Accumulation of reduced Hg^0 as recoverable droplets within the bed matrix was aided by adsorption of Hg^0 at biofilm exopolysaccharides (Wagner-Döbler et al. 2000b). The effluent was further loaded at an activated charcoal column where both adsorption and further microbial reduction reduced the Hg concentration in water below 50 μg l^{-1} (Wagner-Döbler et al. 2000a; von Canstein et al. 2002b).

Metal ions that form insoluble precipitates when reduced, involve namely U(IV), Tc(IV) and Cr(III) (Wall and Krumholz 2006; Lovley and Coates 1997). The capacity of dissimilatory Fe^{3+}-reducing bacteria and SRB to anaerobically reduce UO_2^{2-} or $Tc_2O_8^-$ to U(IV) or Tc(IV) species was recognized soon. Microbial reduction has been proposed as a bioremediation method for uranium—contaminated groundwater (Lovley 1995; Lloyd et al. 2003) as production of U(IV) species results in precipitation of insoluble uraninite mineral (Wall and Krumholz 2006). To promote the activity of UO_2^{2-}-reducing anaerobic species, ethanol or acetate are supplemented to groundwater to establish anoxia and pursue microbial reduction of alternate electron acceptors, including UO_2^{2-} (Anderson et al. 2003). Prerequisite for successful immobilization of uranium in aquifer is association of precipitate bodies with the biomass followed by their aggregation (Bargar et al. 2008). This process should avoid spreading of the precipitate in groundwater and eventual remobilization of uranium at other location.

The CrO_4^{2-}-reducing ability is ubiquitous among heterotrophic bacteria under both aerobic and anaerobic conditions. These bacteria involve members of *Pseudomonas* sp., *Bacillus* sp., *Microbacterium* sp., *Arthrobacter* sp. and SRB (Córdoba et al. 2008; Faisal and Hasnain 2004; Camargo et al. 2003; McLean and Beveridge 2001; Pattanapipitpaisal et al. 2001; Shen and Wang 1995). The cognate enzyme activity is generally localized at the cell surface of anaerobic bacteria and in the cytoplasm of aerobes. Various organic pollutants such as aromatic compounds could be exploited by bacteria as an electron source for chromate reduction, resulting in simultaneous decontamination from organics and chromium. Biofilms have been shown more resistant to CrO_4^{2-} toxicity than suspended growth cells (Gail and Parsek 2003; Chirwa and Wang 1997). Feasibility of biofilm packed bed reactors with CrO_4^{2-}-reducing *Bacillus* sp. (Chirwa and Wang 1997) or *Arthrobacters* CR47 (Córdoba et al. 2008) for the biological reduction of chromate has been demonstrated with bench top bioreactors. Dermou et al. (2005) reported the biological reduction of CrO_4^{2-} with removal rates of up to 530 g CrO_4^{2-} m^{-2} day^{-1} in a pilot-scale tricking filter inoculated with an industrial sludge containing strains from the *Acinetobacter* sp. Other studies have reported the use of anaerobic bacteria and conditions in film fixed bioreactors to reduce CrO_4^{2-} (Battaglia-Brunet et al. 2007; Smith and Gadd 2000). The CrO_4^{2-}-reducing strain of *Enterobacter cloacae* was reported to be resistant to high levels of chromate (10 mM) and both fed-batch and dialysis reactors were suggested for the instrumentation of the process exploiting this bacteria (Fuji et al. 1990; Komori et al. 1990; Ohtake et al. 1990). The fixed-bed column bioreactors inoculated with the SRB *Desulfomicrobium norvegicum* and fed with H_2 could reduce 100 mg l^{-1} of CrO_4^{2-} in the presence of 250 mg l^{-1} of sulphate in synthetic solution and the system was effective also when actual ground waters and waste waters were subjected to CrO_4^{2-} removal (Battaglia-Brunet et al. 2002, 2004a, b, 2006, 2007). The scaled up bioreactors (volume capacity of 2, 20 and 200 l) employing different supports (pozzolana, ceramics, PVC crossfiller) operated under continuous feed between 10 and 35°C for several months. The plants designed for a complete reduction of 190 µg l^{-1} of CrO_4^{2-} in waste water by *Bacterium dechromaticus* (Romanenko) were built-up in Ukraine and Russia in the middle 70's (Serporylov et al. 1981). The reduction rate of the process was reported to be 1 g of potassium chromate per g dry weight of bacteria per 72 h. However, the use of chromate reducing bacteria seems to have a much more favorable economic impact in the case of CrO_4^{2-} reduction and precipitation in soil, than for waste water cleanup.

Bacterial reduction of precious metals was recognized as promising alternative to their recovery by ion-exchange and electrochemical methods. The added value of microbial reduction of Pd^{2+} to Pd^0 is formation of biogenic nanoparticles, a catalyst which could be directly made of waste without pre-purification. Biogenic Pd^0 could directly catalyze various chemical reductions such as dehalogenation of polychlorinated biphenyls (De Windt et al. 2005; Hennebel et al. 2009), reduction of CrO_4^{2-} (Mabbet et al. 2004) or can be used as fuel cell catalyst (Yong et al. 2009). The hydrogenase activity of *Desulfovibrio desulfuricans* catalyzing bioreductive deposition of Pd^0 particles onto cell surfaces was studied in detail (Lloyd et al. 1998; Yong et al. 2002a) and possibility to recover Pd^0 from authentic waste water was demonstrated (Yong et al. 2002b). Pd^0 Nanoparticles with a capacity to reduce CrO_4^{2-} are formed also on the surface of *Bacillus sphaericus* (Pollmann et al. 2006). The initial biosorption of Pd^{2+} on S-layer of this bacterium was followed by reductive step triggered by the addition of H_2 as an electron donor. It is noteworthy that due to crystalline arrangement of bacterial S-layer proteins, these surface structures would be promising templates for fabrication of different inorganic nanocrystal (CdS, Au^0, Pt^0 and Pd^0) arrays (Györvary et al. 2004; Wahl et al. 2001; Dieluweit et al. 1998; Shenton et al. 1997). A wide range of (micro)organisms can reduce Au^{3+} species to Au^0. This phenomenon naturally occurs, e.g., in bacterial biofilms on gold grains (Reith et al. 2007). The capacity of *C. metallidurans* CH34 to produce Au^0 Nanoparticles is attributed to efficient accumulation of Au^{3+} species coupled to the formation of cellular $Au^{(I)}$-S complexes. The later induce metal resistances gene clusters (including a Au-specific operon), which promote efflux, reduction, and possibly methylation of Au-complexes, leading to the formation of $Au^{(I)}$-C-compounds and nanoparticulate Au^0 (Reith et al. 2009). It is expected that recognition of specific genetic responses to Au^{3+} salts and functional characterization of respective genetic determinants would open the way for the development of bioexploration and bioprocessing tools (Southam et al. 2009).

9.3 Mixed-Function Consortia in Bioremediation of Heavy Metals from Water

Many of the above described metal-immobilization mechanisms such as biosorption, metal sulphide precipitation and biotransformation as well as the intracellular metal accumulation operates in the processes that use either natural or artificial biotopes or microbial consortia for decontamination of waste water from metals. The waste streams are applied either directly or mostly after abiotic sewage treatment that can remove up to 40–60% of total metal present.

9.3.1 Activated Sludge in Heavy Metal Removal

The principle of activated sludge process is the lowering of the organic content of the waste (drainage) water by community of microorganisms in a reactor constantly supplied with organic matter (waste) and oxygen. Most of the microorganisms present in the activated sludge are bacteria, but other organisms such as cyanobacteria, fungi, yeasts, algae, protozoa, and some metazoa may also play important roles (Kasan 1993). The principal process involved in the metal removal by activated sludge seems to be biosorption of metal ions and adsorption and entrapment of inorganic precipitates (Kasan 1993; Pujol and Cantler 1992; Yuncu et al. 2006). The biotransformation (namely by reduction) and intracellular metal accumulation are also involved, however, their contribution is assumed to be considerably lower.

The crucial event in activated sludge is formation of flocs of diameters varying from 50 to 500 μm that are composed of wide spectrum of living or dead (i.e., biopolymers of) bacteria, fungi, protozoa and metazoa. The major components of flocs are heterotrophic bacteria of such genera as *Pseudomonas* sp., *Achromobacter* sp., *Flavobacterium* sp., *Alcaligenes* sp., *Citrobacter* sp., and *Zooglea* sp. Hydrophobic interactions between cell surfaces, the formation of highly cross-linked exopolysacharides (e.g., by pseudomonad *Zooglea* sp.) and their cross-linking by lectin-like proteins stabilized by divalent metal ions are thought to be main event in the flocs formation (Novak et al. 1999; Jorand et al. 1998; Keiding and Nielsen 1997; Zita and Hermansson 1997). Yuncu et al. (2006) showed that increasing C/N ratio of nutrients fed to the bioreactor to promote production of exopolysacharides may result in increased sorption of certain heavy metal ions, while biosorption of other was impaired. In an intriguing study Kodukula et al. (1994) showed that biosorption of metals in actiavted sludge consequently led to dissolution of corresponding inorganic precipitates as the equilibrium between soluble metal ion and precipitate is driven by the solubility constant.

The activated sludge has been proposed as an effective tool for remediation of waste water from metal pollution (Arican et al. 2002; Costley and Wallis 2001; Utkigar et al. 2000; Bux et al. 1999; Aksu and Yener 1998; Atkinson et al. 1996; Chang et al. 1995). Still, the heavy metal removal by activated sludge is a consequence of the main process taking place in the sludge—lowering of the biological oxygen demand value of waste water, i.e., removal of organic compounds feeding the consortia. However, industrial metallic effluents are often nutrient limiting and should be supplemeted with carbon and energy sources. Despite encouraging results obtained in laboratory scale (Atkinson et al. 1996), the pilot-plant biosorption process employing activated sludge for cleanup of waste water from the plating plant was considered economically outcompeted with chemical precipitation method (Atkinson et al. 1998a, b). Promising lab-scale system developed Costley and Wallis (2001) with rotating biological contactor (RBC) supporting growth of biofilm formed upon inoculation with sewage activated sludge. The laboratory scale 10 l RBC was used in the treatment of synthetic waste waters contaminated with 100 μM of each of Cd^{2+}, Cu^{2+} and Zn^{2+} with average metal removal efficiencies in one operation cycle of 82%, 50% and 30%, respectively. The RBC was operated in multiple sorption (12 weeks) and desorption (48 h) cycles. Authors reported that biofilm consisting of yeasts, bacteria and filamentous fungi maintained its viability and metal removal eficiency even after quite violent 36-h sequential desorption with 0.1 and 0.5 M HCl.

9.3.2 Heavy Metal Removal Capacity of Artificial Wetlands

The aquatic environments are composed of higher plants that produce the organic matter and remove metals by biosorption and bioaccumulation in tissue, algae that also produce nutrient for other heterotrophic organisms of the environment and together with aerobic microorganisms remove metal from the water column via biosorption and biotransformation. The anaerobic processes (metal sulphide precipitation, biotransformation and sorption and deposition of inorganic precipitates) take place in the sediment.

Natural wetlands have been used for centuries for their capacity to assimilate environmental contaminants at large quantities (Gray et al. 2000). Artificial wetlands emulating the properties of natural wetlands have emerged as a viable option for treatment of a wide range of environmental and water pollution problems (Greenway 1997; Greenway and Simpson 1996). With respect to metal removal, the possible load rates and removal efficiencies are generally greater with constructed wetlands than the natural ones (Mays and Edwards 2001). Wetlands have been recognized particularly suitable approach to treat waste waters from metal and coal mining operations (Johnson and Hallberg 2005a, b; Ye et al. 2001; Cole 1998; Debusk et al. 1996; Brodie et al. 1989a, b; Hammer and Bastian 1989). One of the first examples was the 'Meander channel system' operated from 1980s for the cleanup of the metal-loaded waste water in the Missouri lead mine (referenced in Gadd 1992). At least 99% of input Pb^{2+}, Cu^{2+}, Zn^{2+}, Mn^{2+}, Ni^{2+}, $Fe^{2+/3+}$ and Cd^{2+} is removed from the water column using the complex consorcia of mostly cyanobacteria, algae and higher plants (including pondweed *Potomogeton* and cattalis *Typha* sp.).

The metal removal in the wetland occurs in its three main compartments, soil, hydrology and vegetation, to which various microflora associates. Establishment of anaerobic conditions in hydrology and to some extent also in soil is prerequisite for efficient biogeochemical operation of wetland. Soils are under wetland conditions saturated with water, which during the season allows developing of anaerobic environment in its upper layers. The vegetation, hydrophytic plants inhabiting these soils, is macro and microscopic (algae) with a capacity to take up metals from the water by bisorption and intracellular accumulation (Collins et al. 2005; Williams 2002). As autotrophic organisms, they contribute to wetland fitness by production of reduced carbon compounds serving as nutrients to wetland heterotrophs. The hydrology compartment involves amorphous metal oxyhydroxides, clay, cell surfaces and associated exopolysaccharides, cell debris and heterogeneous polyligands, such as fulvic, humic and tannic acids (Sheoran and Sheoran 2006).

The heavy metal removal from solution in wetland is determined by physical, chemical and biological processes, which are dependent on each other, thus making the overall process very complex (Collins et al. 2005; Dunbabin and Bowmer 1992). The chemical metal removal processes are often closely related to the activity of wetland microflora capable to promote bioprecipitation and reduction of metal ions. Chemical processes are defined as adsorption (to the clay, Fe or Mn oxohydroxides, biomolecules), oxidation and hydrolysis of metal species, metal precipitation (as sulphide, hydroxide, carbonate) and co-precipitation (with Fe oxides) of transition metals (Sheoran and Sheoran 2006). Microbial sulphate and iron reduction support precipitation of metal hydroxides by increasing pH of solution, which is especially beneficial for AMD treatment. Sulphide produced by SRB further precipitates metals as sulphides (Russell et al. 2003; Kalin 2001). The physical removal processes involve namely settling and sedimentation of suspended [metal-containing] solids and precipitates (Johnston 1993). Sedimentation of particles less dense than water is possible only after their flocculation, which is promoted by high concentrations of suspended matter, ionic strength and algae (Sheoran and Sheoran 2006). Flocs may further serve as adsorption or biosorption material for heavy metals.

Plants and phytoplankton play very important role in stabilization and recycling of heavy metals in wetlands (Scholz 2003; Mays and Edwards 2001; Ye et al. 2001; Mashauri et al. 2000; Brix 1993; Pip and Stepaniuk 1992) and during the season can sequester substantial metal concentrations (Matagi et al. 1998; Wood and McAtamney 1994). To net metal uptake contribute, besides biosorption that occurs at cell walls of roots and submerged leafs, namely accumulation of metals in tissues. Some plants species of *Typha* sp., *Desmostachya* sp. or *Saccharum* sp. showed increased capacity to tolerate heavy metals in their tissues (Ye et al. 2001; Greenway 1997; Dunbabin and Bowmer 1992; Skousen et al. 1992). In addition, translocation of oxygen from the shoots to roots followed by leakage of the oxygen from intercellular spaces may impair the reducing power of the rhizosphere. This promotes precipitation of oxyhydroxides of Fe^{3+} and Mn^{2+} followed by additional co-precipitation or adsorption of heavy metal species (Matagi et al. 1998).

Number of reports showed that substantial amount of metal accumulated by examined wetland species is sequestered in roots with a positional benefit of phytostabilization (phytorhizofiltration, see Sect. 9.4.1) of metal in wetland environment (Greenway and Simpson 1996; Greenway 1997; Polprasert et al. 1996; Scholes et al. 1998). Substantial metal deposition in roots was reported for *Cyperus alternifolius* and *Vallarsia exaltata* (Cheng et al. 2002), *Phragmites australis* and *Spartina alternifora* (Weis and Weis 2004; Barley et al. 2005), *Cyperus papyrus* (Denny et al. 1995). Interesting applications would find *Eichhornia crassipes*, water hyacinth, well proliferating in slowly moving rivers and swamps in many countries lying between 40°N and 40°S. The metallophitic nature of water hyacinth with the ability to accumulate heavy metals especially in its roots is well documented (reviewed in Malik 2007). Jayaweera et al. (2008) demonstrated the capacity of *E. crassipes* to remove under nutrient-poor conditions nearly 97% of Fe^{3+} from waste water containing 9.27 of Fe mg l^{-1} during 6 weeks in constructed wetland. The main mechanism remained phytorhizofiltration (with Fe levels in roots reaching 6,500 mg kg^{-1} dry wt), but root-promoted chemical precipitation of Fe_2O_3 and $Fe(OH)_3$ followed by flocculation and sedimentation of Fe-colloidals was also observed. Although the phytostabilisation effect is important, it would be only temporary for most metals. Use of plants with promoted translocation of metals to harvestable shoots would extend the wetland approach with phytoextraction mechanism (see Sect. 9.4.1). For example, water lettuce *Pistia stratiotes* and the small water fern *Azolla pinnata* were shown to reduce Hg concentrations in the coal mining effluent containing 10 μM Hg by 80 and 68%, respectively (Mishra et al. 2009). The Hg levels in roots and shoots were nearly the same and ranged from 400 to 600 μg of Hg g^{-1} dry tissue weight. Another possibility is use of terrestrial plants grown either in soil close proximity wetland or directly in the contaminated aquifer. Roots of many hydroponically grown terrestrial metalophytes such as grasses, sunflower *Helianthus annuus* or Indian mustard *Brassica juncea* were shown to remove toxic metals and radionuclides from aqueous solutions, including genuine contaminated water (see Sect. 9.4.3). To this end, identification and use of suitable hyperaccumulating species would be obviously beneficial.

9.4 Accumulation of Heavy Metals by Different Plant Species

Risk element uptake by plants from soil and water depends on the level of pollution, forms of the element, its mobility and also on plant species (Boruvka and Vacha 2006). Plant growing in metal contaminated environment can accumulate toxic metal ions and efficiently compartmentalize these into various plant parts. Several studies indicated that the partitioning of heavy metals at the whole plant level could be broadly divided into three categories. For instance, Chaney and Giordano classified Mn, Zn, Cd and Mo as metals, which were readily translocated to the plant shoots; Ni, Co and Cu, were intermediate, and Cr, Pb and Hg were translocated to the lowest extent (Alloway 1995). Virtually all plants (and some yeasts) employ phytochelatins (PCs) to detoxify most of toxic heavy metal ions. PCs are small peptides of general structure $(\gamma\text{-Glu-Cys})_n X$ (PCn; n=2–11; X represents Gly, Ser, β-Ala, Glu, Gln or no residue) which are capable of an efficient sequestration of multiple metal ions in metal-thiolate complexes (Clemens 2006; Kotrba et al. 1999). PCs are synthesized enzymatically in a transpeptidation reaction from glutathione (γ-glutamylcysteinylglycine, GSH) or its homologues (*iso*-PCs) by the constitutive PC synthase (PCS) in a metal-dependent manner. The low molecular weight metal-PC complexes of 2–4 kDa formed in cytosolic compartment could be further transported to vacuoles where immobile 6–9 kDa high molecular weight complexes of metal sulphide crystallites covered with PC are formed under the incorporation of S^{2-} (Clemens 2006; Kotrba et al. 1999). The possible redistribution of metal-PC complexes within the plant body via a symplasmic or apoplasmic passage, initiated by their export via ATP-binding cassette (ABC) transporters, has been suggested by Bovet et al. (2005).

Certain plants called hyperaccumulators absorb unusually large amounts of metals in comparison to other plants (e.g. up to 0.1% of chromium, cobalt, copper or nickel, 1% of zinc or manganese, 0.01% of cadmium and 0.0001% of gold in shoot dry weight) (Baker et al. 2000). Such hyperaccumulators are taxonomically widespread throughout the Plant Kingdom (Cunningham et al. 1995).

9.4.1 Phytoremediation Concept

Phytoremediation, remediation using living plants, is of public acceptance and is an aesthetically pleasant, solar-energy driven, passive technique that can be used to clean up sites with shallow, low to moderate levels of contamination (Aken et al. 2010; Vangronsveld et al. 2009; Yang et al. 2005; Macek et al. 2004; Meagher 2000; Salt et al. 1998). Phytoremediation is not only a growing science; it is also a growing industry. This technique can be used along with or, in some cases, in place of physico-chemical cleanup methods (Kayser 1998). Early estimates on the costs for remediating contaminated sites have shown that plants could do that same job as a group of engineers for one tenth of the cost. The soil or water does not need to be gathered in and stored as hazardous waste, requiring large amounts of land, money, and manpower. Plants can be sown, watered, and then harvested with less manpower. The storage of the harvested plants as hazardous waste is seldom required and when needed is less demanding than traditional disposal techniques. However, the main drawback on the use of this novel technology is that it is not applicable to all sites.

Several mechanisms may be involved in the direct and indirect action of phytoremediation in contaminates sites. The phytoremediation of heavy metals can be divided into four groups:

1. Phytoextraction; the use of metal-accumulating plants to remove toxic metals from soil,
2. Phytostabilization; the use of plants to eliminate the bioavailability of toxic metals in soils,
3. Phytorhizofiltration, the use of plant roots to remove toxic metals from polluted waters,
4. Phytovolatization, the conversion of metal *in planta* to volatile forms that evaporate through leaf surface.

9.4.2 Hyperaccumulating Plants

The natural capacity of some plant species to hyperaccumulate heavy metals has sparked the interest of plant physiologists, ecologists, evolutionary biologists for over 50 years. Metal concentrations in the shoots of some known hyperaccumulators can reach really extreme values. Reported examples [and on dry weight basis] involve *Thlaspi caerulescens* accumulating 51,600 mg of Zn or 1,800 mg of Cd kg^{-1}; *Thlaspi rotundifolium* containing up to 8,200 mg of Pb kg^{-1}; *Macadamia neurophylla* with 51,800 mg of Mn kg^{-1}; *Psychotria douarrei* containing 47,500 mg of Ni kg^{-1}; *Ipomoea alpina* containing 12,300 mg of Cu kg^{-1}; and *Haumaniastrum robertii* with shoot Co content of 10,200 mg kg^{-1} (reviewed in Cunningham and Ow 1996).

Main focus was on terrestial plant species, including their potential for remediation of soils contaminated with heavy metals based on mechanisms of phytoextraction and phytostabilization. Phytoextraction, or phytoaccumulation, is referred to as the uptake and translocation of metal contaminants in the soil via the roots into the aboveground portions of the plants. To physically remove metals from the contaminated site the aboveground shoots of the hyperaccumulator plants are to be harvested and subsequently disposed of as hazardous wastes or treated for the recovery of the metals (Evanko et al. 1997). Phytoremediation can be used to remove not only metals (e.g. Ag, Cd, Co, Cr, Cu, Hg, Mn, Mo, Ni, Pb, Zn) but also radionuclides (e.g. 90Sr, 137Cs, 239Pu, 234U, 238U) and certain organic compounds (i.e., petroleum hydrocarbons) (Andrade et al. 2002).

The term hyperaccumulation describing a highly abnormal level of metal accumulation was first applied by Jaffre et al. (1976) in the title of their paper on nickel concentration in the tree *Sebertia acuminata*. The specific use of hyperaccumulation to denote a defined concentration (higher than 1,000 mg of Ni kg^{-1}) was introduced by Brooks et al. (1977) in discussing Ni concentration in species of *Homalium* and *Hybantus* from various parts of the world. To date, more than 440 hyperaccumulator species have been described, three quarters of these being Ni accumulators from extensive occurences of Ni rich ultramafic soils found in many parts of the world (Reeves 2006). The concept of using hyperaccumulator plants to accumulate high quantities of metals in plant biomass to remove heavy metals from contaminated soils was first suggested by Chaney (1983). In addition to the low cost, phytoaccumulation has several other important advantages over the traditional soil removal/replacement remediation methods. For example, it is *in situ*, preserves top soil, reduces the secondary waste stream, is environmentally sustainable and the plant ash may also have economic value (Baker et al. 2000; Chaney et al. 2000). The main attraction of using hyperaccumulators for phytoremediation, that remove and concentrate large amounts of a particular element, is the possibility of employing species that remove and concentrate large amounts of particular element from the soil without significant chemical intervention, other than classical application of fertilisers. It is important that the metal concentration in harvested plant tissue is greater than that in soil. To define, e.g., Ni-hyperaccumulation more precisely Reeves (2006) suggested hyperaccumulator as plant in which a metal concentration is of at least 1,000 mg of Ni kg^{-1} in the dry matter of any above ground tissue in at least one specimen growing in its natural habitat.

Baker and Brooks (1989) have reported about 400 metal-accumulating wild plants that accumulate high concentrations of heavy metals in their shoots. Natural hyperaccumulator plants often grow slowly and have low biomass yield. *T. caerulescens* was reported as a hyperaccumulator of cadmium and zinc. It can accumulate over 3% of zinc and at the same time over 0.1% of cadmium per dry biomass. The practical use of this plant for phytoremediation is restricted by its small biomass yield (Brooks 1998). Metal hyperaccumulators are highly attractive model organisms, because they have overcome major physiological bottlenecks limiting metal accumulation in biomass and metal tolerance.

There are two general approaches to phytoextraction: continuous and chemically enhanced phytoextraction (Salt et al. 1998). The first approach uses naturally hyperaccumulating plants with the ability to accumulate an exceptionally high metal content in the shoots. Hyperaccumulating plants usually hyperaccumulate only a specific metal and metals that are primarily accumulated (Ni, Zn and Cu) are not among the most important environmental pollutants. No plant species has yet been found that demonstrates a wide spectrum of hyperaccumulation (Watanabe 1997). Hyperaccumulators are also mostly slow growing, low biomass-producing species, lacking good agronomic characteristics (Chaney et al. 2005; Cunningham et al. 1995). There is no evidence that natural hyperaccumulator plants can access a less soluble and bio-available pool of metals in soil.

In non-hyperaccumulating plants, factors limiting their potential for phytoextraction include small root uptake and little root-to-shoot translocation of heavy metals. Chemically enhanced phytoextraction has been shown to overcome the above problems (LeCooper et al. 1999; Blaylock et al. 1997; Huang and Cunnigham 1996). Common crop plants with high biomass can be triggered to accumulate high amounts of low bioavailable metals, when their mobility in the soil and translocation from the roots to the green part of plants was enhanced by the addition of mobilizing agents when the crop had reached its maximum biomass. The feasibility of chemically enhanced phytoextraction has been primarily studied for Pb and chelating agents as soil additives, less attention has been given to other metals and radionuclides or their mixtures.

About 360 species worldwide are known to act as Ni hyperaccumulators (Reeves 2006). The plant families most strongly represented are the *Brassicaceae, Euphorbiaceae, Asteraceae, Flacourtiaceae, Buxaceae and Rubiaceae* (Table 9.2). About 90 other species are from more than 30 families, distributed throughout plant kingdom.

Table 9.2 Ni-hyperaccumulating plants. (According to Reeves 2006)

Family and species	Maximum Ni concentration (mg kg^{-1} dry weight basis)	Location
Acanthaceae		
Ruellia geminiflora	3,330	Brazil
Asteraceae		
Berkheya coddii	11,600	South Africa
Pentacalia sp. (ten species)	16,600	Cuba
Senecio sp. (nine species)	11,000	Cuba, S. Africa, Canada
Brassicaceae		
Allyssum sp. (52 taxa)	1,280–29,400	Europe, Turkey, Armenia, Syria, Iraq
Bornmuellera sp. (six taxa)	11,400–31,200	Greece, Albania, Turkey
Cochlearia aucheri	17,600	Turkey
Peltaria emarginata	34,400	Greece
Streptanthus polygaloides	14,800	USA
Thlaspi sp. (23 taxa)	2,000–31,000	Europe, Greece, Turkey, Japan, USA
Buxaceae		
Buxus sp.	1,320–25,420	Cuba
Flacourtiaceae		
Homalium sp. (seven species)	1,160–14,500	New Caledonia
Xylosma sp.	1,000–3,750	New Caledonia
Rubiaceae		
Ariadne shafer	13,070–22,360	Cuba
Euphorbiaceae		
Euphorbia sp.	1,810–9,340	Cuba, Brazil
Leucocroton sp.	2,260–27,240	Cuba
Phyllanthus sp.	1,090–60,170	New Caledonia, Cuba, Philippines, Indonesia

The discovery of zinc accumulation in certain *Viola* sp. and *Thlaspi* sp. in nineteenth century was followed by other species capable to hyperaccumulate more than 10,000 mg of Zn kg^{-1}, notably *Arabidopsis halleri.* This plant is one of the closest relatives of *Arabidopsis thaliana.* It has colonized calamine soils, which are highly contaminated with Zn, Cd, Pb as a consequence of industrial activities. In addition some populations have been reported to contain more than 100 μg g^{-1} of dry biomass of Cd in their leaves. In hydropony *A. halleri* have been shown to tolerate at least 30-fold higher Zn and 10-fold higher Cd concentrations in roots than *A. thaliana* can tolerate (Roosens et al. 2008).

Lead is present in most soils and rocks at concentrations below 50 mg kg^{-1} and generally shows relatively low mobility in soils and into vegetation which typically contains less than 10 mg of Pb kg^{-1}. In cases when Pb does enter the plant roots in larger concentrations from Pb enriched soils, significant translocation to the upper parts of the plant is uncommon. Incresaed concentrations of Pb in aboveground tissue can be caused by entering of that metal bound on dust and soil fine particles directly to leaves through stomata. High lead concentrations in several plant species in their natural habitats are shown in Table 9.3.

Cadmium is a nonessential heavy metal that is widespread in our environment because of contamination by power stations, metal industries and waste incineration. Toxicity to living cells is occurring at very low concentration, with suspected carcinogenic effects in humans. However, the biological effects of this metal and the mechanisms of its toxicity are not yet clearly understood (Suzuki et al. 2001).

Table 9.3 Zn- and Pb-hyperaccumulating plants. (According to Reeves 2006)

Family and species	Maximum metal concentration (mg kg^{-1} dry weight basis)		Location
	Zn	Pb	
Brassiceae			
Arabidopsis halleri	13,620	2,740	Germany
Thlaspi brachypetalum	15,300	1,210	France
Thlaspi caerulescens	43,710	8,200	Europe
Thlaspi ochroleucum	4,130	–	Thasos, Greece
Thlaspi cepaeifolium	18,500	–	Austria, Italy
Thlaspi praecox	15,500	–	Bulgaria
Thlaspi stenopterum	11,500	–	Spain
Thlaspi tatrense	20,100	–	Slovakia
Dichapetalaceae			
Dichapetalum gelonioides	30,000		Sumatra, Mindanao, Sabah
Poaceae			
Agrostis tenuis		13,490	UK
Arrhenatherum elatius		1,500	France
Festuca ovina		11,750	UK
Polygonaceae			
Rumex acetosa	11,000	5,450	
Violaceae			
Viola calaminaria	10,000		Belgium, Germany

Cd is one of the increasingly frequent contaminants of agricultural soils, where it is usually present at 0.1–0.2 mg kg^{-1} but occasionally it has been detected at much higher levels in some regions. Cadmium contamination in agricultural soils is due to either excessive phosphate fertilization, use of sewage sludge as a soil amendment, or to naturally high background levels (Dorlhac de Borne et al. 1998). Cadmium has no essential function in plants and at high concentrations is toxic to plants and animals. Uptake of Cd by plant roots depends on the concentration, the oxidation state of this metal in solution and on the physico-chemical characteristics of the soils such as pH, content of clay minerals and organic matter. Only few plant species have been shown which accumulate more than 100 mg kg^{-1} into their tissue (*T. caerulescens* and *A. halleri*). Recently high accumulation abilities by *Salix* were shown (Dickinson and Pulford 2005).

Normal concentrations of Co and Cu in plants are in the ranges 0.03–2 and 5–25 mg kg^{-1}, respectively. The tupelo or black gum of the southeastern United States (*Nyssa sylvatica*) is remarkable in being able to accumulate as much as 845 mg of Co kg^{-1} from normal soils. However, even on cobalt-enriched soils, such as those derived from ultramafic rocks, plant Cu rarely exceeds 20 mg kg^{-1}.

Extensive screening of many sites of mining and smelting activity throughout Zaire through plant and soil sample collections and analysis, identified 30 hyperaccumulators of cobalt and 32 of copper, with 12 species being common to the two lists. The Co and Cu accumulators have been found in more than dozen families (some are shown in Table 9.4). It should be noted that Co- and Cu-hyperaccumulating plants are not restricted only to metaliferous soils.

Table 9.4 Some examples of Co- and Cu-hyperaccumulating plants. (According to Reeves 2006)

Family species	Maximum Co concentration (mg kg^{-1} dry weight basis)	Maximum Cu concentration (mg kg^{-1} dry weight basis)
Amaranthaceae		
Pandiaka metallorum	2,131	6,270
Asteraceae		
Anisopappus davyi	2,889	3,504
Commelinaceae		
Cyanotis longfolia	4,197	
Cypetraceae		
Ascolepis metallorum	1,118	1,211
Bulbostylis pseudoperennis	2,127	7,783
Euphorbiaceae		
Phyllanthus williamoides	1,140	
Fabaceae		
Crotalaria cobalticola	3,010	
Vigna dolomittica		3,000
Poaceae		
Eragrostis racemosa		2,800
Lamiaceae		
Aeollanthus subacaulis var. *Linearisa*	5,176	13,700
Haumaniastrum robertii	10,230	2,070

Manganese is an essential element activating some of the enzymes involved in citric cycle (tricarboxylic acid cycle) and central role of manganese cluster complexes in oxidation of water to oxygen has been reported. Toxic levels fall in the range of 1,000–12,000 mg kg^{-1}, depending on species. Some species have been found with 1,000–5,000 mg of Mn kg^{-1} on soils with manganese mineralization (more than 1% Mn) as well as on soils with lower Mn concentrations. Concentration of Mn in ultramafic soils may range from 1,000 to 5,000 mg kg^{-1}, which is not regarded as strongly abnormal. Most records of Mn hyperaccumulation come from these areas. Other hyperaccumulators were found on ultramafic soils in New Caledonia with concentrations around 1,000 mg kg^{-1} (Reeves 2006), in six plant species concentrations exceeded 10,000 mg of Mn kg^{-1} and 9 species showed at least one specimen above this level. Mn hyperaccumulators usually belong to Families *Apocynaceae, Celastraceae, Clusiaceae, Myrtaceae, Proteaceae*.

Metal tolerant species and hyperaccumulators represent a valuable and potentially useful biological resource which posses great potential for use in variety of strategies for bioremediation of metals, but some of them have been very rarely collected (Reeves et al. 2006). Seems that there is an urgent need for more exploration of European metalliferous sites, so that more species of hyperaccumulating plants can be found and the distribution and rarity of these species can be better defined. The phytoremediation potential of most known hyperaccumulating species is, however, limited because of their slow growth, low biomass and often tight association with a specific habitant (Chaney et al. 2005). The physiological mechanisms of metal hyperaccumulation and the cognate genetic determinants are being througly investigated to provide solid basis for selection of suitable genes to be (over)expressed in high-biomass plants of well established agriculture to improve their remediation potential.

9.4.3 Potential of Phytorhizofiltration for Waste Water Cleanup from Heavy Metals

Phytorhizofiltration could be defined as the removal of contaminants by plant roots in surface [waste] water through biosorption, intracellular uptake or precipitation by the roots of a plant. The phytorhizofiltration potential of plant roots has been evaluated mainly using hydroponically grown metalophyte terrestrial plants such Indian mustard *B. juncea*, sunflower *H. annuus* and some grasses.

Salt et al. (1997) showed that roots of *B. juncea* concentrated (on dry weight basis) Pb^{2+}, Cd^{2+}, and Ni^{2+} 500–2,000-fold and CrO_4^{2-} 100–250-fold over initial metal concentrations in artificially contaminated water. Similar results were obtained by Dushenkov et al. (1995) in an extensive study involving *B. juncea*, *H. annuus* and various grasses. Roots of these species also effectively removed Cu^{2+}, Cd^{2+}, Cr P, Ni^{2+}, Pb^{2+}, Zn^{2+} and CrO_4^{2-} from aqueous solutions by combined biosorption and tissue uptake. For example roots of *B. juncea* concentrated these metals 131–563-fold above their initial concentration in solutions. Removal of Pb^{2+} was attributed to biosorption and root-mediated precipitation of Pb-phosphate species. Recent analysis of Pb deposition in root tissue suggested that substantial Pb^{2+} uptake by *B. juncea* roots is based on intracellular accumulation at the root tip (Meyers et al. 2008). Grass pea plants *Lathyrus sativus* showed capacity to tolerate 0.5 mM Pb and accumulate in roots up to 153 mg of Pb g^{-1} of dry mass (Brunet et al. 2008). Authors reported that nearly 90% of accumulated metal was intracellular.

H. annuus plants selected for their ability to accumulate in roots UO_2^{2-} 30,000-fold above initial solution concentrations were tested in a pilot scale with water from former uranium processing facility in Ashtabula (OH, USA) (Dushenkov et al. 1997). Phytorhizofiltration system used for bioremediation of contaminated water containing up to 874 µg of U l^{-1} reduced uranium concentration to less than 20 µg l^{-1} meeting the USEPA Water Quality Standard. Similar results were obtaine by Lee and Yang (2010), who further showed that not only *H. annuus*, but also *Phaseolus vulgaris* has the capacity to accumulate uranium in roots, albeit with slightly lower efficiency. Study by Tomé et al. (2008) suggested that roots *H. annuus* acomplish nearly complete removal of natural uranium and ^{226}Ra soluble species by two distinct mechanisms. Uptake in roots represented 50% and 70% of removed uranium and ^{226}Ra, respectively, and remaining radionuclides were found in the precipitate of yet undefined composition.

9.4.4 Improving Metal Accumulation by Plants Through Genetic Modifications

Requirement for plants removing heavy metals is to grow fast in the contaminated environment, to be resistant, able to accumulate toxic metals and transfer cations or oxyanions into the harvestable (above ground) parts, or transform them into less-toxic forms. The most important parameter for selection of suitable plants is not the tolerance of the plant to heavy metals, but the effectiveness in their accumulation. In addition to accumulation capacity, biomass production must be considered in order to determine the total metal uptake.

Enhancement of the metabolic abilities of plants can be achieved by traditional breeding, protoplast fusion, and the direct insertion of novel genes. Genetic engineering methods are widely used for the improvement of different crop plants. The genes introduced usually bring higher resistance against pests or herbicides or improve the technological properties of the plant. This is especially remarkable in the field of herbicide resistance, but also genes responsible for other important traits are being introduced into plant species. A similar approach is expected to improve the plant abilities in the field of environmental detoxification (Kotrba et al. 2009; Doty 2008; Macek et al. 1998, 2000, 2008; Kotrba et al. 1999; Newman et al. 1998; Rugh et al. 1998).

Many attempts have been made to breed willow, poplar, and other plants having properties useful for phytoremediation. Very promising results were obtained by means of genetic engineering. The aim is the formation of plants combining a high ability to accumulate, detoxify, or degrade xenobiotics and pollutants, with resistance toward the toxic compounds present and with suitable agrotechnical characteristics, in other words the improvement of the process by using genetically modified plants specifically tailored for phytoremediation purposes.

At present the main goal is metal-hyperaccumulation traits that might be introduced into fast growing and, high-biomass plants. Plants have developed their own systems for binding heavy metals based largely on the synthesis of phytochelatins, described by Grill et al. (1989). Heavy metal binding in plants is normally achieved, as reviewed by Clemens (2006), Kotrba et al. (1999) or described by Bailey et al. (2003), by phytochelatins and phytosiderophores. Different attempts to improve the heavy metal accumulation capacity of high biomass yielding plants have used genetic engineering (Dorlhac de Borne et al. 1998). The other option is to use traditional breeding to produce, for example, alpine pennycress plants that grow faster and taller.

Many studies have concentrated on hyperaccumulator plants (most studied being cadmium hyperaccumulator, *T. caerulescens*) in order to clarify the mechanisms of hyperaccumulation, transforming them e.g. by *Agrobacterium rhizogenes* to obtain hairy roots (Nedelkoska and Doran 2000). The problem is very complex and seems full understanding and engineering plant metal accumulation is a long way ahead (Clemens et al. 2002). Heavy-metal binding peptides and proteins in plants have been studied already for decades; there is a large choice of such compounds (Kotrba et al. 1999). There is also a growing number of studies on the plant metal transporters (Krämer et al. 2007). Many proteins need metals for their proper function and metal homeostatsis and tolerance to excess metal ions is being primarily etablished by coordinated action of metal transporters, metallothioneins, phytochelatins and glutathione. Different strategies targeting the metal homeostasis and tolerance determinants are used to obtain genetically modified plants with improved properties and have been summarised in reviews (Kotrba et al. 2009; Macek et al. 2008; Cherian and Oliveira 2005; Krämer and Chardonnens 2001; Pilon-Smith and Pilon 2002) and some examples shown in Table 9.5.

Modifications of levels of enzymes normally present, e.g. overexpression of glutathione synthetase in *Brassica juncea* enhances cadmium tolerance and accumulation (Bennett et al. 2003; Pilon-Smits et al. 1999). Similar results have also been obtained in hybrid poplar (*Populus tremula* × *P. alba*), in which overproduction of enzymes involved in glutathione synthetesis promoted accumulation of Cd^{2+} in young leaves (Koprivova et al. 2002; Bittsánszky et al. 2005). Overproduction of PCs due to overexpression of PCS synthase gene in roots of model *A. thaliana* supported the translocation of Cd^{2+} into the shoots and reduced metal accumulation in the roots (Gong et al. 2003). *Nicotiana glaucum* is metalophyte with wide geographic distribution, high biomass, deep roots and resistant to herbivores (Barazani et al. 2004) and thus fulfills many requirements for promising candidate in phytoremediation efforts. Overexpression of PCS in *N. glaucum* further substantially improved accumulation of Cu^{2+}, Zn^{2+}, Pb^{2+} and Cd^{2+} in shoots (Martínez et al. 2006).

A very different way to enhance metal uptake uses cloning of nicotianamine synthase genes (Higuchi et al. 1999), involved in the synthesis of phytosidero-

Table 9.5 Properties of plants engineered for improved phytoremediation capacity. (Acording to Kotrba et al. 2009)

Transgenic plant	Expression of gene encoding:	Metal accumulation phenotype of transgene	Reference
B. juncea	Glutathione synthetase	1.5 times increased Cd and Zn levels in shoots	Bennett et al. (2003)
P. tremula × *P. alba*	γ-glytamylcysteinyl synthetase	Three times increased Cdaccumulation from soil	Koprivova et al. (2002)
N. glauca	Phytochelatin synthase	6.0, 3.3, 4.8, 18.2 and 2.6 times more of Pb, Cd, Zn, Cu and Ni, respectively, accumulated in shoots	Martínez et al. (2006)
A. sinicus	Phytochelatin synthase and MT in symbiotic rhizobia	Colonized roots accumulated 3 times more Cd from soil.	Ike et al. (2007)
N. tabacum	Yeast MT CUP1	Three times higher Cu accumulation from soil	Thomas et al. (2003)
N. tabacum	HisCUP1	Up to 90% higher Cd accumulation from soil, reduced accumulation of Cd in roots	Macek et al. (2002) Pavlíkova et al. (2004)
N. tabacum	Vacuolar transporter CAX2	3.4, 2.3 and 1.9 times increased accumulation of Cd, Mn and Zn, respectively, from solution	Korenkov et al. (2007b)
A. thaliana	Vacuolar transporter YCF1	1.5 times higher levels of Cd and Pb in shoots	Song et al. (2003)
A. thaliana	Mercuric reductase	Hg volatized from solution at rate of 5 μg [g fresh wt]$^{-1}$ min^{-1}	Rugh et al. (1996)
A. thaliana	Mercutic reductase and organomercurial lyase	Hg volatized from solution with 25 μM phenyl-Hg^{+} at rate of 760 ng [g fresh wt]$^{-1}$ min^{-1}	Bizily et al. (2003)
N. tabacum	Mercuric reductase	Amount of Hg volatized from solution with 25 μM Hg^{2+} was per g and day 23.8 μg by roots, 6.9 μg Hg^{0} by leaves and 4.1 μg by stem	He et al. (2001)
L. tulipifera	Mercuric reductase	Hg volatization from hydroponic media with 10 μM Hg^{2+} at rate of 1.2 μg Hg^{0} [g fresh wt]$^{-1}$ day^{-1}	Rugh et al. (1998)

phores. The occurence of the essential metal binding amino acid nicotianamine in plants and microorganisms was described already in early eighties by Rudolph et al. (1985). Secondary metabolites or organic acids seem to have a still underestimated role in metal uptake, for example aluminium tolerance in transgenic plants could be achieved by alteration of citrate synthesis (De la Fuente et al. 1997).

An interesting approach to improve phytorhizofiltration capability of leguminous milkvetch *Astragalus sinicum* by engineering its symbiotic root-associated rhizobia was employed by Ike et al. (2007). Colonization of roots with *Mesorhizobium huakuii* subsp. *rengei* (strain B3) expressing PCS along with mammalian MTs promoted Cd^{2+} uptake in roots by three times.

In the last decade, many attempts were made to prepare transgenic plants bearing genes coding for metallothioneins (MTs) of different origin, including human MT, animal MT, and yeast MT (Bailey et al. 2003; Macek et al. 1998, 2002; Dorlhac de Borne 1998; Hasegawa et al. 1997; Evans et al. 1992). MTs are gene-encoded cysteine-rich peptides capable of high affinity coordination of heavy metal ions via cysteine residues shared along the peptide sequence in Cys-X-Cys or Cys-Cys motifs (Vasák 2005; Kotrba et al. 1999). Mammalian MTs are composed of 61 or 62 amino-acid residues and their 20 cysteine residues form 7 and 12 coordination centres for divalent and monovalent heavy metal ions, respectively. In the yeast S*accharomyces cerevisiae* form 12 cysteine residues of CUP1, a 53 amino-acid MT variant, eight binding centres for monovalent and four binding centres for divalent heavy metal ions. In these organisms, the intracellular sequestration of toxic heavy metal ions via MTs, represents one of the principal mechanisms conferring tolerance to particular heavy metal ions. Cadmium partitioning in transgenic tobacco plants expressing a mammalian metallothionein gene is described in detail by Dorlhac de Borne et al. (1998). The authors found that the Cd content is increased in roots but decreased in leaves This fact would have a beneficial potential for the phytorhizofiltration or transient phytostabilization of heavy metals in soil. Increased Cu^{2+} accumulation was reported also for roots of *A. thaliana,* overexpressing the pea MT (Evans et al. 1992). Thomas et al. (2003) showed that production of CUP1 (MT variant induced in *S. cerevisiae* by Cu^{+}) also significantly promoted the accumulation of Cu^{2+}, but not of Cd^{2+}, in leaves of *N. tabaccum.* Further enhancement of the performance of transgene products can be expected by implanting an additional heavy metal binding site, like a polyhistidine extension, known for its high affinity for heavy metals. On overeression of HisCUP1 gene (CUP1 additionally modified with an N-terminal hexahistidine, His; Macek et al. 1996) showed transgenic *N. tabacum* improved phytoextraction potential for Cd^{2+} both from solutions and sandy and humous soils (Pavlíkova et al. 2004; Macek et al. 2002).

Relatively broad substrate specificity of particular metal transporters (Krämer et al. 2007) makes them a promising tool to improve metal uptake for phytoremediation. In *N. tabaccum*, the overproduction of its NtCBP4 transporter (structurally similar to animal K^{+} channels), suported uptake and translocation of Pb^{2+} to shoots (Sunkar et al. 2000; Bizily et al. 1999; Arazi et al. 1999). Detoxification of accumulated metal at cellular level may involve its transport into vacuoles as the final metabolically-inactive sink. Enhanced accumulation (in vacuoles) and translocation of Cd^{2+} and Pb^{2+} to shoots was achieved with *A. thaliana* on expression of the heterologous yeast vacuolar transporter YCF1 known to be effective on the $(glutathione)_2Cd$ complex (Li et al. 1997). Overproduction of the vacuolar metal ion/H^{+} antiporters CAX2 and CAX4 in *N. tabacum* provided transgene plants capable of efficiently detoxifying Cd^{2+}, Zn^{2+} and Mn^{2+} and improved uptake of metal ions in roots (Korenkov et al. 2007a, b; Hirschi et al. 2000).

Concerning the treatment of heavy metal contaminated sites, an interesting approach is phytoremediation of mercury or methylmercury by genetically modified plants, using genes *merA* encoding bacterial mercuric reductase and *merB* encoding bacterial organomercurial lyase (Table 9.5). These genes, separately or together, were expressed in *A. thaliana* (Yang et al. 2003; Bizily et al. 1999, 2003; Heaton et al. 1998; Rugh et al. 1996), *N. tabacum* (Ruiz et al. 2003; He et al. 2001), rice *Oryza sativa* (Heaton et al. 2003), saltmarsh cordgrass *Spartia alterniflora* (Czakó et al. 2006), yellow poplar *Liriodendron tulipifera* (Rugh et al. 1998), and cottonwood *Populus deltoides* (Che et al. 2003; Lyyra et al. 2007). Production of respective enzymes provided transgenes with Hg^{2+} and organo-Hg^{+} tolerant phenotypes and pronounced capacity to volatize Hg.

9.5 Conclusions

The natural capacity of microbes, algae and plants and their symbiotic associations with each other to alter chemical speciation of metals and radionuclides in waters, soils, sediments (Fig. 9.1) have beneficial consequences for bioremediation of the environment and contaminated waste streams. The achievements reviewed above provide solid basis to state that exploitation of metabolic activities, which directly or indirectly result in immobilization of toxic species, would complement pure biosorptive removal of metals and radionuclides from solutions. Moreover, the use of living cells and of their mixed-function consortia is a viable approach for direct treatments of underground water *in situ* and of some complex industrial and municipal effluents in which degradation of organic pollutants may simultaneously occur. A range of reactor designs and *in situ* bioremediation systems have been proposed, some of which were successful at pilot scale level for treatment of genuine effluents (Table 9.1). Many demonstrations indicated that use of multiple species consortia often provide better performance and stability against physico-chemical fluctuations in effluent composition and less sensitivity to environmental factors. Although many proposed and pilot scale bioreactor processes employing metabolic activities of particular microorganism or microbial consortia showed high metal removal efficiency of nearly 100%, their disadvantage would be need for nutrient supplements to effectively treat nutrient-poor effluents. Efforts are thus needed to identify cheap growth substrates to develop really economically favorable process. The nutrient demand is not an issue in self-maintaining artificial wetlands, in which is the proliferation of biological component supported by photoautotrophic plats and algae. Biological removal of metals in wetland relies namely upon established anaerobic and aerobic environments within, which are important for microbially induced immobilization of metal species and to a large extent upon capacity of plants to accumulate metals. Efforts towards implementation of hyperaccumulating plant species to wetland systems would be desirable. Overall performance of the wetland is affected by abiotic components, including dead biota and their decomposition products, and physico-chemical parameters and processes that determine chemical equilibria between soluble and insoluble metal species. Increasing the understanding of complex mechanisms controlling the metal removal process increases the probability of success of wetland application and will further require multidisciplinary approach.

The growing number of studies demonstrates improved metal and radionuclide removal efficiency by transgenic microorganisms and plants at various bioremediation setups and suggests benefits of genetic engineering approach. However, the potential of engineered organism should be demonstrated in pilot scale trials. The underlying economics and safety issues relevant to use of genetically modified organisms should be carefully evaluated and weighted against pros and cons of established remediation techniques as well as against risks of having the recalcitrant heavy metal or radionuclide species in our environment.

Acknowledgements This chapter was processed with the help of grant Centrum 1M06030, MSM 6046137305 and 1M6837805002 from Czech Ministry of Education, Youth and Sports

References

Aken BV, Correa PA, Schnoor JL (2010) Phytoremediation of polychlorinated biphenyls: new trends and promises. Environ Sci Technol 44:2767–2776

Aksu Z, Yener J (1998) Investigation of the biosorption of phenol and monochlorinated phenols on the dried activated sludge. Process Biochem 33:649–655

Alloway BJ (1995) Heavy metals in soils, 2nd edn. Blackie, Glasgow

Anderson RT, Vrionis HA, Ortiz-Bernad I, Resch CT, Long PE, Dayvault R, Karp K, Marutzky S, Metzler DR, Peacock A, White DC, Lowe M, Lovley DR (2003) Stimulating the *in situ* activity of *Geobacter* species to remove uranium from the groundwater of a uranium-contaminated aquifer. Appl Environ Microbiol 69:5884–5891

Andrade JCM, Mahler CF (2002) Soil phytoremediation. 4th International Conference of Engineering Geotechnology. Rio de Janeiro, Brazil

Arazi T, Sunkar R, Kaplan B, Fromm H (1999) A tobacco plasma membrane calmodulin-binding transporter confers Ni^{2+} tolerance and Pb^{2+} hypersensitivity in transgenic plants. Plant J 20:171–182

Arican B, Gokcay CF, Yetis U (2002) Mechanistics of nickel sorption by activated sludge. Process Biochem 37:1307–1315

Atkinson BW, Bux F, Kasan HC (1996) Bioremediation of metal-contaminated industrial effluents using waste materials. Water Sci Technol 34:9–15

Atkinson BW, Bux F, Kasan HC (1998a) Consideration for application of biosorption technology to remediate metal-contaminated industrial effluents. Water SA 24:129–135

Atkinson BW, Bux F, Kasan HC (1998b) Waste activated sludge remediation of metal-plating effluents. Water SA 24:355–359

Bailey NJC, Oven M, Holmes E, Nicholson JK, Zenk MH (2003) Metabolomic analysis of the consequences of cadmium exposure in *Silene cucubalus* cell cultures via H-1 NMR spectroscopy and chemometrics. Phytochemistry 62:851–858

Baker AJM, Brooks RR (1989) Terrestrial higher plants which hyperaccumulate metallic elements—a review of their distribution, ecology and phytochemistry. Biorecovery 1: 81–126

Baker AJM, McGrath SP, Reeves RD, Smith JAC (2000) Metal hyperaccumulator plants: a review of the ecology and physiology of a biological source for phytoremediation of metal-polluted soil. In: Terry N, Bañuelos GS (eds) Phytoremediation of contaminated soil and water. CRC Press, Boca Raton, pp 85–107

Barazani O, Sathiyamoorthy P, Manandhar U, Vulkan R, Golan-Goldhirsh A (2004) Heavy metal accumulation by *Nicotiana glauca* Graham in a solid waste disposal site. Chemosphere 54:867–872

Bargar JR, Bernier-Latmani R, Giammar DE, Tebo BM (2008) Biogenic uraninite nanoparticles and their importance for uranium remediation. Elements 4:407–412

Barkay T, Miller SM, Summers AO (2003) Bacterial mercury resistance from atoms to ecosystems. FEMS Microbiol Rev 27:355–384

Barley RW, Hutton C, Brown MME, Cusworth JE, Hamilton TJ (2005) Trends in biomass and metal sequestration associated with reeds and algae at Wheel Jane Biorem pilot passive treatment plant. Sci Total Environ 338:107–114

Barnes LJ (1993) Process for their microbial oxidation of sulphidic wastes. United States Patent 5840191

Battaglia-Brunet F, Foucher S, Ignatiadis I, Michel C, Morin D (2002) Reduction of chromate by fixed films of sulphate-reducing bacteria using hydrogen as electron source. J Ind Microbiol Biotechnol 28:154–159

Battaglia-Brunet F, Foucher S, Denamur A, Margraff M, Morin D, Ignatiadis I (2004a) Chromate reduction at low sulphate concentration in hydrogen-fed bioreactors. Environ Technol 25:101–109

Battaglia-Brunet F, Foucher S, Morin D, Ignatiadis I (2004b) Chromate (CrO_4^{2-}) reduction in ground waters by using reductive bacteria in fixed-bed bioreactors. Water Air Soil Pollut: Focus 4:127–135

Battaglia-Brunet F, Touze S, Michel C, Ignatiadis I (2006) Treatment of a chromate-polluted groundwater in a 200-dm^3 pilot bio-reactor fed with hydrogen. J Chem Technol Biotechnol 81:1506–1513

Battaglia-Brunet F, Michel C, Joulian C, Ollivier B, Ignatiadis I (2007) Relationship between sulphate starvation and chromate reduction in a H_2-fed fixed-film bioreactor. Water Air Soil Pollut 183:341–353

Bennett LE, Burkhead JL, Hale KL, Terry N, Pilon M, Pilon-Smits EAH (2003) Analysis of transgenic Indian mustard plants for phytoremediation of metal-contaminated mine tailings. J Environ Q 32:432–440

Bittsánszky A, Kömives T, Gullner G, Gyulai G, Kiss J, Heszky L, Radimszky L, Rennenberg H (2005) Ability of transgenic poplars with elevated glutathione content to tolerate Zn^{2+} stress. Environ Int 31:251–254

Bizily SP, Rugh CL, Summers AO, Meagher RB (1999) Phytoremediation of methylmercury pollution: merB expression in Arabidopsis thaliana confers resistance to organomercurials. Proc Natl Acad Sci U S A 96:6808–6813

Bizily SP, Kim T, Kandasamy MK, Meagher RB (2003) Subcellular targeting of methylmercury lyase enhances its specific activity for organic mercury detoxification in plants. Plant Physiol 131:463–471

Blaylock MJ, Salt DE, Dushenkov S, Zakharova O, Gussman C, Kapulnik Y, Ensley BD, Raskin I (1997) Enhanced accumulation of Pb in Indian mustard by soil-applied chelating agents. Environ Sci Technol 31:860–865

Bonthrone KM, Basnakova G, Lin F, Macaskie LE (1996) Bioaccumulation of nickel by intercalation into polycristalline hydrogen uranyl phosphate deposited via an enzymatic mechanism. Nat Biotechnol 14:635–638

Boruvka L, Vacha R (2006) Litavka river alluvium as a model area heavily polluted with potentially risk elements. In: Morel JL, Echevarria G, Goncharova N (eds) Phytoremediation of metal contaminated soils. IV. Earth and environmental sciences, vol 68. NATO Science Series. Springer, Dordrecht, pp 267–298

Bovet L, Feller U, Martinoia E (2005) Possible involvement of plant ABC transporters in cadmium detoxification: a cDNA sub-microarray approach. Environ Int 31:263–267

Brim H, McFarlan SC, Fredrickson JK, Minton KW, Zhai M, Wackett LP, Daly MJ (2000) Engineering *Deinococcus radiodurans* for metal remediation in radioactive mixed waste environments. Nat Biotechnol 18:85–90

Brix H (1993) Macrophytes-mediated oxygen transfer in wetlands: transport mechanism and rate. In: Moshiri TA (ed) Constructed wetlands for water quality improvement. CRC-Press, Boca Raton

Brodie GA, Hammer DA, Tomljanovich DA (1989a) Treatment of acid drainage with a constructed wetland at the Tennessee Valley Authority 950 coal mine. In: Hammer DA (ed) Constructed wetlands for waste water treatment. Lewis Publishers Inc., Chelsea, pp 201–210

Brodie GA, Hammer DA, Tomljanovich DA (1989b) Constructed wetlands for treatment of an ash pond seepage. In: Hammer DA (ed) Constructed wetlands for waste water treatment. Lewis Publishers Inc., Chelsea, pp 211–220

Brooks RR, Lee J, Reeves RD, Jaffre T (1977) Detection of nickeliferous rocks by analysis of herbarium specimens of indicator plants. J Geochem Explor J 7:49–57

Brooks RR (1998) General introduction. In: Brooks RR (ed) Plants that Hyperaccumulate Heavy Metals, CABI, Wallingford, pp 1–14

Brunet J, Repellin A, Varrault G, Terryn N, Zuily-Fodil Y (2008) Lead accumulation in the roots of grass pea (*Lathyrus sativus* L.): a novel plant for phytoremediation systems? C R Biol 331:859–864

Brunke M, Deckwer W-D, Frischmuth A, Horn JM, Lunsdorf H, Rhode M, Rohricht KN, Timmis KN, Weppen P (1993) Microbial retention of mercury from waste streams in a laboratory column containing *merA* gene of bacteria. FEMS Microbiol Rev 11:145–152

Bux F, Atkinson B, Kasan HC (1999) Zinc biosorption by waste activated and digested sludge. Water Sci Technol 39:127–130

Camargo FAO, Bento FM, Okeke BC, Frankenberger WT (2003) Chromate reduction by chromium-resistant bacteria isolated from soils contaminated with dichromate. J Environ Q 32:1228–1233

Chaney RL (1983) Plant uptake of inorganic waste constituents. In: Parr JF, Marsh PB, Kla JM (eds) Land treatment of hazardous wastes. Noyes Data Corporation, Park Ridge, pp 50–77

Chaney RL, Li YM, Angle JS, Baker AJM, Reeves RD, Brown SL, Homer FA, Malik M, Chin M (2000) Improving metal hyperaccumulator wild plants to develop commercial phytoextraction systems: approaches and progress. In: Terry N, Bañuelos GS (eds) Phytoremediation of contaminated soil and water. CRC Press, Boca Raton, pp 131–160

Chaney RL, Angle JS, McIntosh MS, Reeves RD, Li YM, Brewer EP, Chen K-Y, Rosenberg RJ, Perner H, Synkowski EC, Broadhurst CL, Wang S, Baker AJM (2005) Using hyperaccumulator plants to phytoextract soil Ni and Cd. Z Naturforsch 60:190–198

Chang D, Fukushi K, Ghosh S (1995) Stimulation of activated sludge cultures for enhanced heavy metal removal. Water Environ Res 67:822–827

Che D, Meagher RB, Heaton ACP, Lima A, Rugh CL, Merkle SA (2003) Expression of mercuric ion reductase in eastern cottonwood (Populus deltoides) confers mercuric ion reduction and resistance. Plant Biotechnol J 1:311–319

Cheng S, Grosse W, Karrenbrock F, Thoennessen M (2002) Efficiency of constructed wetlands in decontamination of water polluted by heavy metals. Ecol Eng 18:317–325

Cherian S, Oliveira MM (2005) Transgenic plants in phytoremediation: recent advances and new possibilities. Environ Sci Technol 39:9377–9390

Chirwa EMN, Wang Y (1997) Hexavalent chromium reduction by *Bacilus* sp. in a packed-bed bioreactor. Environ Sci Technol 31:1446–1451

Clemens S (2006) Toxic metal accumulation, responses to exposure and mechanisms of tolerance in plants. Biochimie 88:1707–1719

Clemens S, Palghren MG, Krämer U (2002) A long way ahead: understanding and engineering plant metal accumulation. Trends Plant Sci 7:309–315

Cole S (1998) The emergence of treatment wetlands. J Environ Sci Technol 32:218–223

Collins BS, Sharitz RR, Coughlin DP (2005) Elemental composition of native wetland plants in constructed mesocosm treatment wetlands. Biores Technol 96:937–948

Córdoba A, Vargas P, Dussan J (2008) Chromate reduction by *Arthrobacter* CR47 in biofilm packed bed reactors. J Hazard Mater 151:274–279

Costa MC, Duarte JC (2005) Bioremediation of acid mine drainage using acidic soil and organic wastes for promoting sulphate-reducing bacteria activity on a column reactor. Water Air Soil Pollut 165:325–345

Costa MC, Martins M, Jesus C, Duarte JC (2008) Treatment of acid mine drainage by sulphate-reducing bacteria using low cost matrices. Water Air Soil Pollut 189:149–162

Costley SC, Wallis FM (2001) Bioremediation of heavy metals in a synthetic wastewater using a rotating biological contactor. Water Res 35:3715–3723

Cunningham SD, Ow DW (1996) Promises and prospects of phytoremediation. Plant Physiol 110:715–719

Cunningham SC, Berti WR, Huang JW (1995) Phytoremediation of contaminated soils. Trends Biotechnol 13:393–397

Czakó M, Feng X, He Y, Liang D, Márton L (2006) Transgenic *Spartina alterniflora* for phytoremediation. Environ Geochem Health 28:103–110

D'Souza SF, Sar P, Kazy SK, Kubal BS (2006) Uranium sorption by *Pseudomonas* biomass immobilized in radiation polymerized polyacrylamide bio-beads. J Environ Sci Health A Tox Hazard Subst Environ Eng. 41:487–500

De la Fuente JM, Ramírez-Rodríguez V, Cabrera-Ponce JL, Herrera-Estrella L (1997) Aluminium tolerance in transgenic plants by alteration of citrate synthesis. Science 276:1566–1568

De Windt W, Aelterman P, Verstraete W (2005) Bioreductive deposition of palladium (0) nanoparticles on *Shewanella oneidenis* with catalytic activity towards reductive dechlorination of polychlorinated biphenyls. Environ Microbiol 7:314–325

Debusk AT, Laughlin RB, Schwartz LN (1996) Retention and compartmentalization of lead and cadmium in wetland microcosms. Water Res 30:2707–2716

Denny P, Bailey R, Tukahirwa E, Mafabi P (1995) Heavy metal contamination of Lake George (Uganda) and its wetlands. Hydrobiologia 297:229–239

Dermou E, Velissariou A, Xenos D, Vayenas DV (2005) Biological chromium(VI) reduction using a trickling filter. J Hazard Mater B 126:78–85

Dickinson NM, Pulford ID (2005) Cadmium phytoextraction using short-rotation coppice *Salix*: the evidence trail. Environ Int 31:609–613

Diels L, VanRoy S, Mergeay M, Doyen W, Taghavi S, Leysen R (1993a) Immobilization of bacteria in composite membranes and development of tubular membrane reactors for heavy metal recuperation. In: Paterson R (ed) Effective membrane processes: new perspectives. Kluwer Academic Publishers, The Netherlands, pp 275–293

Diels L, VanRoy S, Taghavi S, Doyen W, Leysen R, Mergeay M (1993b) The use of Alcaligenes eutrophus immobilized in a tubular membrane reactor for heavy metal recuperation. In: Torma AE, Apel ML, Brieley CL (eds) Biohydrometallurgical technologies, vol 2. The Minerals, Metals & Materials Society, Warrendale, pp 133–144

Diels L, Dong Q, van der Lelie D, Baeyens W, Mergeay M (1995) The czc operon of Alcaligenes eutrophus CH34: from resistance mechanism to the removal of heavy metals. J Ind Microbiol 14:142–153

Diels L, De Smet M, Hooyberghs L, Kinaer L, Brox G (2001) Bioremediation of heavy metal contaminated soils with a biometal sludge reactor (BSMR). Process Metall 11B:479–486

Diels L, Spaans PH, Van Roy S, Hooyberghs L, Ryngaert A, Wouters H, Walter E, Winters J, Macaskie L, Finlay J, Pernfuss B, Woebking H, Pümpel T, Tsezos M (2003) Heavy metals removal by sand filters inoculated with metal sorbing and precipitating bacteria. Hydrometallurgy 71:235–241

Dieluweit S, Pum D, Sleytr UB (1998) Formation of a gold superlattice on an S-layer with square lattice symmetry. Supramol Sci 5:15–19

Dorlhac de Borne F, Elmayan T, de Roton C, de Hys L, Tepfer M (1998) Cadmium partitioning in transgenic tobacco plants expressing a mammalian metallothionein gene. Mol Breed 4:83–90

Doty SL (2008) Enhancing phytoremediation through the use of transgenics and endophytes. New Phytol 179:318–333

Dunbabin JS, Bowmer KH (1992) Potential use of constructed wetlands for treatment of industrial waste waters containing metals. Sci Total Environ 3:151–68

Dushenkov V, Nanda Kumar PBA, Motto H, Raskin I (1995) Rhizofiltration: the use of plants to remove heavy metals from aqueous streams. Environ Sci Technol 29:1239–1245

Dushenkov S, Vasudev D, Kapulnik Y, Gleba D, Fleisher D, Ting KC, Ensley B (1997) Removal of uranium from water using terrestrial plants. Environ Sci Technol 31:3468–3474

Evanko CR, Dzombak DA (1997) Remediation of metals-contaminated soils and groundwater. Technology evaluation report. Ground Water Remediation Technology Analysis Center, Pittsburgh, PA, USA

Evans K, Gatehouse J, Lindsay W, Shi J, Tommey A, Robinson N (1992) Expression of the pea metallothionein like gene *PsMTa* in *Escherichia coli* and *Arabidopsis thaliana* and analysis of trace metal ion accumulation: implications for PsMTa function. Plant Mol Biol 20:1019–1028

Faisal M, Hasnain S (2004) Microbial conversion of Cr(VI) in to Cr(III) in industrial effluent. Afr J Biotech 3:610–617

Finlay JA, Allan VJ, Conner A, Callow ME, Basnakova G, Macaskie LE (1999) Phosphate release and heavy metal accumulation biofilm-immobilized and chemicalli-coupled cells of *Citrobacter* sp. pre-grown in continuous culture. Biotechnol Bioeng 63:87–97

Fuji E, Toda K, Ohtake H (1990) Bacterial reduction of toxic hexavalent chromium using a fed batch culture of *Enterobacter cloacae* (strain HO1). J Ferment Bioeng 69:365–367

Gadd GM (1992) Microbial control of heavy metal pollution. In: Fry JC, Gadd GM, Herbert RA, Jones CW, Watson-Craik IA (eds) Microbial control of pollution. Cambridge University Press, Cambridge, pp 59–87

Gadd GM (2000) Bioremediation potential of microbial mechanisms of metal mobilization and immobilization. Curr Opin Biotechnol 11:271–279

Gadd GM (2007) Geomycology: biochemical transformations of rocks, minerals, metals and radionuclides by fungi, bioweathwering and bioremediation. Mycol Res 111:3–49

Gadd GM (2010) Metals, minerals and microbes: geomicrobiology and bioremediation. Microbiology 153:609–643

Gail MT, Parsek MR (2003) Heavy metal resistance of biofilm and planktonic *Pseudomonas aeruginosa.* Appl Environ Microbiol 69:2313–2320

Gonçalves MMM, da Costa ACA, Leite SGF, Sant'Anna GL Jr (2007) Heavy metal removal from synthetic wastewaters in an anaerobic bioreactor using stillage from ethanol distilleries as a carbon source. Chemosphere 69:1815–1820

Gong J, Lee DA, Schroeder JI (2003) Long-distance root-to-shoot transport of phytochelatins and cadmium in *Arabidopsis.* Proc Natl Acad Sci U S A 100:10118–10123

Gray S, Kinross J, Read P, Marland A (2000) The nutrient assimilative capacity of Mareal as a substrate for waste treatment. Water Res 34:2183–2190

Greenway M (1997) Nutrient content of wetland plants in constructed wetlands receiving municipal effluent in tropical Australia. Water Sci Technol 35:135–142

Greenway M, Simpson JS (1996) Artificial wetlands for wastewater treatment, water reuse and wildlife in Queens Land, Australia. Water Sci Technol 33:221–229

Grill E, Loffler S, Winnacker EL, Zenk MH (1989) Phytochelatins, the heavy-metal-binding peptides of plants, are synthesised from glutathione by a specific γ-glutamylcysteine dipeptidyl transpeptidase (phytochelatin synthase). Proc Natl Acad Sci U S A 84:6838–6846

Györvary E, Schroedter A, Talapin DV, Weller H, Pum D, Sleytr UB (2004) Formation of nanoparticle arrays on S-layer protein lattices. J Nanosci Nanotechnol 4:115–120

Hammer DA, Bastian RK (1989) Wetland ecosystems: natural water purifiers? In: Hammer DA (ed) Constructed wetlands for waste water treatment. Lewis Publishers Inc., Chelsea, pp 5–19

Hansen CL, Zwolinski G, Martin D, Williams JW (1984) Bacterial removal of mercury from sewage. Biotechnol Bioeng 26:1330–1333

Hasegawa I, Terada E, Sunairi M, Wakita H, Shinmachi F, Noguchi A, Nakajima M, Yazaki J (1997) Genetic improvement of heavy metal tolerance in plants by transfer of the yeast metallothionein gene (CUP1). Plant Soil 196:277–281

He YK, Sun JG, Feng XZ, Czakó M, Márton L (2001) Differential mercury volatilization by tobacco organs expressing a modified bacterial merA gene. Cell Res 11:231–236

Heaton ACP, Rugh CL, Wang NJ, Meagher RB (1998) Phytoremediation of Hg-polluted soils by genetically engineered plants. J Soil Cont 7:497–509

Heaton ACP, Rugh CL, Kim T, Wang NJ, Meagher RB (2003) Toward detoxifying mercury-polluted aquatic sediments with rice genetically engineered for mercury resistance. Environ Toxicol Chem 22:2940–2947

Hennebel T, Gusseme BD, Verstraete W (2009) Biogenic metals in advanced water treatment. Trends Biotechnol 27:90–98

Higuchi K, Suzuki K, Nakanishi H, Yamaguchi H, Nishizawa N-K, Mori S (1999) Cloning of nicotianamine synthase genes, novel genes involved in the synthesis of phytosiderophores. Plant Physiol 119:471–479

Hirschi KD, Korenkov VD, Wilganowski NL, Wagner GJ (2000) Expression of *Arabidopsis CAX2* in tobacco. Altered metal accumulation and increased manganese tolerance. Plant Physiol 124:125–133

Horn JM, Brunke M, Deckwer WD, Timmis KN (1994) *Pseudomonas putida* strains which constitutively overexpress mercury resistance for biodetoxification of organomercurial pollutants. Appl Environ Microbiol 60:357–362

Huang JW, Cunningham SD (1996) Lead phytoextraction: species variation in lead uptake and translocation. New Phytol 134:75–84

Huisman JL, Schouten G, Schultz C (2006) Biologicaly produced sulphide for purification of process streams, effluent treatment and recovery of the metals in the metal and mining industry. Hydrometallurgy 83:106–113

Ike A, Sriprang R, Ono H, Murooka Y, Yamashita M (2007) Bioremediation of cadmium contaminated soil using symbiosis between leguminous plant and recombinant rhizobia with the MTL4 and the PCS genes. Chemosphere 66:1670–1676

Jaffre T, Brooks RR, Lee J, Reeves RD (1976) *Sebertia acuminata*, a hyperaccumulator of nickel from New Caledonia. Science 193:579–580

Jayaweera MW, Kasturiarachchi JC, Kularatne RK, Wijeyekoon SL (2008) Contribution of water hyacinth (*Eichhornia crassipes* (Mart.) Solms) grown under different nutrient conditions to Fe-removal mechanisms in constructed wetlands. J Environ Manager 87:450–460

Jeong BC, Hawes C, Bonthrone KM, Macaskie LE (1997) Localization of enzymatically enhanced heavy metal accumulation by *Citrobacter* sp. and metal accumulation in vitro by liposomes containing entrapped enzyme. Microbiology 143:2497–2507

Jeong BC, Poole PS, Willis AC, Macaskie LE (1998) Purification and characterization of acid-type phosphatases from a heavy-metal-accumulating *Citrobacter* sp. Arch Microbiol 169:166–173

Johnson DB, Hallberg KB (2005a) Biogeochemistry of the compost bioreactor components of a composite acid mine drainage passive remediation system. Sci Total Environ 338:81–93

Johnson DB, Hallberg KB (2005b) Acid mine drainage remediation options: a review. Sci Total Environ 338:3–14

Johnston CA (1993) Mechanisms of water wetland water quality interaction. In: Moshiri GA (ed) Constructed wetlands for water quality improvement. CRC-Press, Boca Raton, pp 293–300

Jorand F, Boue-Binge F, Block JC, Urbain V (1998) Hydrophobic/hydrophilic properties of activated sludge exopolymeric substances. Water Sci Technol 37:307–315

Kaksonen AH, Puhakka JA (2007) Sulfate reduction based bioprocesses for the treatment of acid mine drainage and recovery of metals. Eng Life Sci 6:541–564

Kalin M (2001) Biogeochemical and ecological consideration in designing wetland treatment systems in post-mining landscapes. Waste Manag 21:191–196

Kalin M, Fyson A, Smith MP (1993) ARUM-acid reduction using microbiology. In: Torma AE, Apel ML, Brierley CL (eds) Biohydrometallurgical technologies. The Minerals, Metals & Materails Society, Warrendale, pp 319–328

Kalin M, Caetano, Chaves WL (2003) Acid reduction using microbiology: treating AMD effluent emerging from an abandoned mine portal. Hydrometallurgy 71:217–225

Kasan HC (1993) The role of waste activated sludge and bacteria in metal ion removal from solution. Crit Rev Environ Sci Technol 23:79–117

Kayser A, Felix HR (1998) Five years of phytoremediation in the field. In: Timmis KN (ed) Book of ext. abstr int workshop–innovative potential of advanced biological systems for remediation. Technical University Hamburg, Harburg, pp 81–86

Kazy SK, D'Souza SFD, Sar P (2009) Uranium and thorium sequestration by a *Pseudomonas* sp.: mechanism and chemical characterization. J Hazard Mater 163:65–72

Keiding K, Nielsen PH (1997) Desorption of organic macromolecules from activated sludge: effect of ionic composition. Water Res 31:1665–1672

Kodukula PS, Patterson JW, Surampalli RY (1994) Sorption and precipitation of metals in activated sludge. Biotechnol Bioeng 43:874–880

Komori K, Rivas A, Toda K, Ohtake H (1990) A method for removal of toxic chromium using dialysis-sac cultures of a chromate-reducing strain of *Enterobacter cloacae*. Appl Microbiol Biotechnol 33:117–119

Koprivova A, Kopriva S, Jäger D, Will D, Jouanin L, Rennenberg H (2002) Evaluation of transgenic poplars over-expressing enzymes of glutathione synthesis for phytoremediation of cadmium. Plant Biol 4:664–670

Korenkov V, Hirschi K, Crutchfield JD, Wagner GJ (2007a) Enhancing tonoplast Cd/H antiport activity increases Cd, Zn, and Mn tolerance, and impacts root/shoot Cd partitioning in *Nicotiana tabacum* L. Planta 226:1379–1387

Korenkov V, Park S, Cheng N, Sreevidya C, Lachmansingh J, Morris J, Hirschi K, Wagner GJ (2007b) Enhanced Cd^{2+} -selective root-tonoplast-transport in tobaccos expressing Arabidopsis cation exchangers. Planta 225:403–411

Kotrba P, Macek T, Ruml T (1999) Heavy-metal binding peptides and proteins in plants. Coll Czech Chem Commun 64:1057–1086

Kotrba P, Najmanova J, Macek T, Ruml T, Mackova M (2009) Genetically modified plants in phytoremediation of heavy metal and metalloid soil and sediment pollution. Biotechnol Adv 27:799–810

Krämer U, Chardonnens A (2001) The use of transgenic plants in the bioremediation of soils contaminated with trace elements. Appl Microbiol Biotechnol 55:661–672

Krämer U, Talke I, Hanikenne M (2007) Transition metal transport. FEBS Lett 581:2263–2272

LeCooper EM, Sims JT, Cunningham SD, Huang JW, Berti WR (1999) Chelate-assisted phytoextraction of lead from contaminated soils. J Environ Qual 28:1709–1719

Lee M, Yang M (2010) Rhizofiltration using sunflower (*Helianthus annuus* L.) and bean (*Phaseolus vulgaris* L. var. *vulgaris*) to remediate uranium contaminated groundwater. J Hazard Mater 173:589–596

Li ZS, Lu YP, Zhen RG, Szczypka M, Thiele DJ, Rea PA (1997) A new pathway for vacuolar cadmium sequestration in *Saccharomyces cerevisiae*: YCF1-catalyzed transport of bis(glutathionato) cadmium. Proc Natl Acad Sci U S A 94:42–47

Lloyd JR, Yong P, Macaskie LE (1998) Enzymatic recovery of elemental palladium by using sulfate-reducing bacteria. Appl Environ Microbiol 64:4607–4609

Lloyd JR, Lovley DR, Macaskie LE (2003) Biotechnological application of metal-reducing microorganisms. Adv Appl Microbiol 53:85–128

Lovley DR (1995) Bioremediation of organic and metal contaminants with dissimilatory metal reduction. J Ind Microbiol 14:85–93

Lovley DR, Coates JD (1997) Bioremediation of metal contamination. Curr Opin Biotechnol 8:285–289

Lyyra S, Meagher RB, Kim T, Heaton A, Montello P, Balish RS, Merkle SA (2007) Coupling two mercury resistance genes in eastern cottonwood enhances the processing of organomercury. Plant Biotechnol J 5:254–262

Mabbet AN, Yong P, Farr JP, Macaskie LE (2004) Reduction of Cr(VI) by "palladized" biomass of *Desulfovibrio desulfuricans* ATCC 29577. Biotechnol Bioeng 87:104–109

Macaskie LE, Dean ACR (1984) Cadmium accumulation by immobilized cells of a *Citrobacter* sp. Biotechnol Lett 5:177–186

Macaskie LE, Dean ACR (1985) Uranium accumulation by immobilized cells of a *Citrobacter* sp. Biotechnol Lett 7:457–462

Macaskie LE, Empson RM, Cheetham AK, Grey CP, Skarnulis AJ (1992) Uranium bioaccumulation by a *Citrobacter* sp. as a result of an ezymatically mediated growth of polycrystalline HUO_2PO_4. Science 257:782–784

Macaskie LE, Bonthrone KM, Yong P, Goddard DT (2000) Enzymically mediated bioprecipitation of uranium by a *Citrobacter* sp.: a concerted role for exocellular lipopolysaccharide and associated phosphatase in biomineral formation. Microbiology 146:1855–1867

Macek T, Macková M, Truksa M, Singh-Cundy A, Kotrba P, Yancey N, Scouten WH (1996) Preparation of transgenic tobacco with a yeast metallothionein combined with a polyhistidine tail. Chem Listy 90:690

Macek T, Macková M, Burkhard J, Demnerová K (1998) Introduction of green plants for the control of metals and organics in remediation. In: Holm FW (ed) Effluents from alternative demilitarization technologies. NATO PS Ser. 1, vol 22. Kluwer Academic Publishers, Dordrecht, pp 71–85

Macek T, Macková M, Káš J (2000) Exploitation of plants for the removal of organics in environmental remediation. Biotechnol Adv 18:23–35

Macek T, Macková M, Pavliková D, Száková J, Truksa M, Singh Cundy A, Kotrba P, Yancey N, Scouten WH (2002) Accumulation of cadmium by transgenic tobacco. Acta Biotechnol 22:101–106

Macek T, Pavlikova D, Mackova M (2004) Phytoremediation of metals and inorganic pollutants. In: Singh A, Ward OP (eds) Applied bioremediation and phytoremediation. Springer, Berlin, pp 135–157

Macek T, Kotrba P, Svatos A, Novakova M, Demnerova K, Mackova M (2008) Novel roles for genetically modified plants in environmental protection. Trends Biotechnol 26:146–152

Malik A (2004) Metal bioremediation through growing cells. Environ Int 30:261–278

Malik A (2007) Environmental challenge *vis a vis* opportunity: the case of water hyacinth. Environ Int 33:122–138

Martínez M, Bernal P, Almela C, Vélez D, García-Agustín P, Serrano R, Navaro-Aviño J (2006) An engineered plant that accumulates higher levels of heavy metals than *Thlaspi caerulescens*, with yields of 100 times more biomass in mine soils. Chemosphere 64:478–485

Martins M, Santos ES, Pires C, Barros RJ, Costa MC (2010) Production of irrigation water from bioremediation of acid mine drainage: comparing the performance of two representative systems. J Clean Prod 18:248–253

Mashauri DA, Mulugu DMM, Abdul Hussein BS (2000) Constructed wetland at the University of Dar Es Salaam. Water Res 24:1135–1144

Matagi SV, Swai D, Mugabe R (1998) A review of heavy metal removal mechanisms in wetlands. Afr J Trop Hydrobiol Fish 8:23–35

Meyers DE, Auchterlonie GJ, Webb RI, Wood B (2008) Uptake and localisation of lead in the root system of *Brassica juncea*. Environ Pollut 153:323–332

Mays PA, Edwards GS (2001) Comparison of heavy metal accumulation in a natural wetland and constructed wetlands receiving acid mine drainage. Ecol Eng 16:487–500

Mishra VK, Tripathi BD, Kim KH (2009) Removal and accumulation of mercury by aquatic macrophytes from an open cast coal mine effluent. J Hazard Mater 172:749–754

McLean J, Beveridge TJ (2001) Chromate reduction by a *Pseudomonad* isolated from a site contaminated with chromated copper arsenate. Appl Environ Microbiol 67:1076–1084

Meagher RB (2000) Phytoremediation of toxic elemental and organic pollutants. Curr Opin Plant Biol 3:153–162

Mergeay M, Monchy S, Vallaeys T, Auquier V, Benotmane A, Bertin P, Taghavi S, Dunn J, van der Lelie D, Wattinez R (2003) *Ralstonia metallidurans*, a bacterium specifically adapted to toxic metals: towards a catalogue of metal-responsive genes. FEMS Microbiol Rev 27:385–410

Nedelkoska TV, Doran PM (2000) Hyperaccumulation of cadmium by hairy roots of *Thlaspi caerulescens*. Biotechnol Bioeng 67:607–615

Newman LA, Doty SL, Gery KL, Heilman PE, Muiznieks I, Shang TQ, Siemieniec ST, Strand SE, Wang X, Wilson AM, Gordon MP (1998) Phytoremediation research at the University of Washington. J Soil Cont 7:531–542

Novak JT, Higgins MJ, Love NG (1999) The effect of cationic salt addition on the settling and dewatering properties of an industrial activated sludge. Water Environ Res 71:252–254

Ohtake H, Fuji E, Toda K (1990) A survey of effective electron-donors for reduction of toxic hexavalent chromium by *Enterobacter cloacae* (strain-HO1). J Gen Appl Microbiol 36:203–208

Pattanapipitpaisal P, Brown NL, Macaskie LE (2001) Chromate reduction by *Microbacterium liquefaciens* immobilized in polyvinyl alcohol. Biotech Lett 23:61–65

Pavlíková D, Macek T, MacKová M, Száková J, Balík J (2004) Cadmium tolerance and accumulation in transgenic tobacco plants with a yeast metallothionein combined with a polyhistidine tail. Int Biodeteior Biodegrad 54:233–237

Pilon-Smits, Hwang, Mel Lytle C, Zhu, Tai, Bravo, Chen Y, Leustek T, Terry N (1999) Overexpression of ATP sulfurylase in Indian mustard leads to increased selenate uptake, reduction, and tolerance. Plant Physiol 119:123–132

Pilon-Smith E, Pilon M (2002) Phytoremediation of metals using transgenic plants. Crit Rev Plant Sci 21:439–456

Pip E, Stepaniuk J (1992) Cadmium, copper and lead in sediments. Arch Hydrobiol 124:337–355

Pol LW, Lens PNL, Weijma J, Stams AJM (2001) New developments in reactor and process technology for sulphate reduction. Water Sci Technol 44:67–76

Pollmann K, Raff J, Merroun M, Fahmy K, Selenska-Pobell S (2006) Metal binding by bacteria from uranium mining waste piles and its technological applications. Biotechnol Adv 24:58–68

Polprasert C, Dan NP, Thayalakumaran N (1996) Application of constructed wetlands to treat some toxic wastewaters under tropical conditions. Water Sci Technol 34:165–171

Pujol R, Cantler JP (1992) Biosorption and dynamics of bacterial populations in activated sludge. Water Res 26:209–212

Pümpel T, Ebner C, Pernfuss B, Schinner F, Diels L, Keszthelyi Z, Stankovic A, Finlay JA, Macaskie LE, Tsezos M, Wouters H (2001) Treatment of rinsing water from electroless nickel plating with a biologically active moving-bed sand filter. Hydrometallurgy 59:383–293

Reeves RD (2006) Hyperaccumulation of trace elements by plants. In: Morel JL, Echevarria G, Goncharova N (eds) Phytoremediation of metal contaminated soils, IV. Earth and Environmental Sciences, vol 68. NATO Science Series, Springer, Dordrecht, pp 25–52

Reith F, Lengke MF, Falconer D, Craw D, Southam G (2007) The geomicrobiology of gold. ISME J 1:567–584

Reith F, Etschmann B, Grosse C, Moors H, Benotmane MA, Monsieurs P, Grass G, Doonan C, Vogt S, Martinez-Criado G, George GN, Nies DH, Mergeay M, Pring A, Southam G, Brugger J (2009) Mechanisms of gold biomineralization in the bacterium *Cupriavidus metallidurans*. Proc Natl Acad Sci U S A 106:17757–17762

Renninger N, McMahon KD, Knopp R, Nitsche H, Clark DS, Keasling JD (2001) Uranyl precipitation by biomass from an enhanced biological phosphorus removal reactor. Biodegradation 12(6):401–410

Renninger N, Knopp R, Nitsche H, Clark DS, Keasling JD (2004) Uranyl precipitation by *Pseudomonas aeruginosa* via controlled polyphosphate metabolism. Appl Environ Microbiol 70:7404–7412

Roosens NHCJ, Willems G, Saumitou-Laprade P (2008) Using Arabidopsis to explore zinc tolerance and hyperaccumulation. Trends Plant Sci 5:208–215

Rudolph A, Becker E, Scholz G, Prochazka Z, Toman J, Macek T, Herout V (1985) The occurence of the amino acid nicotianamine in plants and microorganisms. A reinvestigation. Biochem Physiol Pflanzen 180:557–563

Rugh CL, Wilde HD, Stack NM, Thompson DM, Summers AO, Meagher RB (1996) Mercuric ion reduction and resistance in transgenic *Arabidopsis thaliana* plants expressing a modified bacterial *merA* gene. Proc Natl Acad Sci U S A 93:3182–3187

Rugh CL, Senecoff JF, Meagher RB, Merkle SA (1998) Development of transgenic yellow poplar for mercury phytoremediation. Nat Biotechnol 16:925–928

Ruiz ON, Hussein HS, Terry N, Daniell H (2003) Phytoremediation of organomercurial compounds via chloroplast genetic engineering. Plant Physiol 132:1344–1352

Ruml T, Kotrba P (2003) Microbial control of metal pollution: an overview. In: Fingerman M, Nagabhushanam R (eds) Recent advances in marine biotechnology. Bioremediation, vol 8. Science Publishers, Enfield, pp 81–153

Russell RA, Holden PJ, Wilde KL, Neilan BA (2003) Demonstration of the use of *Scenedesmus* and *Carteria* biomass to drive bacterial sulfate reduction by *Desulfovibrio alcoholovorans* isolated from an artificial wetland. Hydrometallurgy 71:227–234

Salt DE, Pickering IJ, Prince RC, Gleba D, Dushenkov S, Smith RD, Raskin I (1997) Metal accumulation by aquacultured seedlings of Indian mustard. Environ Sci Technol 31:1636–1644

Salt DE, Smith RD, Raskin I (1998) Phytoremediation. Annu Rev Plant Physiol Plant Mol Biol 49:643–68

Sar P, D'Souza SF (2001) Biosorptive uranium uptake by a *Pseudomonas* strain: characterization and equilibrium studies. J Chem Technol Biotechnol 76:1286–1294

Sar P, D'Souza SF (2002) Biosorption of thorium by a *Pseudomonas* biomass. Biotechnol Lett 24:239–243

Scholes L, Shutes RBE, Revitt DM, Forshaw M, Purchase D (1998) The treatment of metals in urban runoff by constructed wetlands. Sci Total Environ 214:211–219

Scholz M (2003) Performance predictions of mature experimental constructed wetlands which treat urban water receiving high loads of lead and copper. Water Res 37:1270–1277

Serporylov NS, Ponomaryev YE, Korenkov VN, Zhukov IM, Sukhoverova LL, Ksandopulo BA (1981) Biochemical removal of chromium (VI) compounds from wastewater under anaerobic conditions in the presence of substrate. IC:CO2F003-28. Appl. SU 81-3306746

Sharma PK, Balkwill DL, Frenkel A, Vairavamurthy MA (2000) A new Klebsiella planticola strain (Cd-1) grows anaerobically at high cadmium concentrations and precipitates cadmium sulfide. Appl Environ Microbiol 66:3083–3087

Shen H, Wang Y (1995) Hexavalent chromium removal in two-stage bioreactor system. J Environ Eng 121:798–804

Shenton W, Pum D, Sleytr UB, Mann S (1997) Synthesis of cadmium sulphide superlattices using self-assembled bacterial S-layers. Nature 389:585–587

Sheoran AS, Sheoran V (2006) Heavy metal removal mechanism of acid mine drainage in wetlands: a critical review. Mineral Eng 19:105–116

Silver S, Phung LT (2005) A bacterial view of periodic table: genes and proteins for toxic inorganic ions. J Ind Microbiol Biotechnol 32:587–605

Skousen J, Sexstone A, Garbutt K, Sencindiver J (1992) Wetlands for treating acid mine drainage. Green Lands 22:31–39

Smith W, Gadd G (2000) Reduction and precipitation of chromate by mixed culture sulphate-reducing bacterial biofilms. J Appl Microbiol 88:983–991

Sobolewski A (1996) Metal species indicate the potential of constructed wetlands for long-term treatment of mine drainage. J Ecol Eng 6:259–271

Song W, Sohn EJ, Martinoia E, Lee YJ, Yang Y, Jasinski M, Forestier C, Hwang I, Lee Y (2003) Engineering tolerance and accumulation of lead and cadmium in transgenic plants. Nat Biotechnol 21:914–919

Southam G, Lengke MF, Fairbrother L, Reith F (2009) The biogeochemistry of gold. Elements 5:303–307

Sunkar R, Kaplan B, Bouché N, Arazi T, Dolev D, Talke IN, Maathuis FJ, Sanders D, Bouchez D, Fromm H (2000) Expression of a truncated tobacco NtCBP4 channel in transgenic plants and disruption of the homologous *Arabidopsis* CNGC1 gene confer Pb^{2+} tolerance. Plant J 24:533–542

Suzuki N, Koizumi N, Sano H (2001) Screening of cadmium-responsive genes in *Arabidopsis thaliana*. Plant Cell Environ 24:1177–1188

Tabak HH, Govind R (2003) Advances in biotreatment of acid mine drainage and biorecovery of metals: 2. Membrane bioreactor system for sulfate reduction. Biodegradation 14:437–452

Tabak HH, Scharp R, Burckle J, Kawahara FK, Govind R (2003) Advances in treatment of acid mine drainage and biorecovery of metals: 1. Metal precipitation for recovery and recycle. Biodegradation 14:423–436

Thomas JC, Davies EC, Malick FK, Endreszl C, Williams CR, Abbas M, Petrella S, Swisher K, Perron M, Edwards R, Osenkowski P, Urbanczyk N, Wiesend WN, Murray KS (2003) Yeast metallothionein in transgenic tobacco promotes copper uptake from contaminated soils. Biotechnol Prog 19:273–280

Tomé FV, Rodríguez PB, Lozano JC (2008) Elimination of natural uranium and 226Ra from contaminated waters by rhizofiltration using *Helianthus annuus* L. Sci Total Environ 393:351–357

USEPA (2005a) Annual report. http://www.epa.gov/ORD/NRMRL/std/mtb/mwt/annual/annual2005/adwt/passivebiotreat.htm

USEPA (2005b) Annual report. http://www.epa.gov/ORD/NRMRL/std/mtb/mwt/annual/annual2005/adwt/sulfatereducingbacteriademo.htm

Utkgikar V, Chen B, Tabak HH, Bishop DF, Govind R (2000) Treatment of acid mine drainage. I. Equilibrium biosorption of zinc and copper on non-viable activated sludge. Int Biodeter Biodegrad 46:19–28

Vangronsveld J, Herzig R, Weyens N, Boulet J, Adriaensen K, Ruttens A, Thewys T, Vassilev A, Meers E, Nehnevajova E, van der Lelie D, Mench M (2009) Phytoremediation of contaminated soils and groundwater: lessons from the field. Environ Sci Pollut Res Int 16:765–794

Vasák M (2005) Advances in metallothionein structure and functions. J Trace Elem Med Biol 19:13–17

Veeken AHM, Rulkens WH (2003) Innovative developments in the selective removal and reuse of heavy metals from wastewaters. Water Sci Technol 47:9–16

von Canstein H, Kelly S, Li Y, Wagner-Dobler I (2002a) Species diversity improves the eficiency of mercury-reducing biofilms under changing environmental conditions. Appl Environ Microbiol 68:2829–2837

von Canstein H, Li Y, Leonhauser J, Haase E, Felske A, Deckwer WD, Wagner-Dobler I (2002b) Spatially oscillating activity and microbial succession of mercury-reducing biofilms in a technical-scale bioremediation system. Appl Environ Microbiol 68:1938–1946

Wagner-Döbler I, van Canstein HF, Li Y, Timmis KN, Deckwer WD (2000a) Removal of mercury from chemical wastewater by microorganisms in technical scale. Environ Sci Technol 34:4628–4634

Wagner-Döbler I, Lunsdorf H, Lubbehusen T, von Canstein HF, Li Y (2000b) Structure and species composition of mercury-reducing biofilms. Appl Environ Microbiol 66:4559–4563

Wahl R, Mertig M, Raff J, Selenska-Pobell S, Pompe W (2001) Electronbeam induced formation of highly ordered palladium and platinum nanoparticle arrays on the S-layer of *Bacillus sphaericus* NCTC 9602. Adv Mater 13:736–740

Wall JD, Krumholz LR (2006) Uranium reduction. Annu Rev Microbiol 60:149–166

Watanabe ME (1997) Phytoremediation on the brink of commercialization. Environ Sci Technol 31:182–186

Wächtershäuser G (2006) From volcanic origins of chemoautotrophic life to Bacteria, Archea and Eukarya. Phil Trans Soc B 361:1787–1808

Weis JS, Weis P (2004) Metal uptake, transport and release by wetland plants: implications for phytoremediation and restoration. Environ Int 30:685–700

White C, Gadd GM (1998) Accumulation and effects of cadmium on sulphate-reducing bacterial biofilms. Microbiology 144:1407–1415

White C, Gadd GM (2000) Copper accumulation by sulphate-reducing bacterial biofilms. FEMS Microbiol Lett 183:313–318

Whitlock JL (1990) Biological detoxification of precious metal processing wastewaters. Geomicrobiol J 8:241–249

Williams JB (2002) Phytoremediation in wetland ecosystems: progress, problems, and potential. Crit Rev Plant Sci 21:607–635

Wood B, McAtamney C (1994) The use of macrophytes in bioremediation. Biotechnol Adv 12:653–662

Yang H, Nairn J, Ozias-Akins P (2003) Transformation of peanut using a modified bacterial mercuric ion reductase gene driven by an actin promoter from *Arabidopsis thaliana*. J Plant Physiol 160:945–952

Yang X, Feng Y, He Z, Stoffella PJ (2005) Molecular mechanisms of heavy metal hypeaccumulation and phytoremediation. J Trace Elem Med Biol 18:339–353

Ye ZH, Whiting SN, Qian JH, Lytle CM, Lin ZQ, Terry N (2001) Wetlands and aquatic processes, trace elements removal from coal ash leachate by a 10 year old constructed wetland. J Environ Qual 30:1710–1719

Yong P, Rowson NA, Farr JP, Harris IR, Macaskie LE (2002a) Bioaccumulation of palladium by *Desulfovibrio desulfuricans*. J Chem Technol Biotechnol 77:593–601

Yong P, Rowson NA, Farr JP, Harris IR, Macaskie LE (2002b) Bioreduction and biocrystallization of palladium by *Desulfovibrio desulfuricans* NCIMB 8307. Biotechnol Bioeng 80:369–379

Yong P, Mikheenko IP, Deplanche K, Sargent F, Macaskie LE (2009) Biorecovery of precious metals from wastes and conversion into fuel cell catalyst for electricity production. Adv Materials Res 71–73:729–732

Yuncu B, Sanin FD, Yetis U (2006) An investigation of heavy metal biosorption in relation to C/N ratio of activated sludge. J Hazard Mater B 137:990–997

Zita A, Hermanso M (1997) Effects of bacterial cell surface structures and hydrophobicity on attachment to activated sludge flocs. Appl Environ Microbiol 63:1168–1170

Chapter 10
Yeast Biosorption and Recycling of Metal Ions by Cell Surface Engineering

Kouichi Kuroda and Mitsuyoshi Ueda

Abstract Metal biosorption by microorganisms contributes to the removal of toxic heavy metal ions and collection of metal resources from waste streams. Cell surface biosorption enhanced by cell surface engineering is a unique and effective approach for the construction of a novel biosorbent. Through biosorption, adsorbed metal ions can be easily recovered without disintegration of cells, and the repeated use of cells as adsorbents for the biosorption and recuperation of metal ions becomes feasible. Thus, engineering of cell surfaces for enhanced biosorption of metal ions has the potential to be more suitable than genetic manipulations that promote intracellular accumulation. Metal-binding proteins and peptides were displayed on the yeast cell surface by an α-agglutinin-based display system, and cell surface-engineered yeasts showed enhanced biosorption of, and tolerance to, heavy metal ions. In addition, a yeast biosorbent for biosorption and recovery of molybdate ions was constructed by cell surface display of molybdate-binding protein (ModE). The metal biosorption ability of cell surface-engineered yeasts relies upon features of the displayed metal-binding proteins and natural properties of particular cell wall. Metal-binding proteins with a capacity to form selective coordination spheres and provide tailored biosorption could be generated by direct screening a mutant library with combinatorial mutations in the metal recognition domain at the yeast cell wall background. Therefore, generation of novel metal-binding proteins and molecular breeding of yeast biosorbents showing selective biosorption can be concurrently achieved by cell surface engineering.

Keywords Cell surface biosorption • Recovery of metal ions • Recycling • Rare metal ions • Yeast cell surface engineering • Combinatorial bioengineering

K. Kuroda (✉)
Division of Applied Life Sciences, Graduate School of Agriculture, Kyoto University, Sakyo-ku, Kyoto 606-8502, Japan
e-mail: k_kuro@kais.kyoto-u.ac.jp

P. Kotrba et al. (eds.), *Microbial Biosorption of Metals*,
DOI 10.1007/978-94-007-0443-5_10, © Springer Science+Business Media B.V. 2011

10.1 Introduction

Metals are very important materials that are closely related to our daily life because of their implications on human health, economy, and diplomacy between countries. Along with the development of a civilized society, mining- and manufacturing-induced pollution of soil and water has emerged as a social issue. Heavy-metal-ion pollution in the hydrosphere has been caused mainly by effluent waste water from factories, mines, and metal refineries. This is a serious environmental issue that needs to be addressed; however, the removal of heavy metal ions from the environment is difficult because, unlike organic pollutants, they cannot be chemically degraded. Industrial depression by unprecedented environmental pollution and depletion of resources is a growing concern. The use of inorganic metal elements such as rare earth metals is essential in the modern industrial society. The demand for rare metals is increasingly growing because they are considered indispensable for high-technology industries to maintain and develop their international competitiveness. A stable supply of rare metals is critical for the development of global science and technology. Disposal of metal ions causes environmental pollution, but collected metal ions could be useful resources. Therefore, innovative technologies for efficient separation, recovery, and recycling of metal ions from discarded high-technology products and wastewater have to be developed. To address this issue, two main approaches for treatment of metal-containing water could be considered: conventional physico-chemical techniques and bioremediation, including approaches using microorganisms as a biosorbent. Conventional methods employing namely various precipitation and ion-exchange techniques are often ineffective and costly when applied to dilute effluents (Marques et al. 1991; Kapoor and Viraraghavan 1995). It prompted many efforts toward bioremediation of heavy metal ions by biosorption (Gadd and White 1993; Lovely and Coates 1997; Eccles 1999).

Substantial efforts have been made in the metal biosorption field which demonstrated feasibility of this approach (reviewed elsewhere in this book) and metal biosorption is also expected to have advantages in individually specific removals and concentration of certain metals that is difficult to achieve by conventional methods. Studies on biosorption aim at development of methods making cleanup of (waste) water to meet regulatory agency limits possible as well as to efficient recovery and recycling of metals from the biosorbent. Therefore, a novel approach using the yeast cell surface protein display system (yeast cell surface engineering) (Ueda and Tanaka 2000; Kondo and Ueda 2004) was designed for the molecular breeding of a biosorbent. In this chapter, yeast biosorbents constructed by displaying metal-binding proteins and peptides on the yeast cell surface and their performance in metallosorption are introduced. Developed system allows not only the removal of toxic heavy metal ions but it was found also suitable for recovery, and recycling of useful metal ions from solutions. Cell surface design for biosorption and recovery of rare metal ions is also introduced, and the advantages and potential of cell surface design in the construction of biosorbents for bioremediation and resource recycling are discussed.

10.2 Metal Uptake and Microbial Biosorption: Why Use Yeasts?

Recovery of metal ions from aqueous solutions first involves the binding of metal ions on appropriate material, followed by an effective release of accumulated metal ions by simple change in the physico-chemical parameters. Various chemical polymeric resin ion-exchangers, which are extensively used as conventional materials for metal remediation from waste water streams, were formulated. During the last decades, biosorption of metal ions using microorganisms and plants has attracted serious attention because of its low cost and efficient metal recovery from effluents with low concentrations of metal ions (Gadd and White 1993). Living organisms are equipped with systems that ensure maintenance of metal homeostasis and keep intracellular metal ions within a defined beneficial concentration range independent of the environmental metal concentration. In this system, functional proteins that recognize metal ions play a role in intracellular sorting and eventually convert toxic metal ions into their nontoxic forms by intracellular sequestration. Attempts to enhance the natural capacity of microorganisms and plants to accumulate heavy metal ions are actively carried out. Biosorption of metal ions in aqueous solutions using living microorganisms is expected to be a prospective method (Malik 2004). The metal uptake phenomenon caused by the interaction between cells and metals is biphasic and two processes could be recognized: (1) metabolism independent biosorption on the cell surface, which is the first cell component that interacts with the external environment and (2) absorption by transport into the cell (i.e., intracellular accumulation), which may take place also with metals of no biological importance through transporters involved in uptake of essential metal ions.

Several approaches have focused on intracellular accumulation, in which the cellular ability to take up and accumulate metal ions has been improved (Malik 2004; Pazirandeh et al. 1995). It should be noted, however, that intracellular accumulation can be insufficient in terms of the recovery of incorporated metal ions. Systems utilizing intracellular accumulation are often limited to single use because disintegration of cells is required for the extraction of the metal ions incorporated and accumulated inside the cells. Unlike, e.g., phytoremediation approaches, that aim at bioremediation of metal polluted soils and uptake of the pollutant in easy-to-harvest shoot tissues is beneficial, extracellular events such as biosorption are desired for bioremediation of water. Biosorption on the cell surface offers significant advantages: biosorption proceeds in a short time; small size of microorganisms provides a high ratio of surface area to volume and thus a large contact area interacting with metal ions from the environment; bound metal ions can be easily recovered and recycled without the need for disruption of cell integrity; repeated use of cells for biosorption and recovery of metal ions is possible; and the biosorption capacity of cells is maintained as long as the metal-binding molecules on the cell surface are functionally active, even when nonliving biomass is used. Yeasts fully conform to these features and are being often proposed suitable sources of the biosorbent material (Wang and Chen 2006). Yeasts, and *Saccharomyces. cerevisiae* in particular, offer many additional advantages involving easy cultivation in inexpensive growth media at large scale; availability of continuous supply from various biotechnology industries as their (waste) by-product; *S. cerevisiae* is generally regarded as safe and biosorbents made from *S. cerevisiae* can be easily accepted by the public; and *S. cerevisiae* is well understood at molecular level and thus particularly attractive for genetic engineering approaches. The mechanisms involved in biosorption and performance of different biosorbent materials formulated from natural yeast biomass are in detail reviewed elsewhere in this book.

10.3 Emerging Strategy of Metal Biosorption: Cell Surface Engineering

Cell surface design can be an effective approach for the construction of a novel biosorbent and the innovation and tailoring of metallosorption properties of cellular surfaces. Since the cell surface engineering technology, in which cell surface properties are designed by anchoring various functional proteins and peptides on the cell surface, has been established, it attracted significant attention because of its diverse applications, including: derivatization: of cell surfaces with novel metal binding sites (Georgiou et al. 1993; Ueda and Tanaka 2000; Kondo and Ueda 2004). A number of cell surface-engineered microorganisms with new functions have already been constructed (see also Chap. 11).

In cell surface engineering, the display of target proteins on the cell surface is achieved by their genetic fusion with cell surface proteins. Cell surface proteins have been identified for anchoring to the cell surface in various microorganisms, including *S. cerevisiae.* Among the yeast cell surface proteins, α-agglutinin is a mannoprotein involved in the sexual adhesion of mating-type **a** and α cells (Lipke and Kurjan 1992). In the yeast display system, many functional heterologous proteins were successfully displayed using the N-terminal signal sequence targeting the fusion to endoplasmatic reticulum and ultimately at cell wall and the C-terminal cell wall-anchoring domain of α-agglutinin. This domain contains the glycosylphosphatidylinositol (GPI) anchor attachment signal (Fig. 10.1). The α-agglutinin–based system allowed successful display of many functional peptides and proteins of relatively large molecular masses (e.g., enzymes or antibodies) in their functional forms (Ueda and Tanaka 2000; Kondo and Ueda 2004). Cell surface designs directed toward improved biosorption and recovery of metal ions from aqueous solution employed several proteins and peptides that can bind metal ions (Fig. 10.2) (Nishitani et al. 2010; Kuroda and Ueda 2003, 2006; Kuroda et al. 2001, 2002).

Biosorbents constructed by cell surface engineering play the dual role of a carrier and a producer of metal-binding proteins. Consequently, the production and conjugation of metal-binding molecules to cells could be achieved by a simple procedure such as cell cultivation. This allows the preparation of large amounts of tailored biosorbents in a relatively short time and the direct application of the cultivated cells to metal biosorption. The biosorption ability of cell surface-engineered yeast

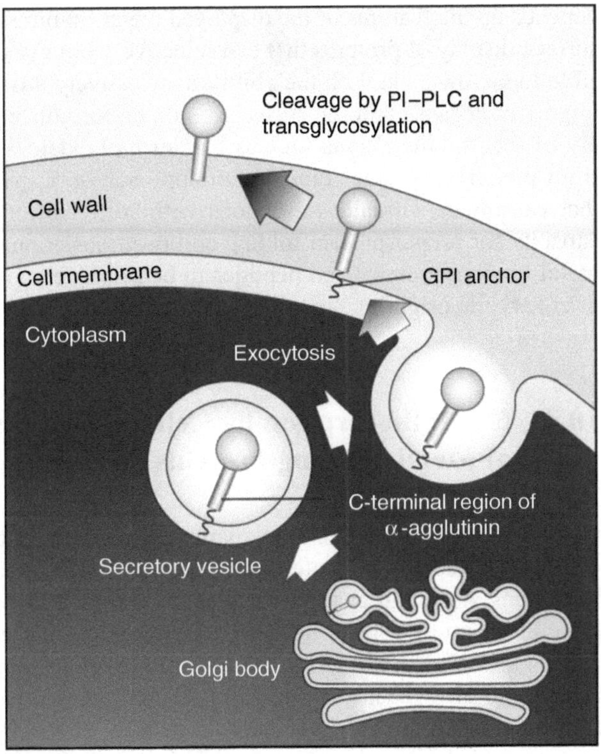

Fig. 10.1 Cell surface display system using the cell wall-anchoring domain of α-agglutinin

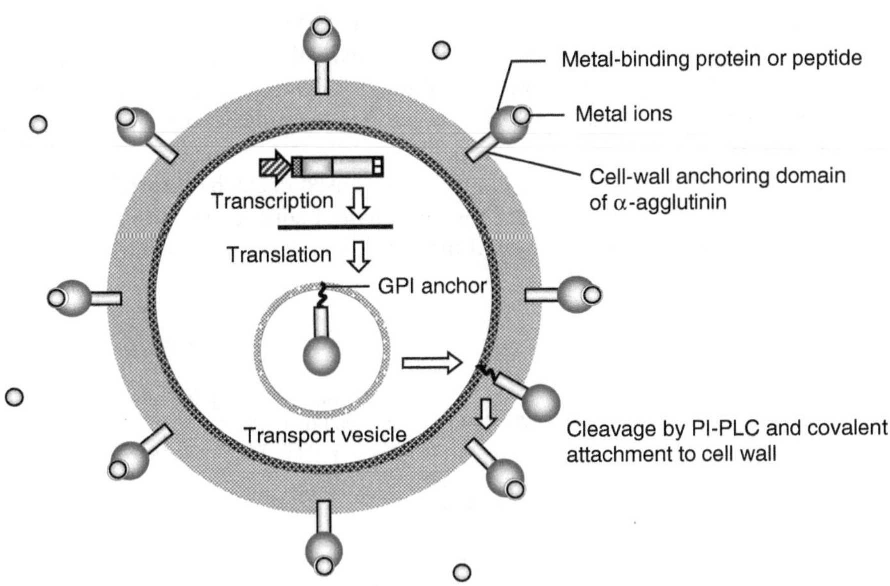

Fig. 10.2 Yeast biosorbent constructed by cell surface engineering

depends on the features of the displayed metal-binding proteins. Therefore, the cell surface display of proteins that can selectively bind target metal ions makes it possible to endow cells with the ability to selectively adsorb metal ions. This is very important because tailored surfaces could be formulated and applied for the recovery of specific metal ions such as Ni^{2+} or molybdate, which are hardly removable from mixtures by other biosorption approaches. *S. cerevisiae* is eukaryotic yeast that can utilize virtually all codons with sufficient efficiency, and possesses the capacity for proper protein folding and posttranslational modifications. Therefore, metal-binding proteins and peptides to be displayed could be of either prokaryotic or eukaryotic origin.

10.4 Metal Biosorption by Cell Surface Display of Metal-Binding Proteins on Yeast

Cell surface design for bioremediation of metal pollution was performed by the cell surface display of proteins and peptides with the ability to bind toxic heavy metal ions. As an initial attempt, divalent heavy metal ions were targeted, and hexahistidine peptide ($[His]_6$; hexa-His) was used as the metal-binding peptide (Kuroda et al. 2001). Hexa-His is widely used as an affinity tag in protein purification (Houchi et al. 1987). For the cell surface display of hexa-His, the prepro leader sequence of the α-factor precursor (Takahashi et al. 1998), hexa-His-encoding sequence, and the 3′ half of the α-agglutinin gene were fused in flame (Fig. 10.3). The constitutive expression of this genetic fusion in *S. cerevisiae* was controlled by glyceraldehyde-3-phosphate dehydrogenase (GAPDH) promoter (Sawai-Hatanaka et al. 1995) (Fig. 10.3). Cell surface display of hexa-His was confirmed by immunofluorescence labeling of transformants (Fig. 10.4). The observed fluorescence on the cell surface indicated that hexa-His was localized and successfully displayed on the cell surface.

The constructed hexa-His-displaying yeast was used in batch biosorption experiments using 100 μM of heavy metal ions in a model aqueous solution. The hexa-His-displaying yeast showed enhanced biosorption and recovery of Cu^{2+} and Ni^{2+}

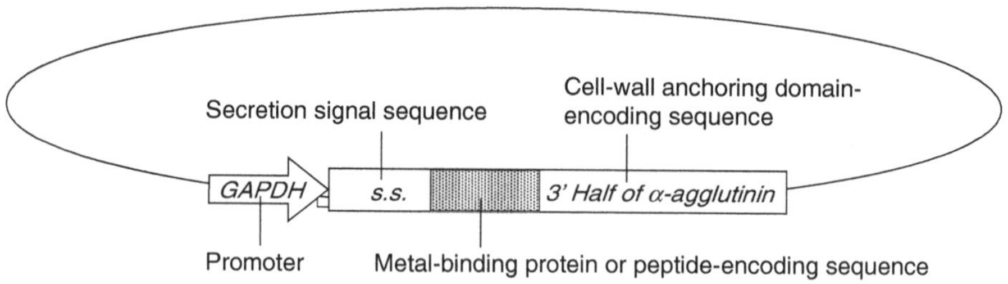

Fig. 10.3 Fusion gene for cell surface display of metal-binding proteins and peptides on yeast cell surface

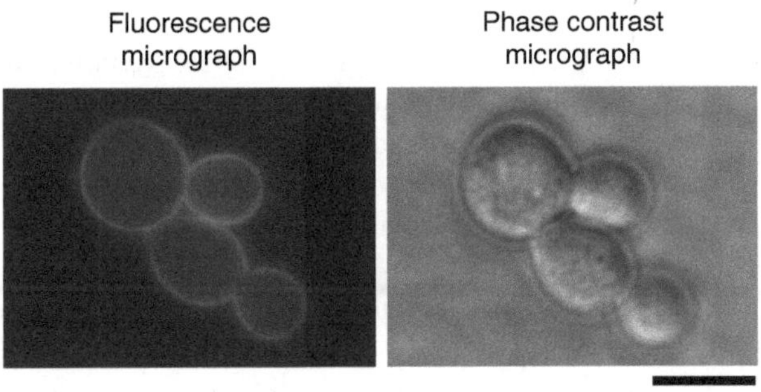

Fig. 10.4 Immunofluorescence labeling of a cell surface-engineered yeast displaying hexa-His. Scale bar, 5 μm

ions compared with control *S. cerevisiae* (Fig. 10.5). Moreover, cell surface-bound metal ions were efficiently recovered with ethylendiaminetetraacetic acid (EDTA) treatment. These results indicated that hexa-His displayed on the cell surface is functional and that the biosorption approach by cell surface design is effective for molecular breeding of biosorbents. Interestingly, hexa-His-displaying yeast proliferated in a medium containing 2 mM Cu^{2+}, in which control strains cannot grow, thereby suggesting that the enhanced biosorption of Cu^{2+} by hexa-His-bearing cell surface also sustained increased tolerance to the toxic levels of this metal ion. Such a phenotype has not been reported in other microorganisms that display hexa-His, such as *Escherichia coli* and *Staphylococcus* species (Samuelson et al. 2000; Sousa et al. 1996).

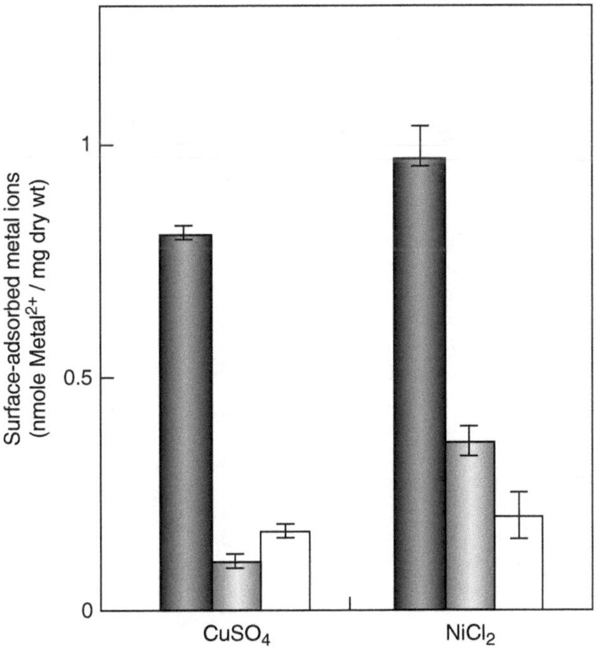

Fig. 10.5 Biosorption and recovery of metal ions by a hexa-His-displaying yeast. *Left columns*, hexa-His-displaying yeast; *middle columns*, control yeast harboring an empty vector; *right columns*, parent strain

Yeast metallothionein (YMT) was used as another metal-binding protein in the second attempt (Kuroda and Ueda 2003). Metallothioneins are cysteine-rich and ubiquitous proteins that sequester intracellular heavy metal ions such as Cu^{+}, Cd^{2+}, Zn^{2+}, Ag^{+} or Hg^{2+}. This protein plays an important role in the detoxification and storage of heavy metal ions for heavy metal homeostasis (Perego and Howell 1997; Butt and Ecker 1987). When displayed on the yeast cell surface using α-agglutinin-based display system, YMT efficiently promoted metallosorption of Cd^{2+} and contributed to enhanced tolerance of modified yeast against in Cd^{2+} medium (Kuroda and Ueda 2003). Tolerance to metal ions enhanced by cell surface design could be beneficial when a living biosorbent is used at a wider range of metal ion concentrations. These results show that enhanced biosorption ability due to cell surface engineering is a useful feature in the molecular breeding of metal-tolerant cells as well as of biosorbents.

In the batch biosorption of metal ions by cell surface-engineered yeast, yeast cells must be separated from treated water after the biosorption. Therefore, the self-aggregation ability in response to the biosorption and accumulation of copper ions was considered an additional development of cell surface-engineered yeasts (Kuroda et al. 2002). Cell aggregation contributes to the low cost and simplicity of the procedure because it causes cells to separate spontaneously from the treated water. To supply the aggregation ability to cells, the fusion gene was constructed and introduced into hexa-His-displaying yeast as follows. The constructed fusion gene consisted of a copper-inducible *CUP1* promoter from the yeast metallothionein gene (Butt and Ecker 1987) and *GTS1* whose overexpression causes constitutive cell aggregation (Bossier et al. 1997). The transformants containing the fusion gene aggregated in response to Cu^{2+} in the culture medium (Fig. 10.6). The copper-inducible cell aggregation occurred at external Cu^{2+} concentration as low as 1 μM with relatively rapid response within 3 h.

To further improve the metallosorption capacity of yeasts, the strains displaying several tandem repetitions of YMT were constructed (Kuroda and Ueda 2006). Four

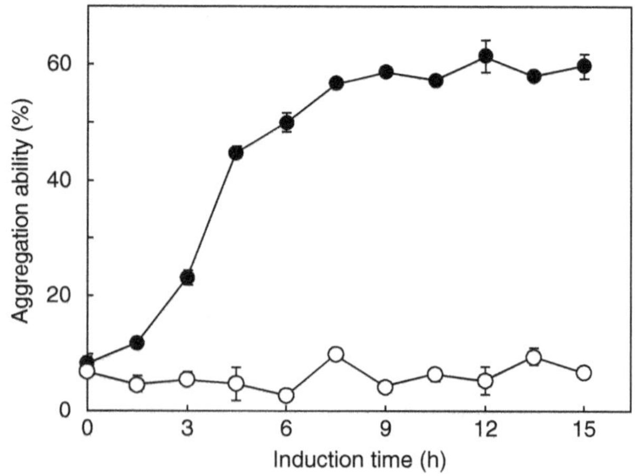

Fig. 10.6 Cell aggregation in response to environmental copper ions by the introduction of the *CUP1* promoter–*GTS1* fusion gene. Aggregation ability after the addition of 100 μM $CuSO_4$ was determined by measuring the concentration of non-aggregating cell just below the meniscus of cultures left to stand for 5 min. Closed circles, yeast cells harboring the *CUP1* promoter–*GTS1* fusion gene; open circles, control yeast cells

and eight repeats as well as one YMT were successfully displayed on the yeast cell surface. The biosorption capacity of modified cells increased proportionally with the number of YMT repetitions.

10.5 Biosorption and Recycling of Rare Metal Ions by Cell Surface-Engineered Yeast

To improve biosorption of rare metal ions by cell surface engineering, biomolecules with the ability to bind rare metal ions are required. Some trace metal ions, including rare metal ions, function as cofactors of metalloproteins and are absolutely necessary for various enzyme reactions. Metalloproteins normally function through specific recognition and binding of essential trace metal ions. Thus, the ability of metalloproteins to recognize and bind metal ions can be utilized in cell surface design for the recovery of rare metals (Nishitani et al. 2010).

Molybdenum is a rare metal that is often used in steel alloys, in electronic components, as a catalyst, and as a lubricant agent, owing to its hardness and heat resistance. Therefore, molybdenum is an increasingly useful metal in industries, and the construction of a system for its biosorption and recovery is required. In living organisms, molybdenum plays an important role in a redox-active center of the molybdopterin cofactor of many enzymes (Kisker et al. 1997). Although most molybdenum compounds are less soluble, the soluble molybdate MoO_4^{2-} is formed upon contact with oxygen and is bioavailable. In *E. coli*, molybdate is transported by ABC transporters consisting of ModA (periplasmic binding protein), ModB (membrane protein), and ModC (ATPase) (Self et al. 2001). The expression of these proteins is regulated by the transcription factor ModE. The ModE polypeptide consists of four structural domains, namely, the N-terminal DNA-binding domain, a linker region, and C-terminal molybdate-binding domain (amino acid residues 120–266) including two mop domains rich in β-structures characteristic to many bacterial molybdate binding proteins (Wagner et al. 2000).

To construct a biosorbent with molybdate-binding ability, the cell surface display of ModE was performed using an α-agglutinin-based display system (Nishitani et al. 2010). In addition to the full-length ModE, the C-terminal domain of ModE was also successfully displayed. *S. cerevisiae* displaying both ModE variants showed enhanced molybdate biosorption from 100 μM molybdate solution (Fig. 10.7). The display of the C-terminal domain of ModE was more effective than that of full-length ModE. Furthermore, the displayed ModE did not show specific binding to other metals except for tungstate, an analog of molybdate (Nishitani et al. 2010). As the next step, desorption of the adsorbed molybdate from the cell surface-engineered yeast is an important process in terms of resource recovery. To this end, the cells displaying C-terminal domain of ModE and preadsorbed with molybdate were washed with buffers of various pH, detergents, and denaturing agents,; however, only a small portion of molybdate was recovered. More than 50% recovery of the adsorbed molybdate was achieved following treatment with protease papain in

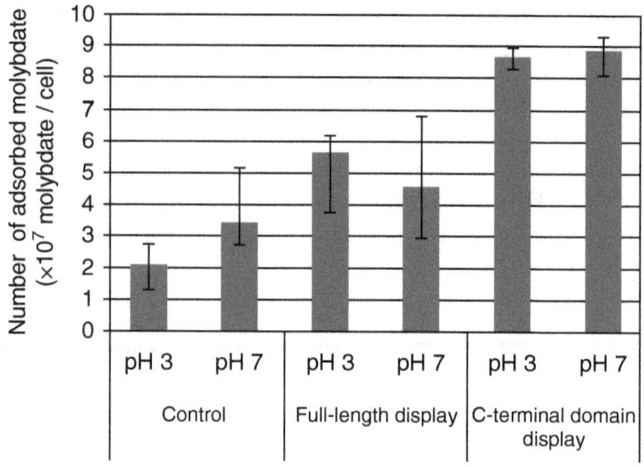

Fig. 10.7 Molybdate biosorption by cell surface-engineered yeasts in 100 μM molybdate solution at pH 3.0 and 7.0

phosphate-buffered saline (PBS; pH 7.4). We demonstrated that papain-treated cells can be recultivated and reused for molybdate biosorption. These data indicate that this biosorption and recovery system is useful not only for the recovery of valuable resources such as rare metal ions but also for the recycling of yeasts with the ability to adsorb molybdate.

In practical biosorption and recovery of rare metal ions, metal ions other than the target rare metal ions can be contained in contaminated streams. Therefore, construction of biosorbents with the ability to specifically adsorb only the target rare metal ions is an important challenge. In the molecular breeding of biosorbents by cell surface design, the metal-binding function of displayed proteins is the determining factor for the metal-binding ability of biosorbents. Generation of proteins with metal-specific binding ability can lead to production of biosorbents that selectively remove rare metal ions. One of the advantages of use of such proteins is the flexibility of response based on the protein design. For the design of proteins by genetic engineering, molecular display on the yeast cell surface represents a potent molecular tool by which the functionality of a displayed protein can be analyzed and adjusted (Ueda 2004). Such an approach employing self-maintaining intact cells does not require protein purification. In other words, cell surface-engineered yeast cells can be treated as microparticles covered with proteins to be modified. This allows a high-throughput screening of protein libraries for desirable functions. Molecular breeding strategy involves altering the metal specificity of metal-binding proteins, creating novel peptides and proteins with specific adsorption ability, and display of these proteins on the yeast cell surface. To this end, libraries consisting of metal-binding proteins that have combinatorial mutations in the region for metal ion recognition and random peptide libraries with random amino acid sequences are to be constructed and displayed on the yeast cell surface. Screening of proteins with selective adsorption abilities from the yeast libraries is a promising tool for the construction of novel metal-binding biomolecules and biosorbents for further studies (Fig. 10.8).

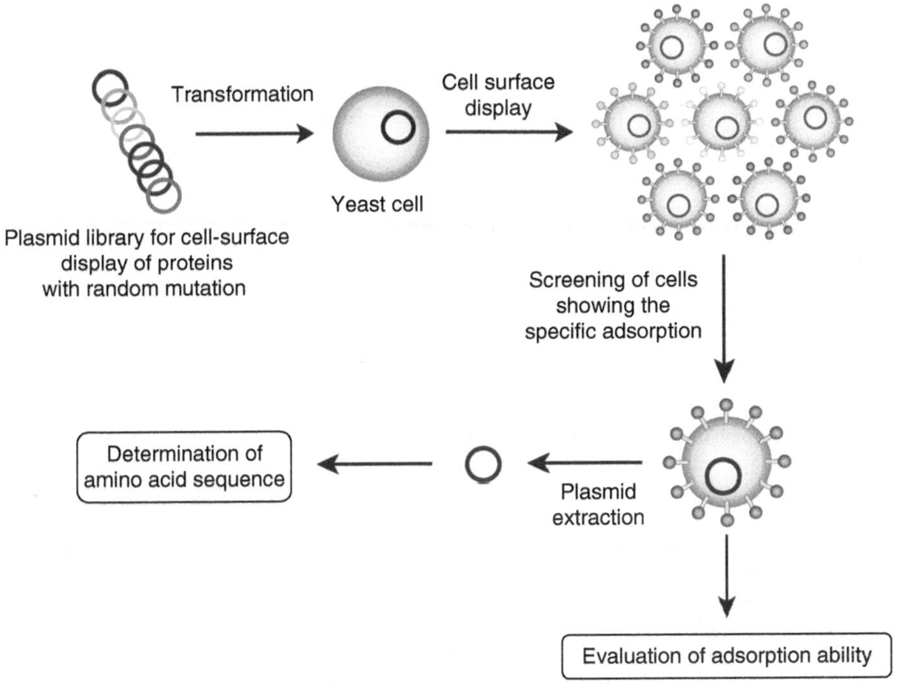

Fig. 10.8 Generation of novel metal-binding biomolecules and molecular breeding of biosorbents for selective biosorption of target metal ions

10.6 Conclusions

Surface design directed toward enhanced biosorption of metal ions by the yeast cell surface was developed as a strategy for molecular breeding of a biosorbent that can find beneficial applications in removing pollutants and recycling resources. As described here, cell surface-engineered yeasts displaying metal-binding peptides or proteins were effective in the biosorption and recovery of metal ions and rare earth elements, and showed several interesting features. The accumulated metals can be desorbed from modified surfaces by using mild chemical manipulation. This allows the recycling of adsorbed metal ions and reuse of the biosorbent material. Furthermore, the enhanced biosorption of metal ions on the cell surface provided cellular tolerance to toxic metal ions. Modified yeast biosorbent can be customized by replacing the proteins displayed on the cell wall according to the target metal ions. Molecular display on the yeast cell surface is also useful for the high-throughput evaluation of protein functions. This technology represents a potent tool for the discovery of novel biomolecules accommodating a wide range of metal ions. Therefore, biosorption, recovery, and recycling of metal ions using cell surface-engineered yeasts need to be further developed through the generation of novel metal-binding proteins and the analysis of metal recognition of proteins.

References

Bossier P, Goethals P, Rodrigues-Pousada C (1997) Constitutive flocculation in *Saccharomyces cerevisiae* through overexpression of the *GTS1* gene, coding for a 'Glo'-type Zn-finger-containing protein. Yeast 13:717–725

Butt TR, Ecker DJ (1987) Yeast metallothionein and applications in biotechnology. Microbiol Rev 51:351–364

Eccles H (1999) Treatment of metal-contaminated wastes: why select a biological process? Trends Biotechnol 17:462–465

Gadd GM, White C (1993) Microbial treatment of metal pollution—a working biotechnology? Trends Biotechnol 11:353–359

Georgiou G, Poetschke HL, Stathopoulos C, Francisco JA (1993) Practical applications of engineering gram-negative bacterial cell surface. Trends Biotechnol 11:6–10

Houchi H, Dobeli H, Schacher A (1987) New metal chelate adsorbent selective for proteins and peptides containing neighbouring histidine residues. J Chromatogr 411:177–184

Kapoor A, Viraraghavan T (1995) Fungal biosorption—an alternative treatment option for heavy metal bearing wastewaters: a review. Bioresour Technol 53:195–206

Kisker C, Schindelin H, Rees DC (1997) Molybdenum-cofactor-containing enzymes: structure and mechanism. Annu Rev Biochem 66:233–267

Kondo A, Ueda M (2004) Yeast cell-surface display—applications of molecular display. Appl Microbiol Biotechnol 64:28–40

Kuroda K, Ueda M (2003) Bioadsorption of cadmium ion by cell surface-engineered yeasts displaying methallothionein and hexa-His. Appl Microbiol Biotechnol 63:182–186

Kuroda K, Ueda M (2006) Effective display of metallothionein tandem repeats on the bioadsorption of cadmium ion. Appl Microbiol Biotechnol 70:458–463

Kuroda K, Shibasaki S, Ueda M, Tanaka A (2001) Cell surface-engineered yeast displaying a histidine oligopeptide (hexa-His) has enhanced adsorption of and tolerance to heavy metal ions. Appl Microbiol Biotechnol 57:697–701

Kuroda K, Ueda M, Shibasaki S, Tanaka A (2002) Cell surface-engineered yeast with ability to bind, and self-aggregate in response to, copper ion. Appl Microbiol Biotechnol 59:259–264

Lipke PN, Kurjan J (1992) Sexual agglutination in budding yeasts: structure, function, and regulation of adhesion glycoproteins. Microbiol Rev 56:180–194

Lovely DR, Coates JD (1997) Bioremediation of metal contamination. Curr Opin Biotechnol 8:285–289

Malik A (2004) Metal bioremediation through growing cells. Environ Int 30:261–278

Marques AM, Roca X, Simon-Pujol MD, Fuste MC, Congregado F (1991) Uranium accumulation by *Pseudomonas* sp. EPS-5028. Appl Microbiol Biotechnol 35:406–410

Nishitani T, Shimada M, Kuroda K, Ueda M (2010) Molecular design of yeast cell surface for adsorption and recovery of molybdenum, one of rare metals. Appl Microbiol Biotechnol 86:641–648

Pazirandeh M, Chrisey LA, Mauro JM, Campbell JR, Gaber BP (1995) Expression of the *Neurospora crassa* methallothionein gene in *Escherichia coli* and its effect on heavy-metal uptake. Appl Microbiol Biotechnol 43:1112–1117

Perego P, Howell SB (1997) Molecular mechanisms controlling sensitivity to toxic metal ions in yeast. Toxicol Appl Pharmacol 147:312–318

Samuelson P, Wernerus H, Svedberg M, Stahl S (2000) Staphylococcal surface display of metal-binding polyhistidyl peptides. Appl Environ Microbiol 66:1243–1248

Sawai-Hatanaka H, Ashikari T, Tanaka Y, Asada Y, Nakayama T, Minakata H, Kunishima N, Fukuyama K, Yamada H, Shibano Y, Amachi T (1995) Cloning, sequencing, and heterologous expression of a gene coding for *Arthromyces ramosus* peroxidase. Biosci Biotechnol Biochem 59:1221–1228

Self WT, Grunden AM, Hasona A, Shanmugam KT (2001) Molybdate transport. Res Microbiol 152:311–321

Sousa C, Cebolla A, de Lorenzo V (1996) Enhanced metalloadsorption of bacterial cells displaying poly-His peptides. Nat Biotechnol 14:1017–1020

Takahashi S, Ueda M, Atomi H, Beer HD, Bornscheuer UT, Schmid RD, Tanaka A (1998) Extracellular production of active *Rhizopus oryzae* lipase by *Saccharomyces cerevisiae.* J Ferment Bioeng 86:164–168

Ueda M (2004) Future direction of molecular display by yeast-cell surface engineering. J Mol Catal B 28:139–143

Ueda M, Tanaka A (2000) Genetic immobilization of proteins on the yeast cell surface. Biotechnol Adv 18:121–140

Wagner UG, Stupperich E, Kratky C (2000) Structure of the molybdate/tungstate binding protein mop from *Sporomusa ovata.* Structure 8:1127–1136

Wang J, Chen C (2006) Biosorption of heavy metals by *Saccharomyces cerevisiae:* a review. Biotechnol Adv 24:427–451

Chapter 11
Bacterial Surface Display of Metal-Binding Sites

Pavel Kotrba, Lubomír Rulíšek and Tomas Ruml

Abstract Biosorption of metal ions is a metabolism-independent metal uptake event at the cell wall polysaccharides, associated molecules, and functional groups. It involves mainly the ion-exchange, chemisorption, adsorption, and, in some cases, also the inorganic microprecipitation of certain heavy metal species. In the search for strategies allowing for enhancements of the biosorption capacity for a specific metal ion, display of particular amino acid sequences with the capacity to form coordination centers for the metal ions at the microbial cell wall has proved to be a promising approach. To anchor particular metal-binding moiety by means of recombinant DNA technology, a wide range of cell-surface display systems is available for both Gram-negative and Gram-positive bacteria. These involve outer membrane proteins, autotransporters, lipoproteins, cell-wall associated and covalently bound proteins, S-layer proteins and subunits of surface appendages. Surface displays of metal-binding oligopeptides, metallothioneins, or metalloproteins has been shown to improve the natural biosorption capacity or even selectivity for particular metal ion in model cells *Escherichia coli.* Furthemore, this approach was successfully extended to environmentally robust *Pseudomonas, Cupravidus or Moraxella* species. This chapter provides overview of available surface display systems and summarizes evidence supporting suitability of cell-surface-display technology in tailoring microbial biosorbents.

Keywords Bioremediation • Biosorption • Cell-surface display • Metal-binding peptide • Heterologous gene expression

---

P. Kotrba (✉)
Department of Biochemistry and Microbiology, Institute of Chemical Technology Prague, Technicka 3, 166 28 Prague, Czech Republic
e-mail: pavel.kotrba@vscht.cz

P. Kotrba et al. (eds.), *Microbial Biosorption of Metals,*
DOI 10.1007/978-94-007-0443-5_11,

11.1 Introduction

Exposure of proteins on the cell surface is a natural property of virtually all cells. Proteins localized at plasma membrane or cell wall serve a wide spectrum of diverse functions. They can be structural components, transporters, porins and appendages involved in conjugation, signal receptors and transducers, enzymes and/or determinants responsible for cell-cell recognition and immunoreactions, surface adhesion and colonization, and substrate binding. An attractive possibility to use these proteins or their engineered parts as vectors for targeting heterologous "passenger" proteins or peptides to the cell surfaces has sparked the interest of molecular biologists and biotechnologists for over 20 years. Bacterial surface display has been proved a viable approach for a wide range of medical, industrial and environmental applications. These involve: biostimulation of specific antibody production by display of antigens at surface of bacterial commensals or attenuated pathogens, an approach used to design life human vaccines or to improve antibody production in animals; biosensor development by anchoring enzymes, receptors or other sensitive elements producing detectable signal in a response to target analyte; high-throughput screening of displayed peptide libraries or of outcome of random protein mutagenesis, often by inspection of binding stringency to target ligand; production of self-maintaining whole-cell biocatalysts by immobilizing enzyme activities useful in industrial substrate conversions and/or environmental remediation of xenobiotics (Lee et al. 2003; Wernérus and Ståhl 2004; Raha et al. 2005; Daugherty 2007; Jose and Mayer 2007). Moreover, the anchoring of metal-binding peptides to the microbial cell wall has proved to be a promising approach in the search for strategies allowing for enhancements of the biosorption capacity for a specific metal ion. Here we provide an overview of approaches and achievements made to improve metallosorption capacity of bacterial surfaces and describe the strategies employed in a search for candidate metal-binding peptides.

11.2 Brief History of Bacterial Cell-Surface Display

The first surface expression system introduced in 1985 employed genetic fusions of short peptides to capsid protein PIII of filamentous bacteriophage M13 without impairing its capacity to infect *Escherichia coli* (Smith 1985). Later utilization of other anchoring proteins of the M13 (e.g. pIV, pVI and pVIII) and λ, T4, T7, MS2 and P4 phage-based surface display systems has been described (Benhar 2001). Although the phage-display remains an effective tool for construction of combinatorial peptide libraries for easy biopanning, the size of the protein to be displayed on their surface is rather limited. Therefore, bacterial surface display systems were developed for the exposure of larger passengers at relatively high surface densities. In 1986 Charbit et al. (1986) and Freudl et al. (1986) demonstrated the capacity of the LamB and OmpA proteins to anchor intramolecular fusions of passenger polypep-

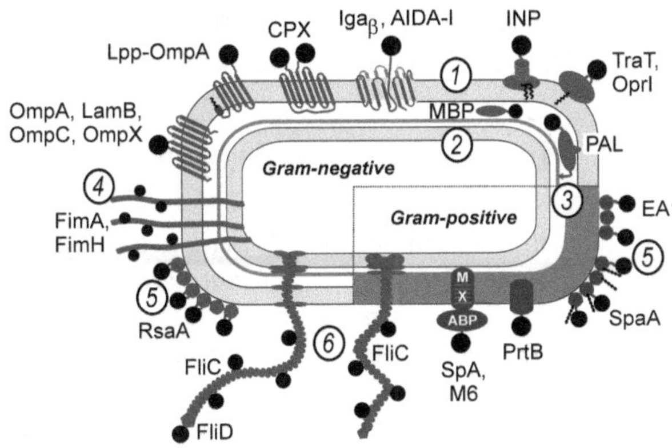

Fig. 11.1 Cell surface display systems in Gram-negative and Gram-positive bacteria. Indicated are names of bacterial proteins used to develop the anchoring carrier protein. Passenger proteins and peptides are shown as *black circles* (not to scale). *1*: outer membrane of Gram-negatiove bacterium; *2*: plasma membrane; *3*: peptidoglycan; *4*: fimbriae; *5*: S-layer; *6*: flagellar filaments

tides at the outer membrane of gram-negative *E. coli.* While the attempts to design artificial carrier proteins were largely unsuccessful (Chen and Georgiou 2002), various gene fusion and display strategies were developed in both Gram-positive and Gram-negative bacteria, which employ outer membrane proteins, autotransporters, lipoproteins, cell-wall associated and covalently bound proteins, S-layer proteins and subunits of surface appendages (Fig. 11.1).

An appropriate anchoring protein should obviously meet several criteria. It should have both the efficient translocation determinants, targeting the protein to the cell surface, and physically powerful anchoring structure, avoiding leakage of displayed fusion from the surface. The anchoring protein should be permissive to insertions of passenger peptides or proteins and the final topology of fusion should ascertain exposure of passenger at the very cell surface. The later is also important to avoid proteolytic cleavage of fused foreign protein "substrate" by surface-associated (e.g. periplasmic) proteases. For example, the first attempt to employ the S-layer protein RsaA of *Caulobacter crescentus* as an anchor revealed that 9 of 11 tested fusion sites resulted in proteolysis of fusion proteins mediated by the S-layer-associated metalloprotease (Bingle et al. 1997; Umelo-Njaka et al. 2002). Not only the size, but also the presence of specific amino acid motifs of the passenger protein is known to significantly affect topology and transportation of fusion product. It has been well documented that passengers prone to form disulphide bridges, containing excess of charged amino acid residues or motifs rich in hydrophobic residues (Nguyen et al. 1995; Jose et al. 1996; Maurer et al. 1997) may impair secretion of recombinant fusion protein. Successful surface display systems for applications in environmental protection should be compatible also with taxons adapted to "outdoor" conditions. Among the popular anchoring proteins, variants of the autotransporter Iga_β and namely of the ice-nucleation lipoprotein (INP) are effective for display of both short and large (up to 45 kDa) passengers at the surface genus *Moraxella, Pseudomonas, Cupravidus* (formerly *Ralstonia*) or *Rhizobium,* thereby extending the effective surface display approach to environmentally robust Gram-negative bacteria (Wu et al. 2008; Saleem et al. 2008).

11.3 Biology and Potential of Useful Bacterial Surface Display Systems

11.3.1 Systems for Gram-Negative Bacteria

The cell wall of Gram-negative bacteria is a two component system composed of outer membrane (OM) and periplasmic space, which extends between inner plasmatic and OM. Located in the 13–25 nm thick periplasm is peptidoglycan cell wall structure with a thickness ranging from 5 to 8 nm. The periplasm can be viewed as a true compartment that fulfills important functions and contains: catabolic enzymes that degrade complex molecules for nutrient acquisition; solute-binding components of ABC transporters or chemotaxis receptors; periplasmic components of serial transport system involved in transport of large substrates which are proteins spanning the periplasm as shuttles connecting outer and inner membranes; detoxifying enzymes; enzymes and chaperones involved in biogenesis of cell envelope structures, including redox enzymes of Dsb family responsible for disulphide bond formation; branched oligoglucans derivatized with phosphoglycerol, phosphoethanolamine and succinic acid, which are induced under hypoosmotic conditions; structural lipoproteins integrating outer membrane with peptidoglycan layer. The OM is an asymmetric lipid bilayer with inner leaflet composed of phospholipids and outer leaflet containing mostly lipopolysaccharides (LPS). Typical LPS consists of three components: membrane forming hydrophobic chain (A region) linked to core oligosaccharide that is extended with hydrophilic oligosaccharide (O antigen) protruding into the medium. The OM prevents entry of noxious compounds and allows permeation of nutrient molecules through the porin channels of a general cut-off of 600 Da. The characteristic of OM channel proteins is that their membrane spanning structure are mainly antiparallel β-strands producing β-barrel structures. As an adaptation to specific environmental and ecological pressures, some Gram-negative bacteria cover their outer membrane with S-layers made of crystalline arrays of 40–170 kDa proteins. The S-layer proteins can be glycosylated and contain substantial amount (40–60%) of hydrophobic residues. The S-layers are the outermost stable cell wall components fixed via their interaction with the sugar moiety of LPS. Two types of surface appendages involve the flagella (organs of locomotion) and fimbriae. Flagellum is a long (3–12 µm) surface appendage of 12–30 nm in diameter constructed of a class of proteins called flagellins, which are assembled to form a cylindrical structure with a hollow core. The filament of flagellum is attached through a hook structure at its end to the basal body, which imparts motion to the flagellum. Fimbriae (or pili) are thin appendages made of protein pilin and have more rigid appearance than flagella. They are distributed profusely over the cell surface (up to around 200 per cell *E. coli* cell), conferring adhesive properties or being involved in bacterial conjugation.

The contribution of native components of Gram-negative cell wall to metallosorption has been studied in detail in *E. coli*. The specific functional groups of the OM participating in metal binding are polar head groups of phospholipids acting mainly at the inner layer of OM, acidic groups of the exposed polypeptides and LPS at the outer half of OM lipopolysaccharides (Hoyle and Beveridge 1983). Although LPS can provide both carboxyl and phosphoryl groups as ligands, only the later is responsible for high affinity binding of divalent metal ions (Ferris and Beveridge 1985). The peptidoglycan layer of *E. coli* binds metal ions via carboxyl group of D-Glu of peptide stem and hydroxyl groups of the glycan backbone (Hoyle and Beveridge 1984). The two-step deposition process was proposed as a mechanism that may be responsible for the increase in the metal-binding capacity of the peptidoglycan (Beveridge and Koval 1981; Hoyle and Beveridge 1984). It would include stoichiometric binding of metal ion generating nucleation site for subsequent precipitation of the metal above stoichiometric amounts.

11.3.1.1 Outer Membrane Proteins for Surface Display

In most outer membrane proteins (OMP), are their N- and C-terminal sequences prerequisites for proper targeting (Boulain et al. 1986; Georgiou et al. 1997). Consequently, the OMPs remain suitable anchoring proteins for intramolecular "sandwich" fusions of passenger proteins and peptides. Gram-negative bacteria share orthologous OmpA protein, which is one of the major proteinaceous components of the OM involved in stabilization the native structure of the OM, through it has very little pore-forming activity (Sugawara et al. 1994). It also participates in conjugation and serves as a receptor for some bacteriophages. The OmpA of *E. coli* is a 325-amino acid protein localizes in OM in the monomeric form. The accepted folding model of the OmpA contains an N-terminal domain (170 residues) consisting of eight antiparallel β-strands (connected by four external and three internal loops), as well as a periplasmic globular C-terminal domain (130 residues) (Sugawara et al. 1996). The OmpA of *E. coli* is able to accept individual or simultaneous oligopeptide insertions in second, third and fourth surface-exposed loops (Freudl et al. 1986; Freudl 1989).

The limitations of passenger size were substantially overcome by constructing Lpp-OmpA chimera, allowing for C-terminal fusions of passenger proteins. This artificial anchor employs the nine amino acids of mature N-terminus of major OM lipoprotein Lpp which provide a signal for fatty acetylation and docking in OM. The second part is derived from OmpA and contains the sequence between amino acid residues 46 and 66 corresponding to third transmembrane β-strand (Georgiou et al. 1996), or between residues 46 and 159 covering the region of third to seventh transmembrane β-strands (Francisco et al. 1992). The OmpA part of Lpp-OmpA is responsible for transportation of heterologous passengers across the outer membrane and ascertains their strong anchoring. The Lpp-OmpA system became extensively used for its versatility and capacity to display relatively large passengers, including enzymes and antibodies, in their active forms (Earhart 2000).

The LamB protein is homotrimeric OMP of porin class of *E. coli* with 421-amino acid monomer forming β-barrel of 18 transmembrane antiparallel β-strands connected by nine external and eight internal loops (Hofnung 1995; Schirmer et al.

1995). It specifically participates in the transport of maltose and maltodextrins across the outer membrane and is also used as a surface receptor by number of coliphages including phage λ. LamB tolerates insertions in a number of sites (Newton et al. 1996). The most popular are insertions in three permissive outer loops (amino acids positions 153, 253 and 374) and one loop faced to periplasm (amino acid position 183). Although these sites can be generally used for displays of heterologous peptides with the upper size limit of 60–70 amino acid residues (Charbit et al. 1988; Su et al. 1992; Sousa et al. 1998), successful display of 232-amino acid staphylococcal protein A inserted at amino acid position 153 was also described (Steidler et al. 1993).

OmpC is another homotrimeric OMP of porin class that allows small hydrophilic molecules to pass through OM of *E. coli.* This 367-amino acid protein consists of 16 consecutive β-strands connected by eight external and seven internal loops. The insertions of polypeptides up to the lengths of 449-amino acids were tolerated at amino acid positions 162 and 298 (Xu and Lee 1999; Cruz et al. 2000; Baek et al. 2010).

OmpX is a small monomeric 148 amino acid residue β-barrel protein with eight antiparalel β-strands. It is present in high number of copies in the OM of *E. coli* (Lai et al. 2004). It possesses four external and three periplasmic loops. Rice et al. (2006) constructed a circularly permuted variant of the OmpX named CPX with changed order of C terminal and N-terminal segments of the protein. In CPX are the transmembrane β-strands five to eight (OmpX numbering) located at the N-terminus and β-strands one to three are fused with a flexible peptide linker. This approach resulted in new N- and C-termini, which corresponds to residues 53 and 54 of OmpX located in native protein in the second external loop. Both termini are efficiently presented at the cell surface of *E. coli* and tolerate (even simultaneous) fusions of heterologous peptide passengers.

11.3.1.2 Autotransporters

Gram-negative bacteria have evolved a simple autotransporter secretion mechanism to transport proteins to the cell surface or secrete them to extracellular environment. Autotransporter precursors are produced as multidomain proteins consisting of three structural and functional features (Jose and Meyer 2007). The N-terminal signal peptide specifies docking of synthesized polypeptide at inner plasma membrane and aids the transport of nascent precursor to periplasm. It is generally accepted that this process employs general Sec-dependent pathway with a signal peptidase that cleaves the N-terminal signal peptide from the precursor. Upon arrival to periplasm, the C-terminal domain associates with OM, folds and forms porin-like β-barrel structure. Finally, the N-terminally attached passenger domain is translocated to the cell surface through this pore. The passenger can either remain anchored in OM (e.g., to provide adhesion function) or released to extracellular milieu and maturated through sequential autoproteolytic cleavages. The later is a case of 106 kDa protease of *Neisseria gonorhoeae,* specific for human IgA1 (Pohlner et al. 1987).

The IgA1 protease is synthesized as a169 kDa autotransporting precursor with the C-terminal Iga_{β} domain that holds the essential pore-forming and translocation functions. The Iga_{β} domain consists of two regions, the transmembrane core (274 amino acid residues) and an essential 60-amino acid peptide linking the β-barrel with the passenger (Klauser et al. 1993a, b). The suitability of IgA1-based autotransporter system for display of heterologous passenger on *E. coli* was first demonstrated by replacing the protease domain with 13 kDa subunit β of cholera toxin (Klauser et al. 1990). N-terminal fusions of various passengers, such as single chain Fv antibody fragments (scFv), leucine zippers as well as short peptides (including cysteine-rich), to Iga_{β} were efficiently displayed on surfaces of *E. coli, P. putida* or *C. metallidurans* (Valls et al. 2000a, b; Veiga et al. 2002, 2003a, b, 2004). However, some other studies indicated that translocation of (presumably unfolded) polypeptide chain through the pore of Iga_{β} could be impaired when the passenger in the periplasm holds stable conformation or is prone to form disulphide bridges (Klauser et al. 1993a; Jose et al. 1996). An efficient autodisplay system, which is less sensitive to inhibition by premature passengers folding, was developed using β-barrel and linker region of AIDA-I, the *E. coli* adhesin involved in the diffuse adhesion to epithelial cells (Maurer et al. 1997, 1999). It is being utilized in display of many peptide and protein passengers, including fully active enzymes such as β-lactamase, esterase A or organophosphate hydrolase (Lattemann et al. 2000; Schultheiss et al. 2002; Li et al. 2008).

11.3.1.3 Lipoproteins

These proteins are anchored to inner or outer leaflet of OM via their covalently attached lipid moiety. The peptidoglycan-associated lipoprotein (PAL) orthologues are widespread in Gram-negative bacteria where they localize in periplasm and form a cross-bridge between the OM and peptidoglycan (Cascales and Lloubes 2004; Godlewska et al. 2009). In *E. coli* forms the N-terminal region of PAL a flexible tail, which binds the protein to the inner leaflet of the OM via lipid moiety attached to the first cysteine residue of the mature protein. The C-terminal region is responsible for interaction with the cell wall peptidoglycan (Lazzaroni and Portalier 1992). It appears that that the majority of contacts are between conserved surface residues of PAL and the peptide region of peptidoglycans. Furthermore, a specific region of the C-terminal domain provides a binding pocket for the m-DAP residue of peptidoglycan (Parsons et al. 2008). The periplasmic display systems employ signal peptide and C-terminal peptidoglycan-binding domain, while the N-terminal fatty acetylated residues are replaced with the heterologous passenger peptide or antibody scFv fragments (Fuchs et al. 1996; Valls et al. 1998; Dhillon et al. 1999).

TraT of *E. coli* and OprI of *Pseudomonas aeruginosa* are lipoproteins useful for display of passenger at the OM surface. TraT is a plasmid-encoded lipoprotein that mediates bacterial conjugation and serum resistance. The mature 220-amino acid protein is anchored to the OM via an N-terminal lipid moiety and two hydrophobic, presumably membrane spanning, moieties located in its central section. TraT allows displays of either intramolecular fusion at positions of residues 125 and 180 (Taylor et al. 1990) or C-terminal fusions of passenger peptides of upper size limit of 100 amino acid residues (Chang et al. 1999; Chang and Lo 2000). C-terminal fusions of up to 334-amino acid protein could be displayed on *E. coli* by using the OprI as an anchor (Cote-Sierra et al. 1998).

Ice nucleation protein (INP) is an atypical OM lipoprotein found in genus *Pseudomonas, Xanthomonas* and *Erwinia* (Kawahara et al. 2002). Its unique feature is the use of glycosylphosphatidylinositol (GPI)-anchoring moiety, which is normally found associated only with some eukaryotic proteins. The amino acid sequence of INP from *Pseudomonas syringae* can be divided into nonrepetitive N-terminal (175 residues) and C-terminal (49 residues) domains and a highly repetitive central domain (Warren and Wolber 1991). The central domain contains 61 repeats of 16 residue fragments (AGYGSTXTAXXXSXLX, where X is a residue that is not conserved in repetitions). These repeats serve as templates for ice nucleation in supercooled water, thereby causing physical injury to plants used by the bacterium for nutrition. C-terminal fusions of heterologous peptide and protein passengers, including enzymes as large as 60 kDa, were efficiently displayed on the surface of *E. coli, Moraxella* sp., *P. putida* and *Stenotrophomonas* sp. (Jung et al. 1998; Shimazu et al. 2001, 2003; Zhang et al. 2004; Yang et al. 2008; Liu et al. 2009). Moreover, the number of repetitions within the central region does not play critical role in proper targeting and anchoring in OM. The INP variants with different lengths of central region could be used, e.g. to prevent steric hindrances, which makes INP one of the most versatile and promising display systems available.

11.3.1.4 S-layer Proteins

Although S-layers appear widespread among bacteria, their exact function remains rather unknown. They likely contribute to pathogenesis as virulence factors, provide the first line of defense against phages, lytic enzymes and may serve as a depository for surface-exposed enzymes (Fernandez and Berenguer 2000). The S-layer of *Caulobacter crescentus* cell is two dimensional crystalline array made of almost 40,000 copies 1026-amino acid protein RsaA, which represents 10–12% of the total cell protein. The N-terminus of RsaA is required for anchoring of the S-layer to the cell surface while the C-terminus contains all the information required for secretion of the protein. The RsaA-based systems employing intramolecular insertions at positions of residues 551 or 723 of RsaA was shown suitable for high-density immunoreactive surface display of 12-amino acid epitopes as well as of larger 61-amino acid IgG-binding domain of streptococcal protein G (Bingle et al. 1997; Umelo-Njaka et al. 2001; Nomellini et al. 2007). Furthemore, the N-terminal fusions to either the last 242 or 336 C-terminal amino acids of RsaA provided an efficient export system for both short peptides and β-glucanase enzyme (Umelo-Njaka et al. 2001; Duncan et al. 2005).

11.3.1.5 Subunits of Surface Appendages

The subunits of *E. coli* flagella and fimbriae are useful carriers for displays of short peptide passengers (Klemm and Schembri 2000; Westerlund-Wikström 2000). High abundance and permissivity to intramolecular insertions of up to 34 residue peptides (Stentebjerg-Olesen et al. 1997) makes the major protein of fimbriae, FimA, an ideal candidate for a high-valency display of heterologous peptide libraries. Fimbriae are adhesive bacterial organelles which enable bacteria to target and to colonize specific host tissues. They are found in up to about 500 copies per cell made up of about 1,000 subunits of the FimA. Minor subunits are FimF, FimG and FimH (Klemm and Schembri 2000) and FimH was also shown to tolerate intramolecular insertions to display up to 56 residues (Pallesen et al. 1995).

The flagellum is the bacterial motility organelle composed of more than 20 different proteins with major FliC subunit present at approximately 20,000 copies per flagellum. Lu et al. (1995) took the advantage of the fact that the central region of FliC is dispensable and in their FLITRX surface display replaced central region with entire thioredoxin of *E. coli*. Resulting surface exposed loop constrained by disulfide bridges of thioredoxin provided an ideal scaffold for a display of 12 amino acid residue peptide libraries. Moreover, the FliC can tolerate intracellular insertions of up to 302 amino acid residue adhesins (Westerlund-Wilkström et al. 1997) and simultaneous display of two different passengers within flagellum were also described (Transkanen et al. 2000). Another flagellum component, 469 amino acid residue FliD, makes a capping structure at the end of flagellum and plays a central role in flagellar polymerization (Ikeda et al. 1987). Its variable central region can tolerate insertion of 39 heterologous amino acid residues, thereby extending the capacity of flagella to simultaneous display of three passengers (Majander et al. 2005).

11.3.2 Systems for Gram-Positive Bacteria

The major component of the cell wall of the Gram-positive bacteria is a peptidoglycan (15–80 nm thick), intimately layered just above the plasma membrane and leaving very limited periplasmic space. Unlike in Gram-negative bacteria, the short peptides crosslinking the polysaccharide component of peptidoglycan in Gram-positive species are covalently linked to teichoic acid (polyol phosphate polymers). Besides these components, the rigid cell wall contains secondary cell wall nonproteinaceous polymers (teichuronic acid, lipoteichoic acid, and other neutral or acidic polysaccharides) in various proportions. Proteins displayed within or on the cell wall fall into four different groups based on the anchoring mechanism: proteins anchored in periplasmic membrane by hydrophobic transmembrane domain(s); lipoproteins posttranslationally attached to the membrane diacylglycerols by a covalent bond; proteins covalently attached to peptidoglycan; proteins anchored noncovalently via cell wall-binding domains (Desvaux et al. 2006).

The covalently attached proteins in the Gram-positive bacteria share a C-terminal cell wall sorting signal, consisting of LPXTG-like motif, a hydrophobic domain of 15–22 amino acid residues and a positively charged six to seven amino acid residue tail, which prevents a release of a polypeptide to the extracellular milieu (Mazmanian et al. 2000). The protein precursor translocated through the plasma membrane remains first attached to the membrane via a hydrophobic domain. Then a membrane-anchored sortase enzyme cleaves the Thr-Gly bond of the LPXTG motif and the carboxyl group of Thr is amide-linked to free amino group of peptidoglycan peptide component. Functions of this group of proteins are very diverse, ranging from adhesins, enzymes receptors and antigens.

Many Gram-positive bacteria (but not, e.g., those of genus *Staphylococcus* and *Listeria*) produce the S-layers. Most S-layers consist of a single type (glyco)protein, but some species produce two species of the S-layer proteins from a single precursor (Calabi et al. 2001) or two transcriptional units (Courture-Tosi et al. 2002). Like the Gram-negative bacteria, many Gram-positive species form flagella and fimbriae (Macnab 2004; Lauer et al. 2005). A specific feature of certain Gram-positive bacteria of families *Lachospiraceae* and *Clostridiaceae* is surface deposition of specific macromolecular structures of cellulosomes (Desvaux 2005). This multienzymatic complex, dedicated to adhesion to plant cell walls and degradation of its component cellulose, is organized around modular integrating non-covalently bound scaffolding protein.

11.3.2.1 Covalently Bound Proteins

Staphylococcal protein A (SpA) is the best characterized cell wall protein as it has been used a model to study anchoring mechanisms of covalently bound proteins. SpA is known to interact with specific mammalian proteins (namely immunoglobulins and macroglubulins) during infection, resulting in adhesion of the bacteria on the host cells and escape from immune system (Foster 2005). SpA consists of N-terminal targeting signal peptide followed by four or five repetitive Z domains (IgG-binding), X domain with repetitive Pro and Gly rich motifs (spans thick peptidoglycan layer, and M domain, which contains the LPXTG motif. The surface display systems employing XM domains of SpA equipped with various signal sequences and spacers (Wernerus and Ståhl 2002), have been successfully used to achieve surface display of N-terminal fusions of undecapeptide libraries and of up to 397 residue passengers in *Staphylococcus xylosus, S. carnosus,* and *Lactobacillus lactis* (Hansson et al. 1992; Samuelson et al. 1995; Gunneriusson et al. 1996; Steidler et al. 1998; Samuelson et al. 2000; Wernerus et al. 2001; Kronqvist et al. 2008; Song and Gu 2009).

The M6 protein, responsible for anti-phagocytic property of group A of staphylococci, is a rod-shaped cell wall protein of α-helical coiled-coil structure. Arming of heterologous peptides and proteins with signal sequences of M6 protein (N-terminal 122 residues and C-terminal 140 residues containing the anchor domain) resulted in efficient display of passengers at the surface of *Streptococcus gordonii*

(Oggioni et al. 1999). This strategy allowed successful surface display viral proteins and their epitopes or domains, tetanus toxin, hemagglutinin and scFv antibodies of sizes ranging from 15 to 515 residues (Oggioni et al. 1999; Giomarelli et al. 2004). While native M6 protein can be surface displayed also in lactobacilli (Piard et al. 1997), the fusion consisting of N-terminal signal 15 residue peptide of lactococcal Usp45 protein, internal passenger (E7 protein of human papillomavirus) and the cell-wall anchor domain of M6 protein was displayed in *Lactobacillus lactis* but not *Lactobacillus plantarum* (Cortes-Perez et al. 2005). Display in *L. plantarum* was achieved when the M6 domain was replaced with cell wall anchoring domain of the native *L. plantarum* cell wall protein.

11.3.2.2 Cell Wall-Associated Proteins and Flagella

Several bacterial proteins are non-covalently anchored to the cell surface via approximately 50 residues long S-layer homology (SLH) domain. The cell surface display mechanism involves a non-covalent interaction between the SLH domain and peptidoglycan-associated polymers (Mesnage et al. 2000). *Bacillus anthracis* synthesizes two S-layer proteins, EA1 and Sap, which account for 5–10% of total protein and are expressed in vivo. Genetic fusions of EA1 and tetanus toxin fragment C and fusions of SLH domains with levansucrase were shown surface-exposed and passengers proved to retained their antigenic properties and enzymatic activity (Mesnage et al. 1999a, b). The mature S-layer of *Paenibacillus alvei* CCM 2051T is composed of 959-amino acid *O*-glycoprotein SpaA. This glycoprotein was recently employed for surface display of C-terminally fused hexahistidine and green fluorescent protein (Zarschler et al. 2010). Production of modified SpaA in *P. alvei* resulted in co-display of fused functional epitopes and an *O*-glycosidically linked glycan in a nanolattice-like fashion. Such arrangements would be of importance especially when mimicking of glycan-mediated clustering effect is desirable. Moreover, the authors constructed the Δ *wsfP* stain, in which non-glycosylated SpaA variants can be displayed.

The cell-surface proteinase PrtB of *Lactobacillus bulgaricus* is composed of five domains. Its N-terminal catalytic domain is followed by A, B, H and W domains (Germond et al. 2003). Although the C-terminal W domain of PrtB contains a LPXTG motif homologue of sequence LPKKT, PrtB does not appear covalently associated with the cell wall. Instead, the high content of positively charged lysine residues within two imperfect 59 residue repeats surrounding the LPXTG motif suggests that PrtB is anchored via salt bridges with negatively charged residues of teichoic acid. The surface display systems, employing secretion signal of amylase AmyA and HW domains of PrtB, have been successfully used to achieve surface display of FliC, a 500 residue flagella antigen of *Salmonella enterica,* on the surface of *Lactococcus lactis* (Kim et al. 2008).

Promising surface display system of biotechnological significance would represent engineered flagella in *Bacillus halodurans* Alk36, an alkaliphilic, thermotolerant bacterium, which continuously over-produces flagellin (Crampton et al. 2007).

The 271-amino acid FliC was shown to tolerate intramolecular fusions of up to 29-amino acid peptide (gp120 epitope of HIV) within its variable region between residues 127–200.

11.4 Sources of Metal-Binding Peptides

The display of proteins on bacterial surfaces enables us to engineer the desired property into the protein (and surface). One of the properties can be the utilized is an enhanced metal binding capacity of the bacterial surface. To achieve this goal, an experience accumulated over several past decades in bioinorganic chemistry can be used. In this section we will review the basic principles governing the metal-ion selectivity and uptake by peptide or protein sequences.

11.4.1 What Can We Learn from the Coordination Preferences of Metal Ions?

11.4.1.1 Metal-Binding Sites in Metalloproteins

It has been estimated that approximately one third of the known proteins contain metal ions (Andreini et al. 2004). They play a key role in many fundamental biochemical processes (e.g., respiration, photosynthesis, and most of other redox processes). The role of metal ions in biomolecules is both functional (catalytic) and structural. The prominent examples of the redox active ions are $Fe^{2+/3+}$, $Mn^{2+/3+}$, and $Cu^{+/2+}$ ions, (Ramirez et al. 1995) whereas the category of structural ions is best represented by the Zn^{2+} ion, (Kuppuraj et al. 2009) the second most abundant metal ion in metalloproteins (though it should be noticed that Zn^{2+} is also quite abundant in catalytic sites). Quite often, the binding of metal ions in metalloproteins is specific (Dudev and Lim 2003). Moreover, the metal-binding sites are tuned to accommodate and/or modify its physicochemical properties (e.g., redox potential) (Han et al. 2002). Therefore, it is a tempting idea to explore the described phenomena in the area of bio/phytoremediation. First, let us describe the basic characteristics of metal-binding sites.

11.4.1.2 The Nature and Essential Characteristics of the Binding of Metal Ions in Proteins

The structural information about the nature and essential characteristics of binding of metal ions in metalloproteins and small-molecule crystals can be drawn from two main sources: Protein Data Bank and Cambridge Structural Database (most of the structures have been collected in the last two decades). The detailed analyses revealed some essential factors influencing the binding and selectivity of metal uptake (Glusker 1991; Rulíšek and Vondrášek 1998; Dudev and Lim 2001; Zheng et al. 2008). Thus, one or two preferred coordination geometries can be assigned to each of the metal ions (e.g., Co^{2+}, Ni^{2+}, Cu^{2+}, Zn^{2+}, Cd^{2+}, Hg^{2+}). Also, a good correlation has been found between the coordination arrangements in small-molecule crystals and the metal-binding sites in metalloproteins (Rulíšek and Vondrášek 1998). It implies that the principles governing the binding of metals in smaller complexes and metalloproteins are similar and that pre-designed binding sites with certain coordination geometry will exhibit greater affinity towards metals that prefer this geometry. The observed abundance of the binding residues is in agreement with the semiquantitative concepts of coordination chemistry, such as hard and soft acids and bases (HSAB) principle of Parr and Pearson (Pearson 1963). Therefore, the occurrence of a particular (or combination of) amino acid side chains in the binding site is another factor determining its specificity.

11.4.2 Natural Metal-Binding Peptides

To enhance metal-binding properties of specific proteins we can either use the natural metal-binding peptides or attempt to design metal-binding peptide sequencies *de novo*. The former approach is described in this section whereas the rational molecular design is mentioned in the next section. The intracellular peptides involved in sequestration of transition heavy metal ions in mammals, yeasts and plants had become popularly used for surface display to improve the metallosorption properties of bacteria and yeasts. These involve metallothioneins (MTs) and phytochelatins (PCs), which combine small size beneficial for efficient surface display and high number of metal-binding centers. Ubiquitous in eukaryotes, MTs are cysteine-rich peptides capable of high affinity coordination of heavy metal ions via cysteine residues shared along the peptide sequence in characteristic Cys-X-Cys or Cys-Cys motifs (Kotrba et al. 1999b; Vasák 2005). Besides the role in detoxification of excess heavy metal ions, MTs are involved namely in Zn and Cu homeostasis, protection against oxidative damage and redox control (Vasák 2005). In mammalian MTs of 61 or 62 amino acid residues form their 20 cysteine residues 7 and 12 coordination centers for divalent and monovalent heavy metal ions, respectively. The mammalian MT molecule is composed of two distinct domains, denoted α- and β-domain. The α-domain consists of first 30 amino acids, contains nine cysteine residues and binds three bivalent metal ions. The β-domain (amino acids 31–61 or 62) contains 11 cysteine residues that bind four bivalent metal ions and the cadmium-binding stability of α-domain is known to be much stronger than that of β-domain and intact wild-type MT (Kotrba et al. 1999b, c).

In the yeast *Saccharomyces cerevisiae* form 12 cysteine residues of CUP1, a 53 amino acid MT variant, eight binding centers for monovalent and four binding centers for divalent heavy metal ions. The shortest MT of 26 amino acids with seven cysteine residues binding six monovalent heavy metal ions is produced by ascomycete *Neurospora crassa* (Muenger et al. 1985). The role of plant MTs is generally attributed to the homeostasis of essential heavy metals (Cobbet and Goldsbrough 2002; Clemens et al. 2006).

Distantly related to MTs are vanabins, quite recently identified cysteine-rich 120-amino acid proteins with most of its 18 cysteine residues organized in Cys-$(X)_3$-Cys motifs (Trivedi et al. 2003; Ueki et al. 2003a). Vanabins were isolated as vanadium-binding proteins from ascidians accumulating extremely high levels of vanadium. Some members of class *Ascidiacea* are able to sequester vanadium in vacuoles of blood cells at concentration reaching 350 mM, i.e., 10^7 times the concentration in seawater (Ueki et al. 2002).

Phytochelatins (PCs) are small peptides of general structure $(\gamma\text{-Glu-Cys})_n X$ (PCn; n=2–11; X represents Gly, Ser, β-Ala, Glu, Gln or no residue) found in virtually all tested plants and in certain yeasts. These peptides are capable of an efficient sequestration of multiple metal and metalloid ions in metal(loid)-thiolate complexes and play a pivotal role in heavy-metal detoxification in plants (Kotrba et al. 1999b; Cobbett and Goldsbrough 2002; Clemens 2006). Unlike gene-encoded MTs, PCs are enzymatically synthesized in a transpeptidation reaction from glutathione (γ-glutamylcysteinylglycine, GSH) or its homologues (*iso*-PCs). Although the presence of γ bond offers a higher degree of structural flexibility than α linkage, γ bond probably does not play a critical role in metal-binding properties of PCs (Dameron and Winge 1990; Bae and Mehra 1997). The presence of Cys-X-Cys motif is of greater importance, making α-linked PC analogues eligible for surface display.

Genes encoding bacterial Hg^{2+} resistance, based on the import of Hg^{2+} into cytoplasm and its subsequent reduction to metallic mercury, are often organized in *mer* operons and virtually all *mer* operons function with homodimeric transcriptional regulator MerR (Barkay et al. 2003; Silver and Phung 2005). In its apoform is MerR transcritional repressor, while upon binding of single Hg MerR acts as a transcriptional activator. MerR shows high specificity, preferring Hg by up to three orders of magnitude over Cd and Zn. MerR consists an N-terminal DNA-binding helix-turn-helix motif, C-terminal Hg^{2+}-binding domain and an interdomain region undefined function. Three conserved cysteine residues localizing within helix 5 of C-terminal domain contribute to the metal-binding center: C117 and C126 of one monomer and C82 from the other (*Tn*21 MerR numbering). Such arrangement results in trigonal thiolate coordination of Hg^{2+}, to which an unusual affinity and specificity of MerR dimer to Hg^{2+} has been attributed (Barkay et al. 2003). MerR is an archetype of transcriptional activators [of MerR family], which use other metal inducers, including Cd^{2+}, Zn^{2+}, Cu^{2+}, Pb^{2+} and Co^{2+} (Brown et al. 2003). Respective regulators may thus represent potential targets for surface display.

Mercuric resistance mechanism involves also small periplasmic Hg^{2+}-binding protein MerP, funneling the metal ion to the uptake transporters (Silver and Phung 2005). The 72-amino acid periplasmic form of MerP is functional as monomer and binds one Hg^{2+} via conserved metal-binding motif GMTCxxC, which is also found singly and as multiple repeats on the N-termini of P-type ATPases involved in transport of transition metal cations (Barkay et al. 2003).

In addition of cysteine residues, many metalloproteins employ also histidines (which is the most abundant metal-binding residue) and carboxylates (Asp, Glu) as potent ligands of their metal-binding centers. As shown below, several artificial short peptide sequences, often following ligand-X-ligand feature of MTs and PC, were designed for surface display and contained cysteine, histidine or both residues as ligands.

11.4.3 Molecular Design In Silico

11.4.3.1 Enhancing the Metal-Binding Properties of the Proteins

There have been many attempts to modify the metal-binding properties of selected proteins in order to change its function (Berry et al. 2002; Kaplan et al. 2004), engineer novel metal-binding sites in natural metalloproteins (Jantz et al. 2004; Reedy and Gibney 2004) or tune particular metalloprotein towards a particular metal ion (Petros et al. 2006; Matzapetakis et al. 2006). One approach is to redesign the metal binding site, i.e., try to incorporate new metal-binding side chains into the site, while not changing the folding of the protein, whose conservation is essential for metal-binding. This, so call, iterative protein redesign has been successfully pursued in the work of Hellinga and coworkers (Benson et al. 2000; Dwyer et al. 2003). By a computational modeling, mutations are introduced into the M^{2+}-binding site and the site further stabilized by the alteration of the second sphere residues. This, rather general, design strategy has been verified experimentally, since the mutated sites have been shown to exhibit the specific metal-binding activity.

A relative success of the above approaches invokes the general question, whether we can design short peptides that would form an arbitrary metal-binding site by folding around the specific metal ion and providing a particular (highly specific) combination of amino acid side chains at the positions of the most favorable coordination polyhedra. In order to decide the optimal combinations of side chains, the extensive screening of the complexation properties (focusing mainly on the metal selectivity) need to be done.

11.4.3.2 Quantum Chemical Studies of the Interactions of Metal Ions with Biologically Relevant Functional Groups

We can provide two examples of the systematic studies dealing with the complexation and selectivity of metal ions in model sites. In the work of Lim and coworkers (Babu et al. 2003; Dudev et al. 2005) the efforts are made to use ab initio/density functional (DFT) or continuous dielectric (CDM) methods (the latter are used in order to obtain a true free energy values including solvation) to predict the equilibrium constants of metal substitutions in model active sites. They have demonstrated that the tendency of a particular metal to bind (interact) in the native binding site is pronounced in the favorable complexation (free) energies, though in few cases, the native metal can be, in principle, substituted by other metal ions (suggested by the higher affinity of the substituting ion). Thus, DFT/CDM studies can sometimes provide insights into mechanisms of e.g, metal poisoning.

Another approach, pursued in the studies of Rulíšek and Havlas (Rulíšek 2000, 2002, 2003) can be viewed as an attempt of the theoretical combinatorial (bioinorganic) chemistry. In the series of papers, the theoretical efforts to devise the most selective combinations of amino acid (AA) side chains (i.e., metal-binding sites) for six studied transition metal (TM) ions—Co^{2+}, Ni^{2+}, Cu^{2+}, Zn^{2+}, Cd^{2+}, Hg^{2+}—are described. Starting from the calculations of the interaction energies of the cations with model functional groups representing amino acid side chains, and defined as the substitution energy of the given functional group for one water molecule in reference perhydrated complex, through the estimation of the so called cooperative effect (i.e., non-additivity of the interation energies upon binding of two functional groups), the complexation energies of these ions have been calculated for each combination of simple functional groups representing amino acid side chains in four coordination geometries (linear, tetrahedral, square planar, octahedral), according to the *approximate* formula. In order to eliminate the errors of 5–15 kcal mol^{-1} inherent in approximate calculations, ten potential candidates for the most specific site for each TM ion have been subject to a rigorous quantum chemical investigation. It resulted in the calculations of complexation energies of six TM ions in 50 different sites and the final assignment of three or four most specific sites for each of them. Moreover, it has been shown that the calculated results well correlate with the abundance of particular combinations of amino acid side chains in the metal-binding sites of metalloproteins and the relative order of the stability constants of the studied ions in two metalloproteins have been reproduced. It gives some confidence about the relevance of the conclusions of these theoretical investigations.

11.4.3.3 Merging the Protein Fragments into One Polypeptide Chain

To accomplish the successful design of novel peptide sequences, one has to link the 'isolated' AA side chains (hypothetically) ligated at the vertices of the particular coordination polyhedron. Such an attempt has been made, using the existing protein fragments as the linking sequences. [r16] To this end, a series of programs has been written that (1) creates the library of AA_i-$(X)_n$–AA_j sequences (n=2–16; AA_i and AA_j is the metal-binding amino acid) from the non-redundant PDB database, (2) merges the fragments into a single AA_1-$(X)_{i1}$–AA_2-$(X)_{i2}$– … -AA_n polypeptide chain, (3) automatically generates the input files for the molecular mechanics calculations, which enable to estimate the strain in the resulting chain. A successful completion of this goal resulted in the design of the short peptide sequences (10–30 amino acids) with the potential of the highly specific binding of TM ions. The peptides designed according to this strategy were experimentally tested (mass spectrometry, NMR, and crystallization) and some of them has exhibited a measurable metal-binding capacity (Kožíšek et al. 2008). The best sequences can be expressed in the surface proteins of plants or bacteria and used for bioaccumulation of selected metal ions.

11.5 Survey of Bacterial Surface Displays for Enhanced Metallosorption

Over last decades, the research directed towards the use of genetically engineered microorganisms or plants for removal or biotransformation of heavy metals or organic xenobiotics from contaminated materials has emerged as a novel strategy in bioremediation (Urgun-Demirtas et al. 2006; Doty 2008; Macek et al. 2008; Saleem et al. 2008; Wu et al. 2008; Abhilash et al. 2009; de Lorenzo 2009; Kotrba et al. 2009; Ruiz and Daniell 2009; Shibasaki et al. 2009; Sylvestre et al. 2009). An approach employing intracellular production of eligible enzyme activities improving the bioremediation potential of microorganism has achieved continuous success in bioremediation of organic xenobiotics *in situ*. A few examples of microbes tailored for removal of metal species *in situ* involve mainly those producing heterologous mercuric reductase, converting Hg^{2+} to Hg^0 that subsequently evaporates to the atmosphere (Brunke et al. 1993; Horn et al. 1994; Brim et al. 2000; Huang et al. 2006).

Unlike the production of intracellular enzymes, the overproduction of metal-binding peptides, such as hexahistidine, metallothionein, phytochelatins or glutathione *S*-transferase, (Kille et al. 1991; Kang et al. 2007) in microorganisms genetically modified for bioremediation of metal ions from liquid streams may encounter several problems. The most serious one is very limited, if any, possibility of recycling of the biomass with accumulated metal species. Intracellular metal ligands, often cysteine-rich peptides such as MTs, may affect the overall status of cellular metallome and tend to be toxic to bacteria, perhaps because of interference with the redox pathways in the cytosol. Heterologous MTs, either as separate MT peptides or as fusion proteins, have been thus often produced in bacteria unstably and at low levels. To achieve stable production of human MT-II, Jacobs et al. (1989) employed, for the first time with metal-binding peptide, the surface display approach. The Lpp-MT-II fusion produced stable MT that retained its metal-binding property. The following sections describe achievements of surface display technology used in construction of model and environmentally robust bacteria for improved metallosorption. Table 11.1 summarizes the data on published metallosorption properties of modified cells.

Table 11.1 Metal uptake with surface-engineered bacteria

Metal-binding peptide or protein	Fusion, localization	Bacterium	Metal	Metal uptake (nmol/mg dry wt)	Reference
Hexahistidine (6His)	LamB-(6His), OM	*E. coli*	Cd	6.3 ± 1.8[a,g]	Sousa et al. (1996)
	LamB-$(6His)_2$, OM		Cd	15.9 ± 1.4[a,g]	
	OmpC-$(6His)_6$, OM		Cd	32.1 ± 2.1[c,m]	Xu and Lee (1999)
	OmpC-$(6His)_2$, OM		Zn	13.8 [b,j]	Cruz et al. (2000)
			Ni	9.9 [b,j]	
			Fe	35.3 [b,j]	
	CstH-(6His), fimbriae		Cd	656 ± 8.5 [b,g]	Saffar et al. (2007)
			Ni	277 ± 12 [b,g]	
	RsaA-(6His), S-layer	*C. crescentus*	Cd	142 ± 31[c,k]	Patel et al. (2010)
$(Gly\text{-}Cys)_2$-Pro-$(Cys\text{-}Gly)_2$ (CP)	LamB-CP, OM	*E. coli*	Cd	9.5 ± 1.2[a,h]	Kotrba et al. (1999a)
$(Gly\text{-}His\text{-}His\text{-}Pro\text{-}His\text{-}Gly)_2$ (HP2)	LamB-HP2, OM		Cd	7.9 ± 1.5[a,h]	Kotrba et al. (1999a)
HPCP fusion peptide	LamB-HPCP, OM		Cd	8.2 ± 1.4[a,h]	Kotrba et al. (1999a)
$(His\text{-}Cys)_4$ (HC peptide)	OprI-HC, OM		Cd	17.8 ± 4.2[a,e]	Bouia et al. (2001)
$(His\text{-}Cys)_4$ (HC peptide)			Cr	24.0 ± 8.1[a,e]	
$(Cys\text{-}Gly\text{-}Cys\text{-}Cys\text{-}Gly)_3$ (CG)	MBP-CG, periplasm		Cd	1.1 ± 0.12[a,e]	Pazirandeh et al. (1998)
			Hg	1.3 ± 0.02[a,e]	
Human MT-II (HMT)	Lpp-HMT, OM		Cd	5.8[a,g]	Jacobs et al. (1989)
Yeast MT CUP1 (YMT)	LamB-YMT, OM		Cd	16.2 ± 0.7[a,g]	Sousa et al. (1998)
Human MT1A (HMT1A)	LamB-HMT1A, OM		Cd	31.9 ± 6.4[a,h]	Sousa et al. (1998)
			Cu	6.7 ± 0.3[a,h]	
			Zn	4.2 ± 0.2[a,h]	
	OmpC-HMT1A, OM		Hg	4.8 ± 0.3[c,i]	Lin et al. (2010)
α-domain of human MT1A	LamB-αHMT1A, OM		Cd	42 ± 6.9[a,h]	Kotrba et al. (1999b)
Mouse MT-I (mMT)	mMT-PAL, PG		Cd	4.6 ± 0.2[a,g]	Valls et al. (1998)
	Lpp-OmpA-mMT, OM		Cd	3.9 ± 0.2[a,g]	Valls et al. (1998)

Table 11.1 (continued)

Metal-binding peptide or protein	Fusion, localization	Bacterium	Metal	Metal uptake (nmol/mg dry wt)	Reference
	mMT-Iga$_{\beta}$, OM	*P. putida*	Cd	28.9±2.5[a,l]	Valls et al. (2000b)
		C. metallidurans	Cd	15.7±0.5[a,h]	Valls et al. (2000a)
		E. coli	Cd	14.8±0.2[a,h]	
	OmpC-mMT, OM		Hg	4.1±0.2[c,i]	Lin et al. (2010)
Tilapiine cichlid fish MT	OmpC-tMT, OM		Hg	5.4±0.2[c,i]	
N. crassa MT	MBP-(NcMT)$_3$, periplasm		Cd	16.2±2.1[a,e]	Pazirandeh and Mauro (2001)
PC-like (Glu-Cys)$_{20}$ (EC20)	Lpp-OmpA-EC20, OM		Cd	61±2.7[a,j]	Bae et al. (2000)
			Hg	19.2±0.8[c,e]	Bae et al. (2001)
	INPNC-EC20, OM	*Moraxella* sp.	Hg	18.8±0.9[a,e]	Bae et al. (2002)
MT-like vanabin1	MBP-vanabin1, periplasm	*E. coli*	Cu	13.5±2.5[c,f]	Ueki et al. (2003b)
MT-like vanabin2	MBP- vanabin2, periplasm		Cu	13.8±3.6[c,f]	Ueki et al. (2003b)
MerR from Tn*21* of *E. coli*	INP-MerR, OM		Hg	119±6[c,d]	Bae et al. (2003)
MerP from *B. cereaus*	selfdisplay in periplasm		Cd	478 (q_{max}/K_d=7.1)[c,d]	Huang et al. (2003)
			Cu	604 (q_{max}/K_d=15.4)[c,d]	
			Pb	413 (q_{max}/K_d=17.6)[c,d]	
			Zn	22,266 (q_{max}/K_d=0.05)[c,d]	Kao et al. (2008)
			Cr	985 (q_{max}/K_d=0.20)[c,d]	
			Ni	723 (q_{max}/K_d=0.27)[c,d]	

OM outer membrane of gram-negative bacterium, *PG* peptidoglycan q_{max} maximum biosorption capacity, K_d Langmuir dissociation constant
[a] Uptake from synthetic minimal media; [b] uptake from rich complete media; [c] batch biosorption experiment in model solution; [d] published as q_{max}; initial metal concentrations: [e] 5 μM; [f] 10 μM; [g] 20 μM; [h] 30 μM; [i] 50 μM; [j] 100 μM; [k] 133 μM; [l] 300 μM; [m] 440 μM

11.5.1 Cadmium Uptake with Surface-Engineered Bacteria

The first study aiming at the improvement of natural metallosorption capacity of bacteria employed intramolecular fusions of single hexahistidine (H1) and two tandem hexahistidine motifs (H2) to position of amino acid 153 within an outer loop of LamB protein (Sousa et al. 1996). The potential of an *E. coli* displaying H1 and H2 to bind Cd^{2+} from the growth medium was increased 5-fold and 11-fold, respectively, thereby demonstrating correlation between the number of novel metal-binding sites and biosorption capacity of the modified cell. Moreover, the surface hexahistidine allowed the cells to adhere reversibly to a Ni^{2+}-containing solid matrix and behave as metalloaffinity adsorbent, a property useful for specific separation of cells. Xu and Lee (1999) showed that OmpC can efficiently display up to 6 hexahistidine moieties inserted at position of residue 162 within carrier's outer loop. It improved the natural capacity of *E. coli* to bind Cd^{2+} in a batch biosorption experiment more than 3-fold. Resting cells of *E. coli* displaying hexahistidine as an intramolecular fusion to the CstH protein of CS3 fimbriae accumulated three times more Cd^{2+} from media than cells producing unmodified CS3 fimbriae (Saffar et al. 2007). High-density hexahistidine display on *C. crescentus* was achieved using RsaA S-layer protein inserted at position of its residue 723 (Patel et al. 2010). Authors showed in batch that resting modified cells (300 mg/l, dry wt basis) can remove virtually all Cd from solution containing 10 μM Cd^{2+} within 15 min of contact time. Hexahistidine was also inserted at position of amino acid Wang et al. (2004) further extended the hexahistidine display approach also to Gram-positive bacteria, demonstrating that the natural Cd^{2+} binding capacity of *B. thuringiensis* could be doubled by anchoring hexahistidine peptide to its S-layer protein.

Fusions of LamB protein with short metal-binding sequences Gly-Cys-Gly-Cys-Pro-Cys-Gly-Cys-Gly (named CP) and Gly-His-His-Pro-His-Gly (named HP) showed the Cd-to-peptide stoichiometry of 3:1 and 1:1 *in vitro,* respectively (Kotrba et al. 1999a). While the sequence of the CP peptide was selected by screening of sub-library of synthetic peptides (Kotrba et al. 1996), the HP peptide represents a natural metal-binding motif of several tandem repetitions in plasma histidine-rich glycoprotein (Morgan 1985). Surface display of two tandem repetitions of HP multiplied the natural ability of *E. coli* to bind Cd^{2+} 3-fold. While the display of CP improved biosorption of Cd^{2+} 4-fold, there was no additive effect of the combination of HP and CP sequences on the total amount of accumulated Cd^{2+}. The performance of CP peptide was not affected by Zn^{2+} in medium, but the uptake of Cd^{2+} was impaired by 25% when an equimolar concentration of Cu^{2+} was present. The metallosorption properties of isolated cell walls were less pronounced. Modification of walls with CP or HP peptides resulted in 1.8-fold increase in Cd^{2+} binding capacity with maximum of 225 nmol Cd/mg of modified dry wall weight (Kotrba et al. 1999a). Saffar et al. (2005) later showed that resting cells of *E. coli* producing CP fusion to CstH protein accumulated 3-fold more Cd^{2+} than those cells which produced CS3 fimbriae alone, and that the amount of Cd accumulated by CstH-CP cells was 50 times higher compared to *E. coli* producing CP fused to LamB protein.

Combination of metal-binding cysteine and histidine residues in a peptide made of 4 His-Cys repetitions was engineered into outer loop of OprI from *P. aeruginosa* and displayed on *E. coli* (Bouia et al. 2001). The capacity of modified cells to accumulate Cd^{2+} from media than increased 4-fold compared to the wild-type cells. Periplasmic display of peptide containing three repetitions of Cys-Gly-Cys-Cys-Gly as a fusion to maltose-binding protein (MBP) improved the Cd^{2+} uptake 12 times, as compared to cell producing MBP alone (Pazirandeh et al. 1998).

The study of the ability of PAL, Lpp-OmpA and LamB (insertion at periplasmic and outer loops) to efficiently display mammalian MTs proved the LamB protein as a suitable carrier of MTs inserted in its permissive outer loop (Sousa et al. 1998; Valls et al. 1998). *E. coli* cells displaying a 66-amino acid sequence obtained from human MT1A (HMT1A) and 65-amino acid sequence from CUP1 of *S. cerevisiae* bound 20- and 15-fold higher amounts of Cd^{2+} respectively compared to the control cells (Sousa et al. 1998). The metal-binding abilities of displayed separate α- and β-domains of HMT1A were also inspected (Kotrba et al. 1999c). The amount of Cd^{2+} accumulated by *E. coli* variants reflected the metal-binding properties of displayed peptide. The cells displaying α-domain accumulated up to 30-times more Cd^{2+} than the control cells, but the contribution of β-domain was ten times less. Moreover, the presence of equimolar concentration of competing Cu^{2+} impaired the biosorption of Cd^{2+} in cells displaying β-domain by 50%, but only by 20% in cells displaying the α-domain. While attempts to produce multiple fusions of α-domain to LamB did not provide viable cells (P. Kotrba and T. Ruml, unpublished results), relatively small size of *N. crassa* NcMT allowed Pazirandeh and Mauro (2001) to produce up to 12 tandem repetitions of this MT to periplasmic MBP. In terms of metal-binding capacity, the best performing was an *E. coli* variant displaying three MT repetition, which accumulated 15-times more Cd^{2+} than cell producing MBP alone.

Success achieved with surface display of MTs on the model *E. coli* prompted Valls et al. (2000a, b) to extend this approach to robust bacteria, which could be concerned more realistic targets to genetic engineering. A predisposition of *P. putida* to surface engineering was demonstrated with cells expressing a genetic fusion of mouse MT to autotransporting Iga_{β} domain. This modification enhanced capacity of *P. putida* to accumulate Cd^{2+} from media 3-fold (Valls et al. 2000b). Although the contribution of the same fusion to metallosorption in *E. coli* was more pronounced and initiated 10-fold increase in Cd^{2+} uptake (Valls et al. 2000a), the *P. putida* cells displaying MT showed two-times higher binding capacity than the respective *E. coli*. Display of mouse MT using the same carrier in *Cupravidus metallidurans* (formerly *Ralstonia eutropha* and *R. metallidurans*) strain CH34, also led in resulting MTB strain to 3-fold increase in natural Cd^{2+} binding capacity (Valls et al. 2000a). Unlike *E. coli* or *P. putida, C. metallidurans* is a heavy metal-resistant soil bacterium, which uses metal ion efflux to detoxify the cell interior and precipitates the surface-bound heavy metal ions, including Cd^{2+}, in a metabolism-dependent manner (Mergeay et al. 2003). The MTB strain was thus able to retain its viability and improved metallosorption properties also at high (300 μM) external Cd^{2+} concentrations. Moreover, MTB showed markedly improved the capacity to immobilize Cd^{2+} in soil and to protect plants from the biological toxicity of the heavy metal (Valls et al. 2000a). This study thus showed that display of MTs may find potential application for generation of biological amendments for *in situ* immobilization of Cd^{2+} in soil.

The potential of gene-encoded phytochelatin analogue (with α linkage) to promote accumulation of Cd^{2+} was demonstrated with periplasmically expressed fusion of 20 repetitive glutamyl-cysteinyl units (EC20) to MBP (Bae et al. 2000). Purified fusion protein showed the Cd-to-EC20 stoichiometry of 10:1 and *E. coli* cell producing this fusion accumulated nearly nine times more Cd^{2+} from media than did the control cells. When fused to Lpp-OmpA anchor, EC20 and EC11 multiplied the natural capacity of *E. coli* to bind Cd^{2+} 15- and 8-fold, which is consistent with the number of presented metal-binding centers (Bae et al. 2000).

Taking advantage of its natural localization of MerP in the periplasm, Huang et al. (2003) tested performance of *E. coli* producing two MerP orthologues from *Pseudomonas* sp. K-62 and *Bacillus cereus*. The cells producing *Pseudomonas* sp. and *B. cereus* orthologues of MerP showed 55 and 84% increase for Cd^{2+} biosorption capacity over the MerP-free cells, respectively in a batch biosorption experiment.

11.5.2 Mercury Uptake with Surface-Engineered Bacteria

Besides their improved Cd^{2+}-binding, *E. coli* displaying Lpp-OmpA-EC20 fusions showed markedly increased capacity to bind Hg^{2+} (Bae et al. 2001). The authors showed the Hg-to-EC20 stoichiometry of 20:1, which is in a good agreement with linear coordination geometry of the metal. The cells of *E. coli* displaying EC20 accumulated more than 18 times more Hg^{2+} than native cells in a batch biosorption experiment. The metal-binding equilibrium was established in less than 5 min suggesting that biosorption was the metal uptake mechanism. The high affinity of Hg^{2+} to sulfhydryl groups and preferred linear coordination geometry presumably contributed to certain specificity of the engineered surface for Hg^{2+}. An administration of 20-fold molar excess of Cd^{2+} reduced the Hg^{2+} binding by only 20%. In another study Bae et al. (2002) inspected the suitability of the ice nucleation protein (INP) based anchor INPNC for display of EC20. The Hg^{2+} uptake capacity of *E. coli* producing INPNC-EC20 was less pronounced compared to the cells producing Lpp-OmpA-EC20. However, the INPNC was proved suitable anchor to display EC20 on the surface of environmentally robust *Moraxellla* sp. The potential of modified *Moraxella* sp. to bind Hg^{2+} from growth medium was increased 6-fold compared to the wild-type cells (Bae et al. 2002). Under the growth conditions, the uptake capacity of *Moraxella* sp. was significantly higher than that of *E. coli* producing Lpp-OmpA-EC20 or EC20-INPNC.

Lin et al. (2010) tested Hg^{2+}-biosorption properties of *E. coli* displaying human HMT1A, mouse mMT-I and tMT from fish *Oreochromis mossambicus* as intramolecular fusions to OmpC. In batch biosorption experiment accumulated resting cells displaying HMT1A, mMT1 and tMT by 80, 55 and 120% more Hg^{2+} than the control cells. It was by 25% more than authors observed with cell producing tested MTs cytoplasmatically.

High degree of specificity for Hg^{2+} was achieved with display of MerR protein. In a batch biosorption experiment, resting cells of *E. coli* with MerR genetically fused to INP accumulated more than 6-fold higher levels of Hg^{2+} than the wild-type cells (Bae et al. 2003). The biosorption equilibrium was achieved within 10 min of contact time, most effectively at pH 5 and the uptake was virtually insensitive to a presence of 400 μM NaCl. The Hg^{2+} binding did not decline in the presence of 100-fold molar excess of Cd^{2+} or Zn^{2+} or even at 2,000-fold excess of competing ligand EDTA. The authors assume that the high specificity and affinity of displayed INP-MerR fusion was due to spontaneous formation of MerR dimers within outer membrane, thereby providing the metal-binding center with trigonal coordination geometry. To mimic the metal-binding site of MerR dimer in a continuous polypeptide, Song et al. (2004) constructed 117-amino acid MBD protein containing a direct tandem duplication of cysteine-residue-carrying helix five of MerR. Fusion of MBP to Lpp-OmpA, allowed for display of 20,000 copies of MBP on the surface of *E. coli,* with Hg^{2+}-to-MBP stoichiometry of 1:1 (Qin et al. 2006). Consequently, the cells producing Lpp-OmpA-MBP accumulated more than 6-fold more Hg^{2+} from the culture media than those cells which did not display MBD. Displayed MBD remained specific for Hg^{2+} also in the presence of 22-fold molar excess of Cd^{2+} or Zn^{2+}.

11.5.3 Uptake of Copper, Zinc, Chromium, Lead and Nickel by Surface-Engineered Bacteria

Improved biosorption of Zn^{2+}, Ni^{2+} and Fe^{3+} was observed with *E. coli* displaying hexahistidine motifs inserted at the position of amino acid 162 of OmpC (Cruz et al. 2000). The display of hexahistidine sequence improved the natural capacity of *E. coli* to take up these metal ions from media nearly 3-fold. The display of two tandem hexahistidine insertions further doubled the amount of accumulated Ni^{2+} and Fe^{3+} compared to cells displaying single hexahistidine insertion. Display of CP peptide and hexahistidine as fusions to CstH protein of *E. coli* CS3 fimbriae enhanced the Ni biosorption capacity of resting cells 10- and 3-fold, respectively compared to the cells producing CS3 fimbriae alone (Saffar et al. 2005, 2007). Besides improved Cd^{2+}-binding, the display of HMT1A as a fusion to LamB multiplied the natural capacity of *E. coli* cells to bind Cu^{2+} and Zn^{2+} from media more than 4-fold and 2.5-fold, respectively (Sousa et al. 1998; Kotrba et al. 1999b).

E. coli producing MerP showed markedly increased capacity to bind not only Cd^{2+}, but also other bivalent metal ions, as demonstrated in batch experiments with living cell biosorbents (Huang et al. 2003; Kao et al. 2008). Overall, better performance showed cells producing MerP from *B. cereus,* which were able to bind by 142, 121, 72, 33 and 23% more Cu^{2+}, Zn^{2+}, Cr^{3+}, Pb^{2+} and Ni^{2+}, respectively.

The absorption preferences for Cd^{2+}, Cu^{2+} and Pb^{2+} evaluated in two metal systems followed an order $Pb^{2+} \geq Cu^{2+} > Cd^{2+}$, which correlated Langmuir dissociation constants determined as 24 μM for Pb^{2+}, 39 μM for Cu and 60 μM (Huang et al. 2003). Intriguingly, periplasmic display of MerP did not contribute to biosorption of Hg^{2+}. Similar loss of natural specificity for metal was observed with periplasmatically produced vanabins originating from *Ascidia sydneiesis samea* (Ueki et al. 2003b). Production of fusions of vanabins to MBP did not improve uptake of VO^{2+}. In contrast, the capacity of *E. coli* producing MBP-vanabin1 and MBP-vanabin2 to accumulate Cu increased 21- and 14-fold, respectively as compared to the cells producing MBP alone.

11.6 Concluding Remarks

Metal binding by biomolecules of structural components or excreted polymers of bacteria is generally fortuitous and relative efficiencies depend on attributes of the metal ion as well as on reactivity of provided ligands. The rationale behind the genetic modifications of bacterial cells described here was expected self derivatization of cell walls with peptides and proteins capable to form stable coordinations spheres for heavy metal ions, thereby improving biosorption capacity of modified walls. As described, a number of surface-display systems are available (Fig. 11.1) and bacterial display of metal binding peptides often resulted in several fold increase in metal uptake by modified cells (Table 11.1), yet, one would question techno-economic perspective and safety. In most cases, living (genetically modified) bacteria were used and metallosorption inspected under growth conditions with media supplemented with one or a few concentrations of the target metal ion. Moreover, research focused more on model *E. coli* of virtually no potential use in bioremediation process. Only limited number of papers described use of resting cells in batch format with evaluated standard biosorption isotherm. Potential of isolated cell-wall envelopes was essentially neglected, although at present the risk of use of living genetically modified microorganisms in metal remediation seems to outweigh potential benefits. The development and proper characterization of suitable biomass, which produce modified cell walls robust enough for formulation of improved biosorbent material is the ultimate challenge to make the surface display approach viable. For ease of manipulation and current availability of various surface display systems, bacteria and yeast (Chap. 10) are primary targets to such efforts.

Intriguingly, several studies showed that the number of metal ions taken up on modified cell walls by biosorption was orders of magnitude higher than the number of displayed novel metal-binding sites (Valls et al. 2000a, b; Kotrba et al. 1999a, b; Sousa et al. 1996, 1998). A possible explanation was that displayed peptides support the interactions of metal with other cell wall components by increasing the local concentration of metals. Based on such cooperation model, it can be hypothesized that display of a peptide forming high-affinity coordination sphere would not favour metal displacement to other cell wall components and thus contributes less to overall metallosorption properties of modified biosorbent. For example, when HP peptide (selected for its high affinity to Zn^{2+} and Cu^{2+}) was displayed on the surface of *E. coli,* it did not contribute to biosorption of Cu^{2+}, but promoted biosorption of Cd^{2+}, which binds to HP less avidly (Kotrba et al. 1999a). Further evidence supporting the cooperation model was obtained with modified *S. cerevisiae.* Display of HP peptide on the cell wall of *S. cerevisiae* increased the amount of Zn adsorbed by modified cells by only 20% (Vinopal et al. 2007), which was consistent with the HP:Zn^{2+} stoichiometry of 1:1 (with an apparent dissociation constant of 120 nM) determined *in vitro.* In contrast, metallosorption of *S. cerevisiae* cell wall was substantially improved on display of NP peptide harboring the Cys-X-X-Glu-Glu (CXXEE) sequence with expected propensity to exchange the metal ion; an event which should occur during metal export by the bacterial metal efflux transporter ATPase to which CXXEE serves as cytoplasmatic metal-fixation motif (Kotrba and Ruml 2009, 2010). The isolated NP-containing cell walls showed up to 3-fold increased capacity for Pb^{2+} biosorption within the equilibrium concentration range of 9–350 μM with maximum sorption capacity of 315 nmol Pb per mg of dry walls. The S-type Pb^{2+} biosorption isotherms, plus the presence of electron-dense deposits (Fig. 11.2) strongly suggested that the improved biosorption potential of NP-displaying cells is due to the onset of microprecipitation of Pb species on the modified cell wall. It appears reasonable to assume that the contribution of the NP peptide involves a gradual increase in the local Pb^{2+} concentration to a certain level that triggers the microprecipitation. Moreover, this process was not impaired by an excess of potentially competing Cd^{2+} and Zn^{2+} ions and virtually

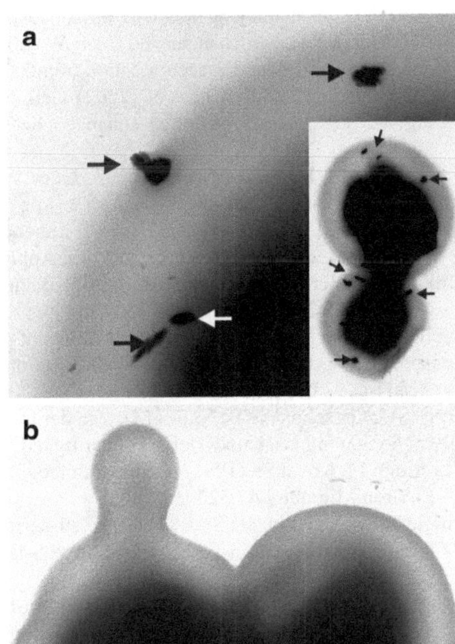

Fig. 11.2 Deposition of Pb microprecipitate at the NP-containing cell walls (**a**) and control native cell walls of *S. cerevisiae* (**b**). The cells preadsorbed in the 150 μM Pb^{2+} solution were directly inspected by TEM without fixation or staining. The electron-dense microprecipitate bodies of averange size of 240 × 80 nm are indicated with *arrows*

all the Pb was recovered from the cell surface by EDTA treatment. This example shows that biosorption mechanism on microbial walls can be specifically upgraded with microprecipitation by the engineering of the biosorbent with an eligible metal-binding peptide. We believe that more detailed studies on the mechanisms involved in biosorption on modified bacterial surfaces would provide valuable information and promote renaissance of surface display concept in biosorption field. As computational chemistry, allowing rational design of specific metal-binding peptides and modulation of their affinity is fast developing discipline, even more successful cell surface modifications may emerge.

Acknowledgements PK and TR are supported from Ministry of Education, Youth and Sports of the Czech Republic (projects MSM 6046137305 and 1M6837805002).

References

Abhilash PC, Jamil S, Singh N (2009) Transgenic plants for enhanced biodegradation and phytoremediation of organic xenobiotics. Biotechnol Adv 27:474–488

Andreini C, Bertini I, Rosato A (2004) A hint to search for metalloproteins in gene banks. Bioinformatics 20:1373–1380

Babu CS, Dudev T, Casareno R, Cowan JA, Lim C (2003) A combined experimental and theoretical study of divalent metal ion selectivity and function in proteins: application to *E. coli* ribonuclease H1. J Am Chem Soc 125:9318–9328

Bae W, Mehra RK (1997) Metal-binding characteristics of phytochelatin analog (Glu-Cys)$_2$-Gly. J Inorg Biochem 68:201–210

Bae W, Chen W, Mulchandani A, Mehra RK (2000) Enhanced bioaccumulation of heavy metals by bacterial cells displaying synthetic phytochelatins. Biotechnol Bioeng 70:518–524

Bae W, Mehra RK, Mulchandani A, Chen W (2001) Genetic engineering of *Escherichi coli* for enhanced uptake and bioaccumulation of mercury. Appl Environ Microbiol 67:5335–5338

Bae W, Mulchandani A, Chen W (2002) Cell surface display of synthetic phytochelatins using ice nucleation protein for enhanced heavy metal bioaccumulation. J Inorg Chem 88: 223–227

Bae W, Wu CH, Kostal J, Mulchandani A, Chen W (2003) Enhanced mercury biosorption by bacterial cells with surface-displayed MerR. Appl Environ Microbiol 69:3176–3180

Baek JH, Han M-J, Lee SH, Lee SY (2010) Enhanced display of lipase on the *Escherichia coli* cell surface, based on transcriptome analysis. Appl Environ Microbiol 76:971–973

Barkay T, Miller SM, Summers AO (2003) Bacterial mercury resistance from atoms to ecosystems. FEMS Microbiol Rev 27:355–384

Benhar I (2001) Biotechnological applications of phage and cell display. Biotechnol Adv 19:1–33

Benson DE, Wisz MS, Hellinga HW (2000) Rational design of nascent metalloenzymes. Proc Natl Acad Sci 97:6292–6297

Berry SM, Gieselman MD, Nilges MJ, Van der Donk WA, Lu Y (2002) An engineered azurin variant containing a selenocysteine copper ligand. J Am Chem Soc 124:2084–2085

Beveridge TJ, Koval SF (1981) Binding of metals to cell envelopes of *Escherichia coli* K-12. Appl Environ Microbiol 42:325–335

Bingle WH, Nomellini JF, Smit J (1997) Cell-surface display of a *Pseudomonas aeruginosa* strain K pilin peptide within the paracrystalline S-layer of *Caulobacter crescentus*. Mol Microbiol 26:277–288

Bouia A, Kholti A, Saghi M, Cornelis P (2001) In-frame fusion of a His-Cys motif into the *Pseudomonas aeruginosa* outer membrane OprI lipoprotein results in increased metal binding capacity by *Escherichia coli*. Res Microbiol 152:799–804

Boulain J, Charbit A, Hofnung M (1986) Mutagenesis by random linker insertion into *lamB* gene of *Escherichia coli* K12. Mol Gen Genet 205:339–348

Brim H, McFarlan SC, Fredrickson JK, Minton KW, Zhai M, Wackett LP, Daly MJ (2000) Engineering *Deinococcus radiodurans* for metal remediation in radioactive mixed waste environments. Nat Biotechnol 18:85–90

Brown NL, Stoyanov JV, Kidd SP, Hobman JL (2003) The MerR family of transcriptional regulators. FEMS Microbiol Rev 27:145–163

Brunke M, Deckwer WD, Frischmuth A, Horn JM, Lünsdorf H, Rhode M, Röhricht M, Timmis KN, Weppen P (1993) Microbial retention of mercury from waste streams in a laboratory column containing merA gene bacteria. FEMS Microbiol Rev 11:145–152

Calabi E, Ward S, Wren B, Paxton T, Panico M, Morris H, Dell A, Doughan G, Fairweather N (2001) Molecular characterization of the surface layer proteins from *Clostridium difficile*. Mol Microbiol 40:1187–1199

Cascales E, Lloubes R (2004) Deletion analyses of the peptidoglycan-associated lipoprotein Pal reveals three independent binding sequences including a TolA box. Mol Microbiol 51:873–885

Chang HH, Lo SJ (2000) Modification with a phosphorylation tag of PKA in the TraT-based display vector of *Escherichia coli*. J Biotechnol 78:115–122

Chang HH, Sheu SY, Lo SJ (1999) Expression of foreign antigens on the surface of *Escherichia coli* by fusion to the outer membrane protein TraT. J Biomed Sci 6:64–70

Charbit A, Boulain JC, Ryter A, Hofnung M (1986) Probing the topology of a bacterial membrane protein by genetic insertion of a foreign epitope; expression at the cell surface. EMBO J 5:3029–3037

Charbit A, Molla A, Saurin W, Hofnung M (1988) Versatility of a vector for expression foreign polypeptides at the surface of Gram-negative bacteria. Gene 70:181–189

Chen W, Georgiou G (2002) Cell-Surface display of heterologous proteins: from high-throughput screening to environmental applications. Biotechnol Bioeng 79:496–503

Clemens S (2006) Toxic metal accumulation, responses to exposure and mechanisms of tolerance in plants. Biochimie 88:1707–1719

Cobbett C, Goldsbrough P (2002) Phytochelatins and metallothioneins: roles in heavy metal detoxification and homeostasis. Annu Rev Plant Biol 53:159–182

Cortes-Perez NG, Azevedo V, Alcocer-González JM, Rodriguez-Padilla C, Tamez-Guerra RS, Corthier G, Gruss A, Langella P, Bermúdez-Humarán LG (2005) Cell-surface display of E7 antigen from human papillomavirus type-16 in *Lactococcus lactis* and in *Lactobacillus plantarum* using a new cell-wall anchor from lactobacilli. J Drug Target 13:89–98

Cote-Sierra J, Jongert E, Bredan A, Gautam D, Parkhouse M, Cornelis P, De Baetselier P, Revets H (1998) A new membrane-bound OprI lipoprotein expression vector. High production of heterologous fusion proteins in gram-negative bacteria and the implications for oral vaccination. Gene 221:25–34

Courture Tosi E, Delacroix H, Mignot T, Mesnage S, Chami M, Fouet A, Mosser G (2002) Structural analysis and evidence for dynamic emergence of *Bacillus anthracis* S-layer networks. J Bacteriol 184:6448–6456

Crampton M, Berger E, Reid S, Louw M (2007) The development of a flagellin surface display expression system in a moderate thermophile, *Bacillus halodurans* Alk36. Appl Microbiol Biotechnol 75:599–607

Cruz N, Le Borgne S, Hernández-Cávez G, Gosset G, Valle F, Bolivar F (2000) Engineering the *Escherichia coli* outer membrane protein OmpC for metal bioadsorption. Biotechnol Lett 22:623–629

Dameron CT, Winge DR (1990) Characterization of peptide-coated cadmium-sulfide srystallites. Inorg Chem 29:1343–1348

Daugherty PS (2007) Protein engineering with bacterial display. Curr Opin Struct Biol 17:474–480

de Lorenzo V (2009) Recombinant bacteria for environmental release: what went wrong and what we have learnt from it. Clin Microbiol Infect 15:63–65

Desvaux M (2005) The cellulosome of *Clostridium cellulolyticum*. Enzymes Microbiol Technol 37:373–385

Desvaux M, Dumas E, Chafsey I, Hébraut M (2006) Protein cell surface display in Gram-positive bacteria: from single protein to macromolecular protein structure. FEMS Microbiol Lett 256:1–15

Dhillon JK, Drew PD, Porter AJ (1999) Bacterial surface display of an anti-pollutant antibody fragment. Lett Appl Microbiol 28:350–354

Doty SL (2008) Enhancing phytoremediation through the use of transgenics and endophytes. New Phytol 179:318–333

Dudev T, Lim C (2001) Metal selectivity in metalloproteins: Zn^{2+} vs Mg^{2+}. J Phys Chem B 105:4446–4452

Dudev T, Lim C (2003) Principles governing Mg, Ca, and Zn binding and selectivity in proteins. Chem Rev 103:773–787

Dudev T, Chang LY, Lim C (2005) Factors governing the substitution of La^{3+} for Ca^{2+} and Mg^{2+} in metalloproteins: a DFT/CDM study. J Am Chem Soc 127:4091–4103

Duncan G, Tarling CA, Bingle WH, Nomellini JF, Yamage M, Dorocicz IR, Withers SG, Smit J (2005) Evaluation of a new system for developing particulate enzymes based on the surface (S)-layer protein (RsaA) of *Caulobacter crescentus:* fusion with the beta-1,4-glycanase (Cex) from the cellulolytic bacterium *Cellulomonas fimi* yields a robust, catalytically active product. Appl Biochem Biotechnol 127:95–110

Dwyer MA, Looger LL, Hellinga HW (2003) Computational Design of a Zn^{2+} receptor that controls bacterial gene expression. Proc Natl Acad Sci 100:11255–11260

Earhart CF (2000) Use of an Lpp-OmpA fusion vehicle for bacterial surface display. Methods Enzymol 326:506–516

Fernandez LA, Berenguer J (2000) Secretion and assembly of regular surface structures in Gram-negative bacteria. FEMS Microbiol Rev 24:21–44

Ferris FG, Beveridge TJ (1985) Site specificity of metallic ion binding in Escherichia coli K-12 lipopolysacharide. Can J Microbiol 32:52–55

Foster TJ (2005) Immune evasion by staphylococci. Nat Rev Microbiol 3:948–958

Francisco JA, Earhart CF, Georgiou G (1992) Transport and anchoring of β-lactamase to the external surface of *Escherichia coli.* Proc Natl Acad Sci U S A 89:2713–2717

Freudl R (1989) Insertion of peptides into cell-surface-exposed areas of the *Escherichia coli* OmpA protein does not interfere with export and membrane assembly. Gene 82:229–236

Freudl R, MacIntyre S, Degen M, Henning U (1986) Cell surface exposure of the outer membrane protein OmpA of *Escherichia coli* K-12. J Mol Biol 188:491–494

Fuchs P, Weichel W, Dubel S, Breitling F, Little M (1996) Separation of *E. coli* expressing functional cell-wall bound antibody fragments by FACS. Immunochemotechnology 2:97–102

Georgiou G, Stephens DL, Stathopoulos C, Poetschke HL, Mendenhall J, Earhart CF (1996) Display of β-lactamase on the *Escherichia coli* surface: outer memrane phenotypes conferred by Lpp'-OmpA'-β-lactamase fusion. Protein Eng 9:239–247

Georgiou G, Stathopoulos C, Daugherty PS, Nayak AR, Iverson BL, Curtiss RI (1997) Display of heterologous proteins on the surface of microorganisms: from the screening of combinatorial libraries to live vaccines. Nat Biotechnol 15:29–34

Germond JE, Delley M, Gilbert C, Atlan D (2003) Determination of the domain of the *Lactobacillus delbrueckii* subsp. *bulgaricus* cell surface proteinase PrtB involved in attachment to the cell wall after heterologous expression of the prtB gene in *Lactococcus lactis*. Appl Environ Microbiol 69:3377–3384

Giomarelli B, Maggi T, Younson J, Kelly C, Pozzi G (2004) Expression of a functional single-chain Fv antibody on the surface of *Streptococcus gordonii.* Mol Biotechnol 28:105–112

Glusker JP (1991) Structural aspects of metal liganding to functional groups in proteins. Adv Protein Chem 42:1–76

Godlewska R, Wiśniewska K, Pietras Z, Jagusztyn-Krynicka EK (2009) Peptidoglycan-associated lipoprotein (Pal) of Gram-negative bacteria: function, structure, role in pathogenesis and potential application in immunoprophylaxis. FEMS Microbiol Lett 298:1–11

Gunneriusson E, Samuelson P, Uhlen M, Nygren PA, Ståhl S (1996) Surface display of a functional single-chain Fv antibody on staphylococci. J Bacteriol 178:1341–1346

Han WG, Lovell T, Noodleman L (2002) Coupled redox potentials in manganese and iron superoxide dismutases from reaction kinetics and density functional/electrostatics calculations. Inorg Chem 41:205–218

Hansson M, Ståhl S, Nguyen TN, Bächi T, Robert A, Binz H, Sjölander A, Uhlén M (1992) Expression of recombinant proteins on the surface of the coagulase-negative bacterium *Staphylococcus xylosus*. J Bacteriol 174:4239–4345

Hofnung M (1995) An intelligent channel (and more). Science 267:473–474

Horn JM, Brunke M, Deckwer WD, Timmis KN (1994) *Pseudomonas putida* strains which constitutively overexpress mercury resistance for biodetoxification of organomercurial pollutants. Appl Environ Microbiol 60:357–362

Hoyle BD, Beveridge TJ (1983) Binding of metallic ions to the outer membrane of *Escherichia coli*. Appl Environ Microbiol 46:749–752

Hoyle BD, Beveridge TJ (1984) Metal binding by the peptidoglycan sacculus of *Escherichia coli* K-12. Can J Microbiol 30:204–211

Huang C-C, Su C-C, Hsieh J-L, Tseng C-P, Lin P-J, Chang J-S (2003) Polypeptides for heavy-metal biosorption: capacity and specificity of two heterologous MerP proteins. Enzyme Microb Technol 33:379–385

Huang CC, Chen MW, Hsieh JL, Lin WH, Chen PC, Chien LF (2006) Expression of mercuric reductase from *Bacillus megaterium* MB1 in eukaryotic microalga *Chlorella* sp. DT: an approach for mercury phytoremediation. Appl Microbiol Biotechnol 72:197–205

Ikeda T, Homma M, Iino T, Asakura S, Kamiya R (1987) Localization and stoichiometry of hook-associated proteins within *Salmonella typhimurium* flagella. J Bacteriol 169:1168–1173

Jacobs FA, Roymer FM, Beauchemin M, Brousseau R (1989) Human metallothionein-II is synthesized as a stable membrane-localized fusion protein in *Escherichia coli*. Gene 83:95–103

Jantz D, Amann BT, Gatto GT Jr, Berg JM (2004) The design of functional DNA-binding proteins based on zinc finger domains. Chem Rev 104:789–799

Jose J, Meyer TF (2007) The autodisplay story, from discovery to biotechnical and biomedical applications. Microbiol Mol Biol Rev 71:600–619

Jose J, Krämer J, Klauser T, Pohlner J, Meyer TF (1996) Absence of periplasmic DsbA oxidoreductase facilitates export of cysteine-containing passenger proteins to the *Escherichia coli* cell surface via the Igaβ autotransporter pathway. Gene 178:107–110

Jung HC, Lebeault JM, Pan JG (1998) Surface display of *Zymomonas mobilis* levansucrase by using the ice-nucleation protein of *Pseudomonas syringae*. Nat Biotechnol 16:576–580

Kang SH, Singh S, Kim JY, Lee W, Mulchandani A, Chen W (2007) Bacteria metabolically engineered for enhanced phytochelatin production and cadmium accumulation. Appl Environ Microbiol 73:6317–6320

Kao W-C, Huang C-C, Chang J-S (2008) Bioadsorption of nickel, chromium and zinc by MerP-expressing recombinant *Escherichia coli*. J Hazard Mater 158:100–106

Kaplan J, DeGrado WF (2004) De novo design of catalytic proteins. Proc Natl Acad Sci U S A 101:11566–11570

Kawahara H (2002) The structures and functions of ice crystal-controlling proteins from bacteria. J Biosci Bioeng 94:492–496

Kille P, Winge DR, Harwood JL, Kay J (1991) A plant metallothionein produced in *E. coli*. FEBS Lett 295:171–175

Kim TW, Igimi S, Kajikawa A, Kim YH (2008) Display of heterologous proteins on the surface of *Lactococcus lactis* using H and W domains of PrtB from *Lactobacillus delburueckii* subsp. *Bulgaricus* as an anchoring matrix. J Appl Microbiol 104:1636–1643

Klauser T, Pohlner J, Meyer TF (1990) Extracellular transport of cholera toxin B subunit using Neisseria IgA protease β-domain: conformation-dependent outer membrane translocation. EMBO J 9:1991–1999

Klauser T, Kramer J, Otzelberger K, Pohlner J, Meyer TF (1993a) Characterization of the *Neisseria* Igaβ-core. The essential unit for outer membrane targeting and extracellular protein secretion. J Mol Biol 234:579–593

Klauser T, Pohlner J, Meyer TF (1993b) The secretion pathway of IgA protease type proteins in gram-negative bacteria. Bioessays 15:368–373
Klemm P, Schembri MA (2000) Fimbrial surface display systems in bacteria: from vaccines to random libraries Microbiology 146:3025–3032
Kotrba P, Ruml T (2009) Method of yeast biosorbent modification resulting in increased biosorption capacity for Pb^{2+} due to induced microprecipitation of Pb^{2+}. Czech Patent no. 301,334, PV 2009-207, 3 Apr/9 Dec 2009
Kotrba P, Ruml T (2010) Surface display of metal fixation motif of bacterial P1-ATPases specifically promotes biosorption of Pb^{2+} by *Saccharomyces cerevisiae.* Appl Environ Microbiol 76:2615–2622
Kotrba P, Dolečková L, Pavlík M, Ruml T (1996) Rapid screening of peptides for heavy metal binding. Biotechnol Tech 10:773–778
Kotrba P, Dolečková L, de Lorenzo V, Ruml T (1999a) Enhanced bioaccumulation of heavy metal ions by bacterial cells due to surface display of short metal binding peptides. App Environ Microbiol 65:1092–1098
Kotrba P, Macek T, Ruml T (1999b) Heavy metal-binding peptides and proteins in plants. a review. Collect Czech Chem Commun 64:1057–1086
Kotrba P, Pospisil P, de Lorenzo V, Ruml T (1999c) Enhanced metallosorption of *Escherichia coli* cell due to surface display of β- and α-domains of mammalizn metallothionein as a fusion to LamB protein. J Receptor Res Signal Transd Res 19:703–715
Kotrba P, Najmanova J, Macek T, Ruml T, Mackova M (2009) Genetically modified plants in phytoremediation of heavy metal and metalloid soil and sediment pollution. Biotechnol Adv 27:799–810
Kožíšek M, Svatoš A, Buděšínský M, Muck A, Bauer M, Kotrba P, Ruml T, Havlas Z, Linse S, Rulíšek L (2008) Molecular design of specific metal-binding peptide sequences from protein fragments. Theory and experiment. Chem-Eur J 14:7836–7846
Kronqvist N, Löfblom J, Jonsson A, Wernérus H, Ståhl S (2008) A novel affinity protein selection system based on staphylococcal cell surface display and flow cytometry. Protein Eng Des Sel 21:247–255
Kuppuraj G, Dudev M, Lim C (2009) Factors governing metal-ligand distances and coordination geometries of metal complexes J Phys Chem B 113:2952–2960
Lai EM, Nair U, Phadke ND, Maddock JR (2004) Proteomic screening and identification of differentially distributed membrane proteins in *Escherichia coli.* Mol Microbiol 52:1029–1044
Lattemann CT, Maurer J, Gerland E, Meyer TF (2000) Autodisplay: functional display of active β-lactamase on the surface of *Escherichia coli* by the AIDA-I autotransporter. J Bacteriol 182:3726–3733
Lauer P, Rinaudo CD, Soriani M, Margarit I, Maione D, Rosini R, Taddei AR, Mora M, Rappuoli R, Grandi G, Telford JL (2005) Genome analysis reveals pili in Group B *Streptococcus.* Science 309:105
Lazzaroni JC, Portalier R (1992) The excC gene of *Escherichia coli* K-12 required for cell envelope integrity encodes the peptidoglycan-associated lipoprotein (PAL). Mol Microbiol 6: 735–742
Lee SY, Choi JH, Xu Z (2003) Microbial cell surface display. Trends Biotechnol 12:45–52
Li C, Zhu Y, Benz I, Schmidt MA, Chen W, Mulchandani A, Qiao C (2008) Presentation of functional organophosphorus hydrolase fusions on the surface of *Escherichia coli* by the AIDA-I autotransporter pathway. Biotechnol Bioeng 99:485–490
Lin K-H, Chien M-F, Hsieh J-L, Huang C-C (2010) Mercury resistance and accumulation in *Escherichia coli* with cell surface expression of fish metallothionein. Appl Microbiol Biotechnol 87:561–569
Liu Z, Yang C, Jiang H, Mulchandani A, Chen W, Qiao C (2009) Simultaneous degradation of organophosphates and 4-substituted phenols by *Stenotrophomonas* species LZ-1 with surface-displayed organophosphorus hydrolase. J Agric Food Chem 57:6171–6177
Lu Z, Murray KS, Van Cleave V, LaVallie ER, Stahl ML, McCoy JM (1995) Expression of thioredoxin random peptide libraries on the *Escherichia coli* cell surface as functional fusions

to flagellin: a system designed for exploring protein-protein interactions. BioTechnology 13:366–372

Macek T, Kotrba P, Svatos A, Novakova M, Demnerova K, Mackova M (2008) Novel roles for genetically modified plants in environmental protection. Trends Biotechnol 26:146–152

Macnab RM (2004) Type III flagelar protein export and flagelar assembly. Biochim Biophys Acta—Mol Cell Res 1694:207–217

Majander K, Korhonen TK, Westerlund-Wilkström B (2005) Simultaneous display of multiple foreign peptides in the FliD capping and FliC filament proteins of the *Escherichia coli* flagella. Appl Environ Microbiol 71:4263–4268

Matzapetakis M, Ghosh D, Weng TC, Penner-Hahn JE, Pecoraro VL (2006) Peptidic models for the binding of Pb(II), Bi(III) and Cd(II) to mononuclear thiolate binding sites. J Biol Inorg Chem 11:876–890

Maurer J, Jose J, Meyer TF (1997) Autodisplay: one-component system for efficient surface display and release of soluble recombinant proteins from *Escherichia coli.* J Bacteriol 179:794–804

Maurer J, Jose J, Meyer TF (1999) Characterization of the essential transport function of the AIDA-I autotransporter and evidence supporting structural predictions. J Bacteriol 181:7014–7020

Mazmanian SK, Liu G, Jensen ER Lenoy E, Schneewind O (2000) *Staphylococcus aureus* sortase mutant defective in the display of surface proteins in the pathogenesis of animal infection. Proc Natl Acad Sci U S A 97:5510–5515

Mergeay M, Monchy S, Vallaeys T, Auquier V, Benotmane A, Bertin P, Taghavi S, Dunn J, van der Lelie D, Wattinez R (2003) *Ralstonia metallidurans,* a bacterium specifically adapted to toxic metals: towards a catalogue of metal-responsive genes. FEMS Microbiol Rev 27:385–410

Mesnage S, Tosi-Couture E, Mock M, Fouet A (1999a) The S-layer homology domain as a means for anchoring heterologous proteins on the cell surface of *Bacillus anthracis.* J Appl Microbiol 87:256–260

Mesnage S, Weber-Levy M, Haustant M, Mock M, Fouet A (1999b) Cell surface-exposed tetanus toxin fragment C produced by recombinant *Bacillus anthracis* protects against tetanus toxin. Infect Immun 67:4847–4850

Mesnage S, Fontaine T, Mignot T, Delepierre M, Mock M, Fouet A (2000) Bacterial SLH domain proteins are non-covalently anchored to the cell surface via a conserved mechanism involving wall polysaccharide pyruvylation. EMBO J 19:4473–4484

Morgan WT (1985) The histidine-rich glycoprotein of serum has a domain rich in histidine, proline, and glycine that binds heme and metals. Biochemistry 24:1496–1501

Muenger K, Germann UA, Lerch K (1985) Isolation and structural organization of the *Neurospora crassa* copper metallothionein gene. EMBO J 4:2665–2668

Newton SM, Klebba PE, Michel V, Hofnung M, Charbit A (1996) Topology of the membrane protein LamB by epitope tagging and a comparison with the x-ray model. J Bacteriol 178:3447–3456

Nguyen TN, Gourdon MH, Hansson M, Robert A, Samuelson P, Libon C, Andréoni C, Nygren PA, Binz H, Uhlén M, Ståhl S (1995) Hydrophobicity engineering to facilitate surface display of heterologous gene products on *Staphylococcus xylosus.* J Biotechnol. 42:207–219

Nomellini JF, Duncan G, Dorocicz IR, Smit J (2007) S-layer-mediated display of the immunoglobulin G-binding domain of streptococcal protein G on the surface of *Caulobacter crescentus*: development of an immunoactive reagent. Appl Environ Microbiol 73:3245–3253

Oggioni MR, Medaglini D, Maggi T, Pozzi G (1999) Engineering the Gram-positive cell surface for construction of bacterial vaccine vectors. Methods 19:163–173

Pallesen L, Poulsen LK, Christiansen G, Klemm P (1995) Chimeric FimH adhesin of type fimbriae: a bacterial surface display system for heterolofous sequences. Microbiology 141:2839–2848

Parsons LM, Grishaev A, Bax A (2008) The periplasmic domain of TolR from *Haemophilus influenzae* forms a dimer with a large hydrophobic groove: NMR solution structure and comparison to SAXS data. Biochemistry 47:3131–3142

Patel J, Zhang Q, McKay RML, Vincent R, Xu Z (2010) Genetic engineering of *Caulobacter crescentus* for removal of cadmium from water. Appl Biochem Biotechnol 160:232–243

Pazirandeh M, Wells BM, Ryan RL (1998) Development of bacterium-based heavy metal biosorbents: enhanced uptake of cadmium and mercury by *Escherichia coli* expressing a metal binding motif. Appl Environ Microbiol 64:4068–4072

Pazirandeh M, Mauro JM (2001) Production and cellular localization of functional oligomeric peptides in *E. coli:* expression of the *N. crassa* polymetallothionein. Colloid Surf Physicochem Eng Asp 177:197–202

Pearson RG (1963) Hard and soft acids and bases. J Am Chem Soc 85:3533–3539

Petros AK, Reddi AR, Kennedy ML, Hyslop AG, Gibney BR (2006) Femtomolar Zn(II) affinity in a peptide-based ligand designed to model thiolate-rich metalloprotein active sites. Inorg Chem 45:9941–9958

Piard JC, Hautefort I, Fischetti VA, Ehrlich SD, Fons M, Gruss A (1997) Cell wall achoring of the *Streptococcus pyogenes* M6 protein in various lactic acid bacteria. J Bacteriol 179:3068–3072

Pohlner J, Halter R, Beyreuther K, Meyer TF (1987) Gene structure and extracellular secretion of *Neisseria gonorrhoeae* IgA protease. Nature 325:458–462

Qin J, Song L, Brim H, Daly MJ, Summers AO (2006) Hg(II) sequestration and protection by the MerR metal-binding domain (MBD). Microbiology 152:709–719

Raha AR, Varma NR, Yusoff K, Ross E, Foo HL (2005) Cell surface display system for *Lactococcus lactis:* a novel development for oral vaccine. Appl Microbiol Biotechnol 68:75–81

Ramirez BE, Malmström BG, Winkler JR, Gray HB (1995) The currents of life: the terminal electron-transfer complex of respiration. Proc Natl Acad Sci 92:11949–11951

Reedy CJ, Gibney BR (2004) Heme protein assemblies. Chem Rev 104:617–649

Rice JJ, Schohn A, Bessette PH, Boulware KT, Daugherty PS (2006) Bacterial display using circularly permuted outer membrane protein OmpX yields high affinity peptide ligands. Protein Sci 15:825–836

Ruiz ON, Daniell H (2009) Genetic engineering to enhance mercury phytoremediation. Curr Opin Biotechnol 20:213–219

Rulíšek L, Havlas Z (2000) Theoretical studies of metal ion selectivity. 1. DFT calculations of interaction energies of amino acid side chains with selected transition metal ions (Co^{2+}, Ni^{2+}, Cu^{2+}, Zn^{2+}, Cd^{2+}, Hg^{2+}). J Am Chem Soc 122:10428–10439

Rulíšek L, Havlas Z (2002) Theoretical studies of metal ion selectivity. 2. DFT calculations of the affinity of metal-binding sites in metalloproteins for selected transition metal ions (Co^{2+}, Ni^{2+}, Cu^{2+}, Zn^{2+}, Cd^{2+}, Hg^{2+}). J Phys Chem A 106:3855–3866

Rulíšek L, Havlas Z (2003) Theoretical studies of metal ion selectivity. 3. A theoretical design of the most specific combinations of functional groups representing amino acid side chains for the selected metal ions (Co^{2+}, Ni^{2+}, Cu^{2+}, Zn^{2+}, Cd^{2+}, and Hg^{2+}). J Phys Chem B 107:2376–2385

Rulíšek L, Vondrášek J (1998) Coordination geometries of selected transition metal ions (Co^{2+}, Ni^{2+}, Cu^{2+}, Zn^{2+}, Cd^{2+}, Hg^{2+}) in metalloproteins. J Inorg Biochem 71:115–127

Saffar B, Yakhchali B, Arbabi M (2005) Enhanced bioadsorption of cadmium and nickel by *E. coli* displaying a metal-binding motif using CS3 fimbriae. Iran J Biotechnol 3:180–185

Saffar B, Yakhchali B, Arbabi M (2007) Development of bacterial surface display of hexahistidine peptide using CS3 pili for bioaccumulation of heavy metals. Curr Microbiol 55:273–277

Saleem M, Brim H, Hussain S, Arshad M, Leigh MB, Zia-ul-hassan (2008) Perspectives on microbial surface display in bioremediation. Biotechnol Adv 26:151–161

Samuelson P, Hansson M, Ahlborg N, Andréoni C, Götz F, Bächi T, Nguyen TN, Binz H, Uhlén M, Ståhl S (1995) Cell surface display of recombinant proteins on *Staphylococcus carnosus.* J Bacteriol 177:1470–1476

Samuelson P, Wernérus H, Svenberg M, Ståhl S (2000) Staphylococcal surface display metal-binding polyhistidyl peptides. Appl Environ Microbiol 66:1243–1248

Schirmer T, Keller TA, Wang Y-F, Rosenbusch JP (1995) Structural basis for sugar translocation through maltoporin channels at 3.1 Å resolution. Science 267:512–514

Schultheiss E, Paar C, Schwab H, Jose J (2002) Functional esterase surface display by the autotransporter pathway in *Escherichia coli.* J Mol Catal 18:89–97

Shibasaki S, Maeda H, Ueda M (2009) Molecular display technology using yeast—arming technology. Anal Sci 25:41–49

Shimazu M, Mulchandani A, Chen W (2001) Simultaneous degradation of organophosphorus pesticides and p-nitrophenol by a genetically engineered *Moraxella* sp. with surface-expressed organophosphorus hydrolase. Biotechnol Bioeng 76:318–324

Shimazu M, Nguyen A, Mulchandani A, Chen W (2003) Cell surface display of organophosphorus hydrolase in *Pseudomonas putida* using an ice-nucleation protein anchor. Biotechnol Prog 19:1612–1614

Silver S, Phung LT (2005) A bacterial view of the periodic table: genes and proteins for toxic metal ions. J Ind Microbiol Biotechnol 32:587–605

Smith GP (1985) Filamentous fusion phage: novel expression vectors that display cloned antigens on virion surface. Science 228:1315–1317

Song D, Gu Q (2009) Surface expression of *Helicobacter pylori* urease subunit B gene E fragment on *Lactococcus lactis* by means of the cell wall anchor of *Staphylococcus aureus* protein A. Biotechnol Lett 31:985–989

Song L, Caguiat J, Li Z, Shokes J, Scott RA, Olliff L, Summers AO (2004) Engineered single-chain, antiparalel coiled coil mimics the MerR metal binding site. J Bacteriol 186:1861–1868

Sousa C, Cebolla A, de Lorenzo V (1996) Enhanced metallosorption of bacterial cells displaying poly-His peptides. Nat Biotechnol 14:1017–1020

Sousa C, Kotrba P, Ruml T, Cebola A, de Lorenzo V (1998) Metallosorption by *Escherichia coli* cells displaying yeast and mammalian metallothioneins anchored to the outer membrane protein LamB. J Bacteriol 180:2280–2284

Steidler L, Remaut E, Fiers W (1993) LamB as a carrier molecule for the functional exposition of IgG-binding domains of the *Staphylococcus aureus* protein A at the surface of *Escherichia coli* K12. Mol Gen Genet 236:187–192

Steidler L, Viaene J, Fiers W, Remaut E (1998) Functional display of a heterologous protein on the surface of *Lactococcus lactis* by means of the cell wall anchor of *Staphylococcus aureus* protein A. Appl Environ Microbiol 64:342–345

Stentebjerg-Olesen B, Pallesen L, Jensen LB, Christiansen G, Klemm P (1997) Authentic display of a cholera toxin epitope by chimeric type 1 fimbriae: effects of insert positions and host background. Microbiology 143:2027–2038

Su GT, Brahmbhatt H, Wehland J, Rohde M, Timmis KN (1992) Construction of stable LamB-shiga toxin B subunit hybrids: analysis of the expression in *Salmonella typhymurium* aroA strains and stimulation of B subunit-specific mucosal and serum antibody responses. Infect Immun 60:3345–3359

Sugawara E, Nikaido H (1994) OmpA protein of *Escherichia coli* outer membrane occurs in open and closed channel forms. J Biol Chem 269:17981–17987

Sugawara E, Steiert M, Rouhani S, Nikaido H (1994) Secondary structure of the outer membrane proteins OmpA of *Escherichia coli* and OprF of *Pseudomonas aeruginosa.* J Bacteriol 178:3037–6069

Sugawara E, Steiert M, Rouhani S, Nikaido H (1996) Secondary structure of the outer membrane proteins OmpA of *Escherichia coli* and OprF of *Pseudomonas aeruginosa.* J Bacteriol 178:6067–6069

Sylvestre M, Macek T, Mackova M (2009) Transgenic plants to improve rhizoremediation of polychlorinated biphenyls (PCBs). Curr Opin Biotechnol 20:242–247

Taylor IM, Harrison JL, Timmis KN, O'Connor CD (1990) The TraT lipoprotein as a vehicle for the transport of foreign antigenic determinants to the cell surface of *Escherichia coli* K12: structure--function relationships in the TraT protein. Mol Microbiol 4:1259–1268

Transkanen J, Korhonen TK, Westerlund-Wilkström B (2000) Flagellar filaments carying two foreign adhesive peptides. Appl Environ Microbiol 66:4152–4156

Trivedi S, Ueki T, Yamaguchi N, Michibata H (2003) Novel vanadium-binding proteins (vanabins) identified in cDNA libraries and the genome of the ascidian *Ciona intestinalis.* Biochim Biophys Acta 1630:64–70

Ueki T, Takemoto K, Fayard B, Salomé M, Yamamoto A, Kohara H, Susini J, Scippa T, Uyama T, Michibata H (2002) Scanning x-ray microscopy of living and freeze-dried blood cells in two vanadium-rich ascidian species, *Phallusia mammillata* and *Ascidia sydneiesis samea.* Zool Sci 19:27–35

Ueki T, Adachi T, Kawano S, Aoshima M, Yamaguchi N, Kanamori K, Michibata H (2003a) Vanadium-binding proteins (vanabins) from a vanadium-rich ascidian *Ascidia sydneiesis samea.* Biochim Biophys Acta 1626:43–50

Ueki T, Sakamoto Y, Yamaguchi N, Michibata H (2003b) Bioaccumulation of copper ions by *Escherichia coli* expressing vamabin genes from the vanadium-rich ascidian *Ascidia sydneiensis samea.* Appl Environ Microbiol 69:6442–6446

Umelo-Njaka E, Nomellini JF, Bingle WH, Glasier LG, Irvin RT, Smit J (2001) Expression and testing of Pseudomonas aeruginosa vaccine candidate proteins prepared with the *Caulobacter crescentus* S-layer protein expression system. Vaccine 19:1406–1415

Umelo-Njaka E, Bingle WH, Borchani F, Le KD, Awram P, Blake T, Nomellini JF, Smit J (2002) *Caulobacter crescentus* synthesizes an S-layer-editing metalloprotease possessing a domain sharing sequence similarity with its paracrystalline S-layer protein. J Bacteriol 184:2709–2718

Urgun-Demirtas M, Stark B, Pagilla K (2006) Use of genetically engineered microorganisms (GEMs) for the bioremediation of contaminants. Crit Rev Biotechnol 26:145–164

Valls M, Gonzales-Duarte R, Atrian S, de Lorenzo V (1998) Bioaccumulation of heavy metals with protein fusions of metallothionein to bacterial OMPs. Biochimie 80:588–861

Valls M, Atrian S, de Lorenzo V, Fernández LA (2000a) Engineering a mouse metallothionein on the surface of *Ralstonia eutropha* CH34 for immobilization of heavy metals in soil. Nat Biotechnol 18:661–665

Valls M, de Lorenzo V, Gonzáles-Duarte R, Atrian S (2000b) Engineering outer-membrane proteins in *Pseudomonas putida* for enhanced heavy-metal bioadsorption. J Inorg Chem 79:219–223

Vasák M (2005) Advances in metallothionein structure and functions. J Trace Elem Med Biol 19:13–17

Veiga E, Sugawara E, Nikaido H, de Lorenzo V, Fernandez LA (2002) Export of autotransported proteins proceeds through an oligomeric ring shaped by C-terminal domains. EMBO J 21:2122–2131

Veiga E, de Lorenzo V, Fernandez LA (2003a) Autotransporters as scaffolds for novel bacterial adhesins: surface properties of *Escherichia coli* cells displaying Jun/Fos dimerization domains. J Bacteriol 185:5585–5590

Veiga E, de Lorenzo V, Fernandez LA (2003b) Neutralization of enteric coronaviruses with *Escherichia coli* cells expressing single-chain Fv-autotransporter fusions. J Virol 77:13396–13398

Veiga E, de Lorenzo V, Fernandez LA (2004) Structural tolerance of bacterial autotransporters for folded passenger protein domains. Mol Microbiol 52:1069–1080

Vinopal S, Ruml T, Kotrba P (2007) Biosorption of $Cd^{2+}$and Zn^{2+} by cell surface-engineered *Saccharomyces cerevisiae.* Int Biodeter Biodegr 60:96–102

Wang L, Su M, Yu Z (2004) Capacity of *Bacillus thuringiensis* S-layer protein displaying polyhistidine peptides on the cell surface. Appl Biochem Biotechnol 119:133–143

Warren G, Wolber P (1991) Molecular aspects of microbial ice nucleation. Mol Microbiol 5:239–243

Wernerus H, Ståhl S (2002) Vector engineering to improve a staphylococcal surface display system. FEMS Microbiol Lett 212:47–54

Wernerus H, Ståhl S (2004) Biotechnological applications for surface-engineered bacteria. Biotechnol Appl Biochem 40:209–228

Wernerus H, Lehtiö J, Teeri T, Nygren P-A, Ståhl S (2001) Generation of metal-binding staphylococci through surface display of combinatorially engineered cellulose-binding domains. Appl Environ Microbiol 67:4678–4684

Westerlund-Wikström B (2000) Peptide display on bacterial flagella: principles and applications. Int J Med Microbiol 290:223–230

Westerlund-Wilkström B, Transkanen J, Virkola R, Hacker J, Lindberg M, Skurnik M, Korhonen TK (1997) Functional expression of adhesive peptides as fusions to *Escherichia coli* flagellin. Protein Eng 10:1319–1326

Wu C, Mulchandani A, Chen W (2008) Versatile microbial surface-display for environmental remediation and biofuels production. Trends Microbiol 16:181–188

Xu Z, Lee SY (1999) Display of polyhistidine peptides on the *Escherichia coli* cell surface bu using outer membrane protein C as an anchoring motif. App Environ Microbiol 65:5142–5147

Yang C, Cai N, Dong M, Jiang H, Li J, Qiao C, Mulchandani A, Chen W (2008) Surface display of MPH on *Pseudomonas putida* JS444 using ice nucleation protein and its application in detoxification of organophosphates. Biotechnol Bioeng 99:30–37

Zarschler K, Janesch B, Kainz B, Rist R, Messner P, Schäffer C (2010) Cell surface display of chimeric glycoproteins via the S-layer of *Paenibacillus alvei*. Carbohydr Res 345:1422–1431. doi:10.1016/j.carres.2010.04.010

Zhang J, Lan W, Qiao C, Jiang H, Mulchandani A, Chen W (2004) Bioremediation of organophosphorus pesticides by surface-expressed carboxylesterase from mosquito on *Escherichia coli*. Biotechnol Prog 20:1567–1571

Zheng H, Chruszcz M, Lasota P, Lebioda L, Minor W (2008) Data mining of metal ion environments present in protein structures. J Inorg Biochem 102:1765–1776

Chapter 12
Immobilized Biosorbents for Bioreactors and Commercial Biosorbents

Pavel Dostálek

Abstract Biosorbents are materials derived from specific kind of biomass which are cheaper and more effective alternatives for the removal of metal ions from aqueous solution. In this chapter, based on the literatures and our research results, the biosorbents widely used for heavy metal removal were reviewed, mainly focusing on their pretreatment, modification, immobilisation, regeneration, modelling of column bioadsorber and potential applications. Part or chapter is dedicated to application of biosorbents in reactors for removal of metals from aqueous solutions and some large scale applications of biosorbents were reviewed too.

Keywords Biosorbent • Biosorption • Packed bed column • Immobilisation by cross-linking • Immobilisation by entrapment

12.1 Introduction: Requirements for Industrial Biosorbents

Heavy metals are widespread pollutants of great environmental concern as they are non-degradable and thus persistent and contribute to pollution of aquatic system. The degree of treatment may range from a main process for seriously polluted industrial waste to a polishing process for removing the trace concentrations which remain after the main treatment. The conventional processes used for effluent treatment are precipitation as hydroxides/sulphides, oxidation/reduction and ion exchange (Gupta et al. 2000). Microbial or plant biomass on the other hand can passively bind large amounts of metals and biomass can be alternative to conventional treatment of polluted water. A phenomenon commonly referred to as biosorption (Gadd 2001), thus providing a cost-effective solution for industrial wastewater management (Volesky and Holan 1995). However, on prolonged contact with the

P. Dostálek (✉)
Department of Fermentation Chemistry and Bioengineering, Institute of Chemical Technology Prague, Technicka 3, 166 28 Prague, Czech Republic
e-mail: pavel.dostalek@vscht.cz

P. Kotrba et al. (eds.), *Microbial Biosorption of Metals*,
DOI 10.1007/978-94-007-0443-5_12, © Springer Science+Business Media B.V. 2011

metalbearing solution, the living biomass is also able to sequester metal intracellularly by an active process called bioaccumulation. Biosorption is possible by both living and nonliving biomass; however, bioaccumulation is mediated only by living biomass (Gupta et al. 2000). Biosorption by different part of the cell can occur via: complexation, coordination, chelatation of metals, ion exchange, adsorption and inorganic microprecipitation (Volesky 1990). An early reference to this mechanism was cited by Rothstein et al. in 1948 (Volesky 1990). They presented evidence that uranium is sorbed at the yeast cell surface and this mechanism is not dependent on glucose metabolism (Rothstein and Larrabee 1948; Rothstein et al. 1948).

Biomaterials which are able to remove metals from water by uptake are called after treatments biosorbents. These biosorbents are produced from biomass microbial or plant origin. The right choice the biomass for formulation of biosorbent is essential. As the biosorption process involves mainly cell surface sequesteration, cell wall modification can greatly alter the binding of metal ions. A number of methods have been employed for cell wall modification of microbial cells in order to enhance the metal binding capacity of biomass and to elucidate the mechanism of biosorption. These modifications can be introduced either during the growth of a microorganism or in the pregrown biomass. The condition in which micro-organisms grow affects its cell surface phenotype which in turn affects its biosorption potential (Gadd 1990). Work has been done on the effect of culture conditions of cells on their biosorptive capacity. Dostalek et al. (2004) investigated biosorption of Cd^{2+}, Cu^{2+} and Ag^{+} ions by C-, N-, P-, S-, Mg- and K-limited cells of *S. cerevisiae*. The binding capacity of yeast cells for cadmium decreases in the order: K-limited ≥ Mg-limited ≅ C-limited > N-limited ≅ S-limited > P-limited. For Ag^{+} ion: P-limited > K-limited > C-limited ≥ N-limited ≅ Mg-limited > S-limited. For copper ion: K-limited > Mg-limited ≥ C-limited > N-limited ≅ P-limited > S-limited (Dostalek et al. 2004). Pictures of individual yeast cells grown under specific elemental limits which were captured by light microscope are shown in Fig. 12.1. Yeast cells cultivated in chemostat under nitrogen limit are elongated in shape. This shape of yeast was detected also by other authors during cultivation in media poor in nitrogen (Gimeno et al. 1992). Change in shape of yeast is reaction of yeast to environment conditions (nitrogen limit) and increase of surface cell area to their volume ratio. This increase of cell surface improves of cell uptake nutrition (Gimeno et al. 1992). Pregrown biomass could be given several physical and chemical treatments to tailor the metal-binding properties of biomass to specific requirements. The physical treatments include heating/boiling, freezing/thawing, drying and lyophilisation. Drying of the biosorbent is mentioned in many related works as a stage preceding the biosorption/desorption cycle (Crist et al. 1981, 1990, 1994). In the case of the algae, drying of the fresh biomass allows for proper storage and enhances its performance for the sorption of heavy metals (Rocha et al. 2006). The results of the analysis evidenced structural changes in the *Sargassum* sp. marine algae like shrinkage and porosity reduction, due to the drying process. Morphological analysis of *Sargassum* sp. algae showed pore cavities filled with cellular material, which tends to be removed during drying. After the cellular material removal, some cavi-

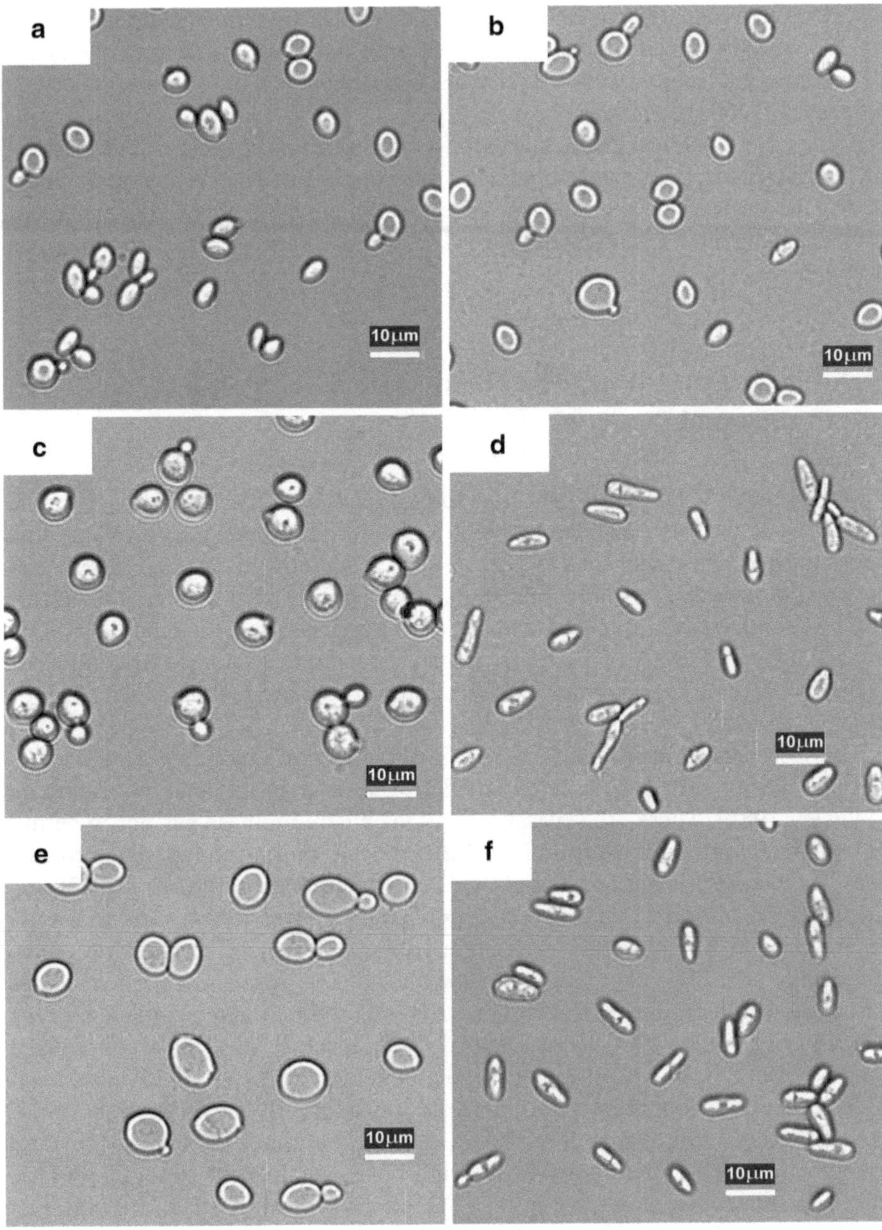

Fig. 12.1 Microphotographs of carbon (**a**), phosphorus (**b**), potassium (**c**), nitrogen (**d**), magnesium (**e**) and sulphur (**f**) limited cells *Saccharomyces cerevisiae* captured by light microscope

ties adhered to each other and collapsed, which evidences shrinkage during drying. For the mild drying conditions the cavities were not altered. The results indicate that volume and consequently area changes are independent of conditions of drying (Rocha et al. 2006).

The various chemical treatments used for biomass modification include washing the biomass with detergents and alkali or acid treatment. The pretreatments could modify the surface characteristics/groups either by removing or masking the groups or by exposing more metalbinding sites (Vieira and Volesky 2000; Wang and Chen 2009).

12.2 Development of Biosorbent Materials: Methods of Immobilisation

The main requirement of an industrial sorption system is that the sorbent can be utilized as a fixed or expanded bed and it should not cause much pressure drop across the bed. This will necessitate some degree of pretreatment, sizing, pelleting, chemical modification or immobilization. These are aimed at obtaining a suitable structure for use in a bed reactor and may enhance metal-specific binding sites. In order to retain the ability of microbial biomass to sorb metals during the continuous industrial process, it is important to utilize an appropriate immobilization technique (Gupta et al. 2000).

The free cells can provide valuable information in laboratory experimentation but are not suited for column packing in industrial applications (Ross and Townsley 1986). There are several exceptions where biomass has not to be immobilized and use for biosorption only after drying, milling and grading process. The first example is sorbent based on *Sargassum* biomass (Volesky 2003) and the second is *Azolla filiculoides* biomass (Fogarty et al. 1999). The free cells generally have low mechanical strength and small particle size and excessive hydrostatic pressures are required to generate suitable flow rates. High pressures can cause disintegration of free biomass. These problems can be avoided by the use of immobilized cell systems (Gadd 1990; Leusch et al. 1995). Immobilized biomass offers many advantages including better reusability, high biomass loading and minimal clogging in continuous flow systems (Holan and Volesky 1994; Gourdon et al. 1990; Gupta et al. 2000). Immobilisation of the biomass in solid structures creates a material with the right size, mechanical strength, rigidity and porosity necessary for use in unit operations typical of chemical engineering. Various techniques are used for the biomass immobilisation. The principal techniques found in the literature for the application of biosorption are based on adsorption on inert supports, on entrapment in a polymeric matrix, on covalent bonds to vector compounds, or on cells cross-linking (Veglio and Beolchini 1997; Wang and Chen 2006, 2009).

The first application of biosorption process was focused to removal of ^{239}Pu from water by activated sludge (Ruchhoft 1949; Volesky 1990). One from the first real biosorbents was constructed by scientists of Czech origin (Nemec et al. 1977). Reinforced fungal mycelia for use as sorbents to separation metal ions, especially U from solution, were prepared by crosslinking mycelium biomass with a formaldehyde or high-mol-weight material. For example, mycelium of *Penicillium chrysogenum* was crosslinked with HCHO, heat cured, and crushed to a grain size of 0.3–0.75 mm; the capacity of the sorbent for U was 95.7 mg/g.

Applicability of biosorbents in practice may also depend on the ease of recovery of bound metals from biosorbent and its possible recovery for further use. The choice of appropriate methods for desorption of the metal is strongly influenced by the mechanism of accumulation. Metabolism independent biosorption is often reversible and the bound metal can be desorbed suitable non-destructive method and the biomass can then be used multiple times. The main desorption agents are diluted inorganic acids and bases, chelating agents and strong salt solutions. However, when the metal is accumulated by the participation of intracellular metabolism, it is usually necessary to use more drastic methods that lead to the destruction of biosorbent (burning or elution using strong acid or alkali). This path can sometimes be economically feasible in the case of obtaining metals from waste cheap biomass (Gadd 2001; Tsezos 1984).

Dilute inorganic acids (e.g. hydrochloric acid, nitric acid) are commonly used agents for desorption of cadmium, copper and other metals on biomass. Metals are eluted with high yield, but using a higher concentration of acid or a longer period of time can damage the biomass, thereby worsening the accumulation of biomass in successive sorption (Tsezos 1984; De Rome and Gadd 1987). With dilute hydrochloric acid (0.1 M) can be removed from the mycelium fungus *Trichoderma viride* over 99% of bound copper, but subsequent accumulation of copper is reduced by about 40% (Townsley et al. 1986). For Cu desorption from live yeast *Saccharomyces cerevisiae* can be used hydrochloric acid (1 M) or a mixture of acids (acetic, lactic, nitric). In both cases, copper is almost completely removed from the biomass, but there is always a subsequent reduction in binding capacity of cells (Junghans and Straube 1991). Silver ions linked to a modified biomass of *Aspergillus niger* can be completely removed with dilute nitric acid (0.1 M) and regenerated biosorbent washing solution containing calcium and magnesium ions. In subsequent uses virtually no loss of binding capacity of biomass (Naseem et al. 1995). Sorbed metals (like copper, chromium, nickel, zinc, cadmium and cobalt) in alga *Chlorella vulgaris* can be quantitatively removed from the algae solution by acidification to pH 2 (Volesky 1990).

Very good and frequently used elution agents are various organic chelating agents. Using ethylenediaminetetraacetic acid (EDTA) can very effectively remove lead and zinc ions from the dead biomass *Streptoverticillium cinnamoneum,* but the recovery is reduced the subsequent sorption of about 20–30% (Puranik and Paknikar 1997). Quite successful, this reagent can also be used for the metals desorption from the activated sludge, where only intracellularly bound metals are not removed (Lawson et al. 1984). When this reagent was used for the desorption of copper from live yeast *Saccharomyces cerevisiae,* only about 60% of bound copper was released (Junghans and Straube 1991). The good chelating agents are further nitrilotriacetic acid (NTA) and less common diethylenetriaminopentaacetic acid (DTPA) (Gadd 2001). NTA is suitable for desorption of cadmium, zinc and copper, but for other metals is not so suitable agent. NTA recovered 70–95% of bound cadmium from immobilized bacteria *Zoogloea ramigera* depending on the amount of accumulated (Kuhn and Pfister 1989). Merkaptoethanol was successfully applied for the selective desorption of gold and mercury from the alga *Chlorella vulgaris* (Darnall et al. 1986). This substance at pH 9 or higher can also be used for the elution of silver from biosorbent based on *Spirulina platensis, Chlorella vulgaris* or *Chlorella pyronoidosa* (Volesky 1990).

Salt solutions should be used in many cases for non-destructive desorption of metal. Carbonates and bicarbonates (ammonium and sodium) or their mixtures are very effective agents for desorption of uranium from *Penicillium* sp., *Chlorella vulgaris, Rhizopus arrhizus* or *Saccharomyces cerevisiae* (Strandberg et al. 1981; Siegel et al. 1990; De Rome and Gadd 1991). Sometimes even regenerated biomass (such as *Penicillium* sp.) exhibits increased binding capacity for accumulated metals (Strandberg et al. 1981). Calcium chloride solution (0.1 M) can be very successfully applied for desorption of cobalt in bound algae *Sargassum natans*. The metal is eluted with around 98% effectiveness and the loss of binding capacity of the biomass is almost negligible (Volesky 1990). We can use for desorption some other salts such as sulfates (Gadd 2001).

If we are suming all claims which are applicable for the biosorbent used for the recovery of metals from solution we can stress the following criteria (Volesky 1990):

- The uptake and release of the metal should be efficient and rapid.
- The active biosorbent agent should be produced at low cost and should be reusable.
- The biosorbent material should have desirable particle size, shape and mechanical properties for use in continuous flow (mixed, packed bed, fluidized bed) systems.
- Removal of the biosorbent from solution should be cheap, efficient and rapid.
- The sorbent should be metal-selective, if desirable, in order to separate specific single metals from a solution containing various metallic species.
- Separation of metal from the sorbent should be metal-selective (if desirable) and economically feasible and loss of the sorbent should be minimal.

12.2.1 Creation of Biofilm

The simplest form of biomass immobilisation is creation of biofilms on the surface of inert materials. Biofiltration represents one of the most important steps in drinking water treatment. Biofiltration enhances biological substrate removal by providing a surface for the attachment and growth of indigenous water microorganisms; this is common for all kinds of biofilters used in drinking water treatment such as slow and rapid sand filters and granular activated carbon (GAC) filters. All of these processes depend on biofilms which have to mature on the filter surfaces (Flemming 2000). The structure of a mature biofilm in a drinking water biofilter is complex and characterized by strong heterogeneity. It is maintained by the EPS (extracellular polymeric substances) which keep the microorganisms together and attach them to the filter material surface. EPS can represent the major part of the organic biomass. Also, they sorb metal ions as well as organic and in particular humic substances which contribute to the brown or black color of the biofilm (Flemming 2000).

There are several applications directly for treatment of waste waters. New system which was developed is called MERESAFIN (The Metal Removal by Sand Filter Inoculation). The system is based on a moving bed sand filter. A biofilm is formed on the sand grains after inoculation with heavy metalresistant bacteria able to biosorb or to bioprecipitate heavy metals. Passage of the wastewater over these biofilms leads to the binding of the metals to the biofilm and consequently the removal of the metals from the wastewater. The metal-laden biofilm is removed from the sand grains in a sand washer created by an airlift for the continuous movement of the filter bed. The metalloaded biomass is separated from the sand in a labyrinth on the top of the sand washer. Nutrients and a carbon source are provided continuously in the system in order to promote the regrowth of the biofilm on the sand grains (Diels et al. 2004; Finlay et al. 2003; Puempel et al. 2001, 2003).

12.2.2 Development of Biosorbent Particles

Type of technique applied for production of biosorption particles depends on type of biomass which we used. Generally some plants and their parts have not so asks for immobilisation. Small single cells on the other hand have to immobilize for their purpose in contactor. There are two techniques use the most frequent for formulation of biosorbents: cross-linking and entrapment in polymeric matrices.

The granulation technology of choice must also poses a number of characteristics which results in a stable, active biosorbent at reasonable cost. The desirable properties for a granulated biosorbent include (Brierley 1990):

- Maximum amount of biomass with minimal quantities of binding agents present
- Excellent porosity—binding material should not diminish porosity or adversely affect biomass metal-binding sites that reside within the granule
- Chemical resistance—the binding technology should produce a granule which remains stable in a wastewater that fluctuates from acidic to alkaline pH extremes.
- Structural integrity—the granule must possess the physical stability for prolonged use in fluid bed contactors and passage through pumps
- Cost—the binding technology must be accomplished at low cost to keep the process competitive with other available wastewater treatment systems.

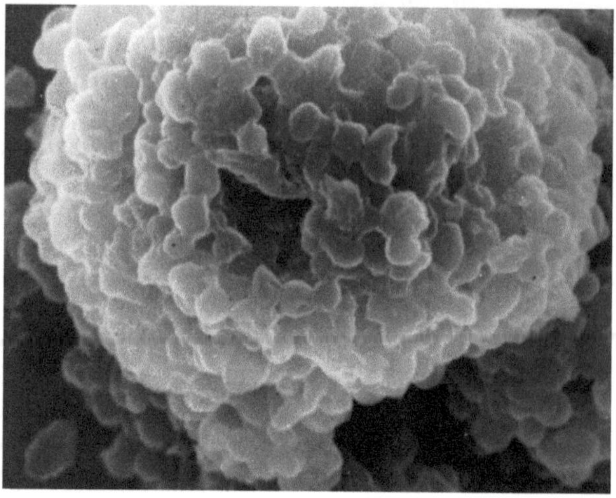

Fig. 12.2 Microphotograph yeast cells immobilised by glutaric dialdehyde (scanning electron microscopy)

12.2.2.1 Immobilisation by Cross-Linkage

The addition of a cross-linker leads to the formation of stable cellular aggregates. This technique was found to be used above all to immobilise algae (Veglio and Beolchini 1997). The most common cross-linkers are: formaldehyde, glutaric dialdehyde (Fig. 12.2), divinylsulfone and formaldehyde-urea mixtures (Veglio and Beolchini 1997). Two types of magnetic biosorbent were prepared by novel protocols from epichlorhydrin-cross-linked *Saccharomyces cerevisiae* cell walls and their biosorption characteristics were compared to those of nonmagnetic cell walls. The magnetic biosorbents I and II were capable of binding Cu^{2+} maximally to 225 and 50 m mol/g, Cd^{2+} to 90 and 25 m mol/g and Ag^{+} to 80 and 45 m mol/g respectively. These values compare with 400, 125 and 75 m mol/g, respectively, for non-magnetic cell walls (Patzak et al. 1997). Cross-linking almost always decreases sorption capacity of used materials, this fact is applicable also for immobilized *Azolla filiculoides* by epichlorhydrin (Fogarty et al. 1999).

12.2.2.2 Immobilisation by Entrapment Method

A number of matrices have been employed for immobilization of cells. One of the matrices that has been used in metal recovery by both viable and non-viable cells is the entrapment in the matrix of insoluble Ca-alginate. The other polymers used are: polyacrolylamide, polysulfone, polyethylenimine and polyhydroxoethylmethyacrylate (Gupta et al. 2000). The materials obtained from the immobilisation in calcium alginate and polyacrylamide are in soft form of gel particles. Those obtained from immobilization in polysulfone and polyethylenimine proved be strongest (Gupta et al. 2000). Several new biosorbents were developed (Fig. 12.3) base on immobilised yeast cell wall envelopes in silica matrix by a sol-gel process (Szilva et al. 1998; Marseaut et al. 2004).

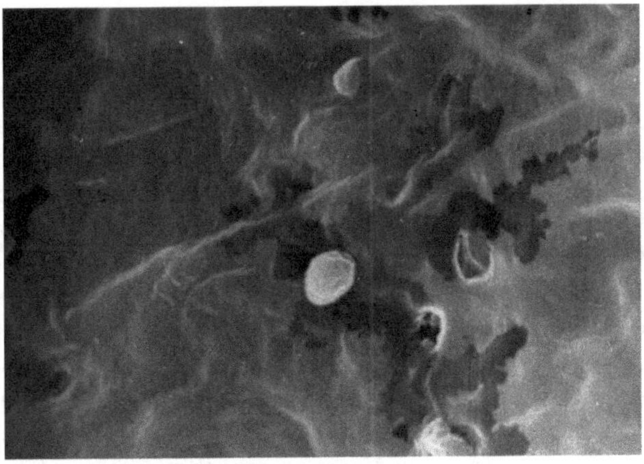

Fig. 12.3 Microphotograph of yeast cell wall envelopes immobilized in silica matrix (scanning electron microscopy)

12.3 Reactors for Biosorption and Desorption Process

The process of metal recovery using biosorbent derived from biomass is basically a solid-liquid contact process consisting of the metal uptake (sequestering) cycle and the metal desorption (elution) cycle. In its technological configuration, it would be very similar to that used in the ion-exchange process or activated carbon applications. The metal solution is contacted with the solid sorbent phase in a batch, semicontinuous, or continuous-flow arrangement. Appropriate contact between the solution and the solid phase can be accomplished routinely in any modification of the following apparatuses (Volesky 1990):

- Batch-stirred tank contactor
- Continuous-flow stirred-tank contactor

Both to be coupled or combined with a solid-liquid separation operation:

- Fixed packed bed contactor
- Pulsating-bed contactor
- Fluidized-bed contactor
- Multiple-bed contact arrangement

12.3.1 Packed Bed Columns

Sorption systems which used fixed packed bed columns have certain advantages in comparison with batch arrangements, because rate a velocity of sorption depends on actual concentration of metal in cleaning solution. If we are use column, biosorbent is continuously in contact with fresh solution. That is why concentration of metal in solution, which is in contact with defined lawyer of biosorbent in column, is relatively stable. In the batch system concentration of metal in solution, which is in contact with certain amount of biosorbent, continuously drop due to metal sorption and thus drop both velocity and effectiveness of biosorption.

Purified solution flows through packed bed of biosorbent, in which take place sorption of metal ions. The conventional flow of sorption or adsorption for fixed bed column is shown in Fig. 12.4. During first phase of process metal is very quickly sorbed by top layers of biosorbent. These layers are in contact with solution which has the highest possible concentration—initial C_0. Small amount of metal which was not bounded inside of first several layers of biosorbent granules is removed from solution in following layers of biosorbent. It is evident, that at the start stage of process almost no metal ions are present in column outflow. In these moments it is called primary adsorption zone located near the top of the column. As purified solution flows through the column, the upper layer of the adsorbent is saturated with metal and gradually become ineffective for further metal binding.

Thus, the primary adsorption zone through the column inches down to where the sorbent is not bound. Shift the primary adsorption zone is generally much slower than the flow of purified solution column. As the primary adsorption zone moves downward toward the end of the biosorbent layer, more and more metal tends to escape out of the eluate (as shown in Fig. 12.4). With increasing time or runoff volume of solution increases the ratio c/C_0 until we get the curve to the "breakthrough point. That in all cases is the stage of practical operation, in which the column is in

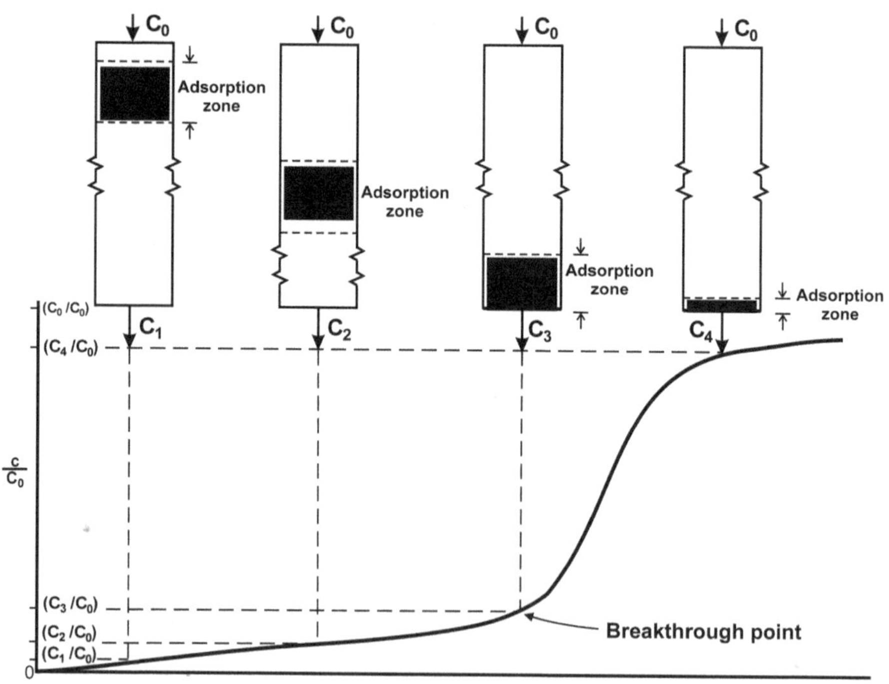

Fig. 12.4 Course of biosorption in packed bed column (x-axis: concentration, y-axis: volume or time)

equilibrium with the purified solution. After this point, it leads to rapid saturation of the remaining metal biosorbents and effluent from the column. Therefore, at this point is mostly regenerated sorbent or change for new one.

Among the factors that influence the course of sorption in packed column belong type of metal, metal concentrations in the treated solution, pH, adsorbent particle size, type of sorbent, sorption kinetics, equilibrium conditions, the flow hydrodynamics solution through the column, the column diameter, height of the adsorbent layer and the flow of purified solution column. Generally, the time needed to reach the breakthrough point is reduced by increased particle size biosorbents, elevated concentrations of metals in cleaning solution, other than optimum pH for biosorption, increasing flow, reducing the height of the layer biosorbents.

If the total height of layer biosorbent in the column is less than the length of primary adsorption zone necessary for the effective removal of metal from the solution, then metal concentration in effluent from the column will increase sharply from the moment when the first eluate leaving the column. For each system column-biosorbent-metal so there is a critical minimum value of the height of layers in the column (Weber 1972).

There are many mathematical models used for computer simulation of column experiments (Volesky 2003). Here is example from our work (Marseaut et al. 2004) which is possible to use for prediction of break-through point. The kinetic and equilibrium data obtained in batch experiments were used for formulation of a mathematical model and computer simulation of adsorption in column. The mathematical model describing the adsorption column dynamics is based on differential balances of metal in solid and liquid phase. We assume the laminar flow regime with axial dispersion controlled by molecular diffusion in a column contractor. It is suggested (Levenspiel 1962) that the effect of axial mixing on the packed bed reactor dynamic could characterize the dimensionless mass Peclet number thus the mathematical model of a fixed bed adsorber can be written in the following form (Marseaut et al. 2004):

Metal balance in liquid phase including interphase mass transfer and axial mixing:

$$E\frac{\partial C}{\partial t} = \frac{F}{V}\left(\frac{d_p}{L}\frac{1}{Pe_p}\frac{\partial C^2}{\partial \xi^2} - \frac{\partial C}{\partial \xi}\right) - \frac{G}{V}\beta(C - C_f)$$

Metal balance in solid phase:

$$\frac{\partial q}{\partial t} = \beta(C - C_f)$$

Initial conditions: $t = 0$: $\xi \in \langle 0,1 \rangle$: $C = 0; q = 0$.

Danckwerts boundary conditions describe the behaviour of liquid phase for the time $t>0$ at the input and entrance of packed bed:

$$\xi = 0: \quad \frac{L}{d_p}Pe_p\left[C^0 - C(0)\right] = -\frac{\partial C(0)}{\partial \xi}$$

$$\xi = 1: \quad \frac{\partial C}{\partial \xi} = 0$$

where

C	actual metal concentration in solution (kg/m^3)
C(0)	concentration of metal at the column entrance influenced by axial diffusion (kg/m^3)
C^0	input metal concentration in solution (kg/m^3)
C_f	metal equilibrium concentration in solution (kg/m^3)
D	coefficient of axial dispersion (m^2/s)
d_p	particle diameter (m)
F	volume flow rate (m^3/s)
G	total mass of the cell walls in an adsorber (silica matrix does not exhibited a sorption activity) (kg)
l	axial length coordinate in fixed bed (m)
L	length of the column (m)
n	number of discretisation points in PDE solver
Pe_p	Peclet number based on particle diameter $Pe_p = \frac{u \cdot d_p}{D}(-)$
q	concentration of metal based on cell wall mass q=G/V (Co−Cf) (−)
Q	ratio of metal uptake of immobilized and free cell walls Q=q immobilized./q free (−)
t	time (s)
u	axial velocity of liquid phase based on empty tube (m/s)
V	total volume of packed bed or suspension in batch experiments (m^3)
β	interphase mass transfer coefficient (s^{-1})
ε	void fraction (−)
ξ	<0,1> axial coordinate (−)

Using the backward difference schema as recommended by literature (Sincovec and Madsen 1975), the set of above created partial differential equations were transformed to a set of nx2 ordinary differential equations (ODE). The resulting ODE's were solved numerically by a standard explicit Runge-Kutta method of forth order with a control of integration step using half step estimation of local integration error (Volesky and Votruba 1992). The numerical test of the algorithm has shown that the selection of ten equidistant knots on the length coordinate granted the desired accuracy of simulation results.

Using this mathematical model of packed bed adsorber, taking into account axial dispersion and mass transfer resistance, the break-through curve is possible predict with high fidelity for crushed and moulded biosorbent.

12.3.2 *Fluidized Bed Columns*

The particles of biosorbent are fluidized in the column bed by upwards flowing liquid. The main advantage of this arrangement is that the feed stream does not need to be completely particle free as required for the paced-bed column which would collect the suspended impurities as a filter (Volesky 2003).

The major disadvantage of the fluidized bed system is that it cannot utilize the biosorbent charge to its maximum potential because its content is being mixed. This way the sorption driving force of the metal concentration gradient between the solid and liquid phases is always lower and it is more difficult to achieve polished effluent (Volesky 2003).

In practice we can in a three-phase fluidized bed bioreactor (Keerthy and Hossain 2008) or sequential stage reactors for heavy metal removal (Lee and Yang 2005).

12.4 Application of Biosorption: Commercial Biosorbents

A large amount of researches on metal biosorption have been published to elucidate the principles of this effective metal-concentration phenomenon during the past 30 years. Biosorption is regarded as a potential cost-effective biotechnology for the treatment of high volume low-concentration complex wastewaters containing heavy metals (Wang and Chen 2009). In despite of mentioned facts biosorption is not still present in large scale in practise yet. There are many reasons for this situation. The situation was described very well by Volesky and Naja (2007).

Pilot installations and few commercial scale units were constructed in the USA, Canada and Israel during 1980s and 1990s. In Table 12.1 are summing all known biosorbents which were used in large scale for industrial applications.

Table 12.1 Biosorbents developed for metal bearing wastewater treatment application

Name	Microorganism/plant	Immobilisation matrix	Particle size	References
BIO-FIX U.S. Bureau of Mines (Golden, Colorado)	Cyanobacteria (*Spirulina*) Yeast Algae Plants (*Lemna* sp., *Sphagnum* sp.)	Polyethylene or Polypropylane or Polysulfone in dimethylformide	0.5–2.5 mm	Bennett et al. (1991)
AMT-Bioclaim	*Bacillus subtilis*			Brierley (1990)
AlgaSORB™ Bio-recovery system, Inc. (Las Cruces, New Mexico)	*Chlorella vulgaris*	Silica or polyacrylamide gels		Darnall et al. (1986); Bedell and Darnal (1990)
B.V. Sorbex, Inc. (Montreal, Canada)	*Sargassum natans* *Ascophyllum nodosum* *Halimeda opuntia* *Palmyra pamata* *Chondrus crispus* *Chlorella vulgaris*			Volesky (1990)
Tsezos	*Rhizopus arrhizus* Activated Sludge *P. chrysogenum*	Polymer coating	0.5–1.0 mm	Tsezos et al. (1987)
Azolla Biofilter	*Azolla filiculoides*			Sela and Tel-Or (1989)

References

Bedell GW, Darnall DW (1990) Immobilization of nonviable biosorbent algal biomass for the recovery of metal ions. In: Volesky B (ed) Biosorption of heavy metals. CRC Press, Boca Raton

Bennett PG, Ferguson CR, Jeffers TH (1991) Biosorption of metal contaminants from acidic mine waters. In: Hausen DM, Petruk W, Hagni RD, Vassiliou A (eds) Process mineralogy XI—characterization of metallurgical and recyclable products. The Minerals, Metals & Materials Society, Warrendale, pp 213–222

Brierley JA (1990) Production and application of a Bacillus-based product for use in metals biosorption. In: Volesky B (ed) Biosorption of heavy metals. CRC Press, Boca Raton

Crist RH, Oberholser K, Saank N, Nguyen M (1981) Nature of bonding between metallic ions and algae cell walls. Environ Sci Technol 15:1212–1217

Crist RH, Martin JR, Guptill PW, Eslinger JM (1990) Interaction of metals and protons with algae. 2. Ion exchange in adsorption and metal displacement by proton. Environ Sci Technol 24:337–342

Crist RH, Martin JR, Carr D, Watson JR, Clarke H J (1994) Interaction of metals and protons with algae. 4. Ion exchange vs. adsorption models and a reassessment of scatchard plots; ion-exchange rates and equilibrium compared with calcium alginate. Environ Sci Technol 28:1859–1866

Darnall DW, Greene B, Henzel MT, Hosea JM, McPherson RA, Sneddon J, Alexander MD (1986) Selective recovery of gold and other metal ions from an algal biomass. Environ Sci Technol 20:206–215

De Rome L, Gadd GM (1987) Copper adsorption by *Rhizopus arrhizus, Cladosporium resinae* and *Penicillium italicum.* Appl Microbiol Biotechnol 26:84–90

De Rome L, Gadd GM (1991) Use of pelleted and immobilized yeast and fungal biomass for heavy metal and radionuclide recovery. J Ind Microbiol 7:97–104

Diels L, Spaans PH, Van Roy S, Hooyberghs L, Ryngaert A, Wouters H, Walter E, Winters J, Macaskie L, Dostálek P, Patzak M, Matějka P (2004) Influence of specific growth limitation on biosorption of heavy metals by *Saccharomyces cerevisiae.* Int Biodeterior Biodegrad 54:203–207

Dostalek P, Patzak M, Matejka P (2004) Influence of specific growth limitation on biosorption of heavy metals by *Saccharomyces cerevisiae*. Int Biodeterior Biodegrad 54:203–207

Finlay J, Pernfuss B, Woebking H, Pumpel T, Tsezos M (2003) Heavy metals removal by sand filters inoculated with metal sorbing and precipitating bacteria. Hydrometallurgy 71:235–241

Flemming HC (2000) Biofims in Biofiltration. In: Rehm HJ (ed) Biotechnology, vol 11c. Wiley, Singapore

Fogarty R, Patzak M, Dostálek P, Votruba J, Tel-Or E, Tobin JM (1999) Column studies of Cu^{2+} removal by immobilized and non imobilized Azolla filiculoides. Biotechnol Tech 13:533–538

Gadd GM (1990) Heavy metal accumulation by bacteria and other microorganisms. Experientia 46:834–840

Gadd GM (2001) Accumulation and transformation of metals by microorganisms. In: Rehm HJ (ed) Biotechnology, vol 10. Wiley, Singapore

Gimeno C, Ljungdahl P, Styles CA, Fink GR (1992) Unipolar cell divisions in the yeast *S. cerevisiae* lead to filamentous growth: regulation by starvation and RAS. Cell 68:1077–1090

Gourdon R, Bhende S, Rus E, Sofer SS (1990) Comparison of cadmium biosorption by Gram-positive and Gram-negative bacteria from activated sludge. Biotechnol Lett 12:839–842

Gupta R, Ahuja P, Khan S, Saxena RK, Mohapatra H (2000) Microbial biosorbents: meeting challenges of heavy metal pollution in aqueous solutions. Curr Sci 78:967–973

Holan ZR, Volesky B (1994) Biosorption of lead and nickel by biomass of marine algae. Biotechnol Bioeng 43:1001–1009

Junghans K, Straube G (1991) Biosorption of copper by yeasts. Biol Met 4:233–237

Keerthy G, Hossain SM (2008) Bacterial metal reclamation of ground water in a three-phase fluidized bed bioreactor. Indian J Environ Prot 28:840–845

Kuhn SP, Pfister RM (1989) Adsorption of mixed metals and cadmium by calcium-alginate immobilized *Zoogloea ramigera.* Appl Microbiol Biotechnol 31:613–618

Lawson PS, Sterritt RM, Lester JN (1984) Adsorption and complexation mechanisms of heavy metal uptake in activated sludge. J Chem Technol Biotechnol 34B:253–261

Lee CI, Yang WF (2005) Heavy metal removal from aqueous solution in sequential fluidized—bed reactors. Environ Technol 26:1345–1353

Leusch A, Holan Z, Volesky B (1995) Biosorption of heavy metal metals (Cd, Cu, Ni, Pb, Zn) by chemically-reinforced biomass of marine algae. J Chem Technol Biotechnol 62:279–288

Levenspiel O (1962) Chemical reaction engineering. An introduction to the design of chemical reactors. Wiley, New York

Marseaut S, Debourg A, Dostálek P, Votruba J, Kuncová G, Tobin JM (2004) A silica matrix biosorbent of cadmium. Int Biodeterior Biodegrad 54:209–214

Naseem AM, Sivarama SK, Maruthi MP (1995) Biosorption of silver ions by processed Aspergillus niger biomass. Biotechnol Lett 17:551–556

Nemec P, Prochazka H, Stamberg K, Katzer J, Stamberg J, Jilek R, Hulak P (1977) Process of treating mycelia of fungi for retention of metals. US Patent 4021368.

Patzak M, Dostálek P, Fogarty RV, Šafařík I, Tobin JM (1997) Development of magnetic biosorbents for metal uptake. Biotechnol Tech 11:483–487

Puempel T, Ebner C, Pernfuss B, Schinner F, Diels L, Keszthelyi Z, Stankovic A, Finlay JA, Macaskie LE, Tzesos M, Wouters H (2001) Treatment of rinsing water from electroless nickel plating with a biologically active moving bed sandfilter. Hydrometallurgy 59:383–393

Puempel T, Macaskie L, Finlay JA, Diels L, Tsezos M (2003) Nickel removal from nickel plating waste water using a biologically active moving-bed sand filter. Biometals 16:567–581

Puranik PR, Paknikar KM (1997) Biosorption of lead and zinc from solutions using *Streptoverticillium cinnamoneum* waste biomass. J Biotechnol 55:113–124

Rocha SCS, Cavalcante JA, da Silva MGC, Pinho CG (2006) Influence of the drying conditions of *Sargassum* sp. *Alga* on the bioadsorption of hexavalent chromium. Environ Technol 27:979–990

Ross IS, Townsley CC (1986) The uptake of heavy metals by filamentous fungi. In: Eccles H, Hunt S (eds) Immobilization of ions by biosorption. IRL Press, Chichester

Rothstein A, Larrabee C (1948) The relationship of the cell surface to metabolism. II. The cell surface of yeast as the site of inhibition of glucose metabolism by uranium. J Cell Comp Physiol 32:247–259

Rothstein A, Frenkel A, Larrabee C (1948) The relationship of the cell surface to metabolism. III. Certain characteristics of the uranium complex with cell surface groups of yeast. J Cell Comp Physiol 32:261–274

Ruchhoft CC (1949) The possibilities of disposal of radioactive wastes by biological treatment methods. Sewage Works J 21:877–883

Sela M, Tel-Or E (1989) Process for the removal of metal ions from aqueous systems. EP 333218

Siegel, SM, Galun M, Siegel BZ (1990) Filamentous fungi as metal biosorbents: a review. Water, Air, Soil Pollut 53:335–344

Sincovec RF, Madsen NK (1975) Algorithm 494 PDONE. Solution of system of partial differential equations. ACM Trans Math Softw 1:261–263

Strandberg GW, Shumate SE, Parrott JR (1981) Microbial cells as biosorbents for heavy metals: accumulation of uranium by *Saccharomyces cerevisiae* and *Pseudomonas aeruginosa.* Appl Environ Microbiol 41:237–245

Szilva J, Kuncova G, Patzak M, Dostálek P (1998) The application of a sol-gel technique to preparation of a heavy metal biosorbent from yeast cells. J Sol-Gel Sci Technol 13:289–294

Townsley CC, Ross IS, Atkins AS (1986) Copper removal from a simulated leach effluent using the filamentous fungus *Trichoderma viride.* In: Eccles H, Hunt S (eds) Immobilisation of ions by bio-sorption. Ellis Horwood, Chichester

Tsezos M (1984) Recovery of uranium from biological adsorbents—desorption equilibrium. Biotech Bioeng 26:973–981

Tsezos M, Baird MHI, Shemilt LW (1987) The use of immobilised biomass to remove and recover radium from Elliot lake uranium tailings streams. Hydrometallurgy 17:357–368
Veglio F, Beolchini F (1997) Removal of metals by biosorption: a review. Hydrometallurgy 44:301–316
Vieira RHSF, Volesky B (2000) Biosorption: a solution to pollution? Int Microbiol 3:17–24
Volesky B (1990) Biosorption and biosorbents. In: Volesky B (ed) Biosorption of heavy metals. CRC Press, Boca Raton
Volesky B (2003) Sorption and biosorption. BV Sorbex, Montreal
Volesky B, Holan ZR (1995) Biosorption of heavy metals. Biotechnol Prog 11:235–250
Volesky B, Naja G (2007) Biosorption technology: starting up an enterprise. Int J Technol Commerc 6:194–211
Volesky B, Votruba J (1992) Modeling and optimization of fermentation processes. Elsevier, Amsterdam
Wang JL, Chen C (2006) Biosorption of heavy metals by Saccharomyces cerevisiae: a review. Biotechnol Adv 24:427–451
Wang J, Chen C (2009) Biosorbents for heavy metals removal and their future. Biotechnol Adv 27:195–226
Weber WJ (1972) Adsorption. Wiley, New York

Chapter 13
Magnetically Responsive Biocomposites for Inorganic and Organic Xenobiotics Removal

Ivo Safarik, Katerina Horska and Mirka Safarikova

Abstract Biosorption of both inorganic and organic xenobiotics using a variety of biological materials represents emerging possibility for the reduction of environmental pollution. In order to improve manipulation of biosorbents, their magnetic derivatives can be prepared. This short chapter provides an overview of magnetically responsive biocomposite materials, both in the form of nanoparticles and microparticles, composed of mainly by individual polysaccharides, complex polysaccharides of the plant origin and microbial and algae cells, and their potential applications in xenobiotics removal. An extensive list of described magnetic biocomposites clearly documents both large variability of biomaterials tested and the procedures used to convert them into magnetic form.

Keywords Biosorbents • Magnetic modification • Magnetic separation • Magnetically responsive biocomposites • Xenobiotics removal

13.1 Introduction

Heavy metal pollution is one of the most important environmental problems today. Various industries (*e.g.*, mining, metallurgy, surface finishing, iron and steel, electroplating, electrolysis, photography, metal surface treating, leatherworking and many others) produce and discharge wastes containing different heavy metals. Such a human activity brings about serious environmental pollution, threatening both hu-

I. Safarik (✉)
Department of Nanobiotechnology, Institute of Systems Biology and Ecology, Academy of Sciences, Na Sadkach 7, 370 05 Ceske Budejovice, Czech Republic
e-mail: ivosaf@yahoo.com

I. Safarik
Regional Centre of Advanced Technologies and Materials, Palacky University, Slechtitelu 11, 783 71 Olomouc, Czech Republic

P. Kotrba et al. (eds.), *Microbial Biosorption of Metals*,
DOI 10.1007/978-94-007-0443-5_13,

man health and ecosystem. Three main types of metal contaminants are of concern, including toxic metals (such as Hg, Cr, Pb, Zn, Cu, Ni, Cd, As, Co, Sn, etc.), precious metals (such as Pd, Pt, Ag, Au, Ru etc.) and radionuclides (such as U, Th, Ra, Am, etc.). The metals cannot be degraded to harmless products and hence persist in the environment indefinitely (Wang and Chen 2009).

In addition to inorganic xenobiotics described above, enormous amount of organic xenobiotics can be found in the environment. A great number of industries such as textile, paper and pulp, printing, iron-steel, coke, petroleum, pesticide, paint, solvent, pharmaceutics, wood preserving etc. consume large volumes of organic based chemicals. Such compounds show a great difference in chemical composition, molecular weight, toxicity, etc. Effluents of the above mentioned industries often contain undesired quantities of these pollutants and need to be treated. Especially important is the group of synthetic dyestuffs, used extensively in textile, paper, printing industries and dyehouses. Dyeing industry effluents constitute one of the most problematic wastewaters not only for their high chemical and biological oxygen demands, suspended solids and content in toxic compounds but also for their color. Dyes may significantly affect photosynthetic activity in aquatic life due to reduced light penetration and may also be toxic to some aquatic life due to the presence of aromatics, metals, chlorides, etc., in them (Aksu 2005).

Contamination of water resources by both inorganic and organic xenobiotics, microbial pathogens and parasites is a major problem in the global context, being one of the leading worldwide causes of deaths and diseases. Therefore an enormous interest is being paid to the development of cost effective and efficient materials and processes for the elimination of main pollutants (Safarik and Safarikova 2010b).

Contaminated water is usually treated by physical or chemical treatment processes. These include flocculation combined with flotation, electroflocculation, membrane filtration, electrokinetic coagulation, electrochemical destruction, ion-exchange, irradiation, precipitation, ozonation, and treatment methods involving the use of activated carbon. However, these technologies are often expensive and can exhibit rather low efficiency of xenobiotics removal (Srinivasan and Viraraghavan 2010).

Adsorption processes have often been used for the removal of various xenobiotics from wastewater. Often activated carbon has been found to be effective, but rather expensive. Many studies have been undertaken to investigate the possible use of low-cost adsorbents such as bentonite, steel-plant slag, fly ash, china clay and silica for xenobiotics removal. However, these low-cost materials have generally low adsorption capacities. Therefore, there is a need to find new, economical, easily available and highly effective adsorbents (Srinivasan and Viraraghavan 2010).

Recently, a large number of studies has appeared in the scientific literature describing the applications of various materials of biological origin that can serve as biosorbents for xenobiotics removal from contaminated water resources. Biological materials such as peat, chitosan, alginate, plant gums, yeast, fungi, bacterial and algae biomass, sawdust, spent coffee grounds, peanut husks etc. have been already used as biosorbents to remove different types of organic and inorganic xenobiotics from solutions and suspensions.

Magnetic separation processes have been shown to be very useful for the selective separation and removal of magnetic materials from difficult-to-handle aqueous systems including suspensions containing particulate impurities. In order to simplify the process of biosorbent separation from the treated aqueous solution and/or suspension, the adsorbent can be converted into magnetic form. In this short chapter we provide an overview of magnetically responsive biocomposite materials and their application for the removal of inorganic and organic xenobiotics from water.

13.2 Biosorption

During the 1970s the search for new techniques and materials capable of inexpensive treatment of polluted waters contaminated with organic and inorganic xenobiotics started. This activity resulted in new technologies involving the removal of toxic compounds from treated water due to biosorption. Biosorption may be defined as the removal of target substances from solutions by biological materials. Substances to be removed can be of both organic and inorganic nature, and in gaseous, soluble or insoluble forms. Biosorption is a physico-chemical process independent of metabolism and includes different types of mechanisms such as absorption, adsorption, ion exchange, surface complexation and precipitation. Biosorption is a property of both living and dead organisms, other biological materials, and their components and has been shown as a promising biotechnology for xenobiotics (pollutants) removal from solutions, and/or for pollutants recovery. Most biosorption studies have been carried out using microbial cells (bacteria, microalgae and fungi), however, practically all biological materials such as macroalgae (seaweeds), plant and animal biomass, waste organic sludges, and many other biological wastes or derived bio-products have an affinity for appropriate xenobiotic(s). In many cases the binding capacities of certain biosorbents are fully comparable with those of the commercial synthetic cation exchange resins. However, despite continuing dramatic increases in published research on biosorption and biosorbents, there has been little or no exploitation in an industrial context (Gadd 2009).

The biosorption process involves a solid phase (appropriate type of biosorbent) and a liquid phase (usually treated water in the case of environmental technology applications) containing dissolved or dispersed species to be sorbed (adsorbate, inorganic/organic xenobiotics). Due to the affinity of the adsorbent for the adsorbate species, the latter is attracted and bound there by different mechanisms. The process continues till equilibrium is established between the amount of solid-bound adsorbate species and its portion remaining in the solution or suspension. The degree of adsorbent affinity for the adsorbate determines its distribution between the solid and liquid phases.

There is an enormous amount of possible biosorbents available for the xenobiotics removal. The renewable character of biomass makes it an inexhaustible pool of biosorbents of all kinds. Biosorption has several advantages compared with conventional techniques, namely low cost of the biosorbents (majority of them is made from abundant or even waste material). In addition, large variability of biological materials and their pretreatment enables to select an appropriate biosorbent exhibiting at least partial selectivity in xenobiotics adsorption (*e.g.*, the proper biosorbent can exhibit preference for specific metal ion). In many cases it is possible to regenerate the biosorbents, enabling their reuse. During the biosorbent regeneration the metal recovery is possible. In many cases biosorption is capable of a performance comparable to the most similar technique, ion exchange treatment; in fact, ion exchange is rather costly, making the low cost of biosorption a major competing factor.

Biosorption of xenobiotics on different biosorbents is not based on only one mechanism. It consists of several mechanisms that quantitatively and qualitatively differ according to the separated compound, the origin of the biomass, and its processing. The complexity of biological materials used as biosorbents means that many functional groups, such as carboxyl, phosphate, sulphate, hydroxyl, amino, amido, thiol, acetamido, imidazole etc. are able to interact with individual xenobiotics. Biosorption often employs standard mechanisms of adsorption, ion exchange and complexation/coordination; in such cases, biosorption can be as rapid and reversible as in the case of standard ion exchange resins applications. Precipitation, where bound metal/radionuclide species can act as loci for their subsequent deposition can lead to very high uptake capacities, but the subsequent desorption can be inhibited. The presence of some functional groups, however, does not guarantee their accessibility for biosorption, perhaps due to steric, conformational, or other barriers (Gadd 2009).

13.3 Magnetic Biocomposite Materials and Magnetic Separation Processes

Magnetically responsive materials (usually in the form of nano- and microparticles) can be used either in their original form (*e.g.*, magnetite, maghemite, different types of ferrites) or mostly in the form of composite materials. In many cases magnetically responsive composites consist of small ferromagnetic, ferrimagnetic or superparamagnetic particles (usually in the nanometer to micrometer range), dispersed in a polymer, biopolymer or inorganic matrix; alternatively magnetic particles can be adsorbed on the outer surface or within the pores of diamagnetic particles or plain magnetic particles can be surface modified (*e.g.*, by silanization). Both native and composite magnetic materials can be classified as "smart materials" or "stimuli responsive materials", because they exhibit response to external magnetic field. Such materials are of special interest both from the scientific point of view but also due to their interesting applications in various areas of biosciences, medicine, biotechnology, environmental technology etc. The wide applicability of magnetic nano- and microparticles is caused by the existence of the following important properties (Safarik and Safarikova 2009):

- Magnetic nano- and microparticles can be selectively separated (removed) from the complex samples using an external magnetic field (*e.g.*, using an appropriate magnetic separator, permanent magnet, or electromagnet). This ability is the basis of the most often used applications of magnetic particles in biosciences and environmental technologies due to the fact that absolute majority of biological materials have diamagnetic properties, which enable efficient selective separation of magnetic and magnetically modified materials.

- Magnetic particles can be targeted to the desired place and kept there using an external magnetic field. These properties can be used, *e.g.*, in the course of magnetic drug targeting or during the construction of magnetic fluid seals.
- Magnetic particles can generate heat when subjected to an alternating magnetic field; this phenomenon is employed especially during magnetic fluid hyperthermia experimentally used for cancer treatment.
- Magnetic iron oxides nanoparticles generate a negative T2 contrast during magnetic resonance imaging thus serving as efficient contrast agents.
- Magnetic particles can be used as labels (*e.g.*, in combination with giant magnetoresistance sensors) enabling sensitive detection of target biologically active compounds.
- Magnetic nano- and microparticles can be used for magnetic modification of diamagnetic biological, synthetic and inorganic materials and for magnetic labeling of biologically active compounds (*e.g.*, antibodies).

There are several possibilities for the magnetic adsorbents separation and/or removal from the treated aqueous solutions and suspensions, depending both on the character of magnetic particles and the volume of the suspension. In most cases small-scale experiments with model aqueous solutions of the tested xenobiotic(s) have been performed and simple commercially available magnetic separators (or even just a piece of strong permanent magnet, see Fig. 13.1) have been used for

Fig. 13.1 Magnetically modified spruce sawdust before (*right*) and after (*left*) magnetic separation. (Reproduced, with permission, from Safarik and Safarikova 2010b)

magnetic separation of micrometer-sized magnetic (bio)composites. Separation of nanometer-sized magnetic materials usually requires more efficient separations systems, such as High gradient magnetic separator (HGMS). Drum magnetic separators can be used for large scale magnetic separation processes employing magnetic microparticles. An overview of commercially available magnetic separators for laboratory applications can be found in a recent paper (Safarik and Safarikova 2004).

13.4 Magnetic Biocomposite Materials for Xenobiotics Removal

Magnetic derivatives of different biological materials have been prepared and studied recently. For the purpose of this chapter the term "magnetically responsive biocomposites" will describe the following groups of materials, where the magnetic properties are caused by the presence of ferro-, ferri- or superparamagnetic materials, while diamagnetic component of the composite is formed by

- individual biopolymers (*e.g.*, chitosan, chitin, alginate, plant gums etc.),
- complex biopolymers, usually of plant origin (*e.g.*, sawdust, spent coffee grounds, peanut husks, spent grain etc.),
- microbial and algae cells (biomass).

Currently no comprehensive list of already described magnetically responsive biocomposites for potential environmental technology applications exists. On that account we have compiled basic information about various types of developed magnetically responsive biocomposites, their applications and other important details in the tabular form (Tables 13.1–13.7).

It can be clearly seen from the tables that there is an enormous variability of the procedures used to prepare different types of magnetic biocomposites. Both biocomposite nanoparticles and microparticles have been successfully prepared and used in model experiments. At present various magnetic derivatives of chitosan are of special interest, both for inorganic and organic xenobiotics removal. On the contrary, magnetic derivatives of plant waste materials represent a new promising type of biocomposites still waiting for further detailed research.

13.4.1 Magnetic Biopolymers

Chitosan has become a very attractive biopolymer for preparation of both non-magnetic and magnetic biocomposite adsorbents (Tables 13.1 and 13.2). For the sorption of metal ions, the amino groups of chitosan serve as the major sorption sites though the hydroxyl groups may also play some part. Under near neutral condition, nitrogen atoms of amino groups hold free electrons which react with metal cations, thus inducing the chelation process. Under acidic conditions, the protonated amino

Table 13.1 Magnetic biopolymer biocomposites for inorganic xenobiotics removal

Biopolymer	Type of composite	Adsorbed metals	Matrix	Additional information	Reference
Chitosan	Iron(III) loaded chitosan-magnetite nanocomposite	Cr^{6+}	Aqueous solution	Freundlich isotherm was used for fitting experimental data	Bajpai and Armo (2009)
	Magnetic chitosan resin	Au^{3+} Ag^{+}	Aqueous solution	Batch and column procedure; thiourea/glutaraldehyde cross-linking	Donia et al. (2007)
	Magnetic chitosan resin	Hg^{2+}	Aqueous solution	Eluted with 0.1 M KI; thiourea/glutaraldehyde cross-linking	Donia et al. (2008)
	Magnetic chitosan/amino resin; magnetic chitosan/quarternary amine resin	Cr^{6+}	Alkaline aqueous solution	Glutaraldehyde cross-linking; regeneration with 2 M NaCl and 0.5 M NaOH	Elwakeel (2010)
	Magnetic chitosan/amino resin; magnetic chitosan/quarternary amine resin	Mo^{6+}	Aqueous solution	Glutaraldehyde cross-linking; elution with 0.1 M ammonia buffer, pH 10	Elwakeel et al. (2009)
	Cross-linked magnetic chitosan beads	Cr^{6+}	Aqueous solution	Epichlorhydrin cross-linking; elution with 0.1 M HCl	Huang et al. (2009)
	magnetic chitosan nanoparticles	Co^{2+}	Aqueous solution	Carboxymethylchitosan covalently bound to magnetite nanoparticles after carbodimide activation	Chang et al. (2006)
	Monodisperse chitosan-bound Fe_3O_4 nanoparticles	Cu^{2+}	Aqueous solution	Carboxymethylchitosan covalently bound to magnetite nanoparticles after carbodimide activation	Chang and Chen (2005b)
	Monodisperse chitosan-coated Fe_3O_4 nanoparticles	Au^{3+}	Aqueous solution	Carboxymethylchitosan covalently bound to magnetite nanoparticles after carbodimide activation; optimal adsorption at pH 2	Chang and Chen (2006)
	Amine-functionalized magnetite nanoparticles—chitosan nanocomposite	Pb^{2+} Cu^{2+} Cd^{2+}	Water solution	Regeneration with weak acidic solution and ultrasound	Liu et al. (2009)

Table 13.1 (continued)

Biopolymer	Type of composite	Adsorbed metals	Matrix	Additional information	Reference
Chitosan (continued)	Cross-linked chitosan-diacetylmonoxime Shiff's base resin	Co^{2+} Cu^{2+} Ni^{2+}	Aqueous solution	Glyoxal cross-linking; regeneration with 0.01–0.1 M EDTA	Monier et al. (2010b)
	Cross-linked chitosan-isatin Shiff's base resin	Co^{2+} Cu^{2+} Ni^{2+}	Aqueous solution	Glyoxal cross-linking; regeneration with 0.01–0.1 M EDTA	Monier et al. (2010a)
	Chitosan-coated magnetite nanoparticles	Fe^{3+}	Blood serum	possible treatment of.β-thalasemia	Namdeo and Bajpai (2007)
	Chitosan-magnetite nanocomposites	Fe^{3+}	Aqueous solution	Cu^{2+} ions decrease uptake of Fe^{3+}	Namdeo and Bajpai (2008)
	Porous magnetic chitosan beads	Cd^{2+}	Waste water	Glutaraldehyde cross-linking and freeze-drying	Rorrer et al. (1993)
	Saccharomyces cerevisiae—magnetic chitosan particles	Cu^{2+}	Aqueous solution	Cells immobilized on chitosan-coated magnetic nanoparticles	Peng et al. (2010)
	Magnetic chitosan microspheres	Cd^{2+} Cu^{2+} Ni^{2+}	Aqueous solution	Chemical modification with ethylenediamine; maximum adsorption capacities were 54.3 mg/g for Cu^{2+}, 20.4 mg/g for Cd^{2+}, and 12.4 mg/g for Ni^{2+}	Zhou et al. (2007)
	Cu(II) ion imprinted composite adsorbent (waste fungal mycelium + chitosan + Fe_3O_4 nanoparticles)	Cu^{2+}	Aqueous solution	metal ion affinity: Cu(II) > Zn(II) > Co(II) > Ni(II); adsorbent could be reused five times	Ren et al. (2008)
	Carboxymethyl chitosan-Fe_3O_4 nanoparticles	Zn^{2+}	Aqueous solution	Maximum adsorption capacity was 20.4 mg/g	Zhou et al. (2006)
	Chitosan-coated magnetic nanoparticles	Cu^{2+}	Aqueous solution	Chitosan modification with α-ketoglutaric acid	Zhou et al. (2009b)

Table 13.1 (continued)

Biopolymer	Type of composite	Adsorbed metals	Matrix	Additional information	Reference
Chitosan (continued)	Thiourea-modified magnetic chitosan microspheres	Hg^{2+} Cu^{2+} Ni^{2+}	Aqueous solution	Maximum adsorption capacities were 625.2, 66.7, and 15.3 mg/g for Hg^{2+}, Cu^{2+}, and Ni^{2+} ions, resp.	Zhou et al. (2009a)
	Chitosan-coated magnetic nanoparticles	Cu^{2+}	Aqueous solution	Chitosan modification with α-ketoglutaric acid; maximum adsorption capacity was 96.15 mg/g	Zhou et al. (2009c)
Chitin	Magnetite-immobilized chitin	Cu^{2+}	Aqueous solution	Maximum adsorption capacity was 53.19 mg/g	Wong et al. (2007)
Alginate	Calcium-alginate encapsulated magnetic sorbent	AsO_4^{3-}	Aqueous solution	Arsenate is reduced to arsenite after its adsorption onto the sorbent	Lim et al. (2009b)
	Magnetic alginate microcapsules	Ni^{2+}	Aqueous solution	Microcapsules contain Cyanex 272	Ngomsik et al. (2006)
	Magnetic alginate beads	Co^{2+}	Aqueous solution	Beads contain Cyanex 272; desorption by nitric acid, pH 1	Ngomsik et al. (2009)
	Calcium alginate encapsulated magnetic sorbent	Cu^{2+}	Aqueous solution	Maximum adsorption capacity was 60 mg/g; optimal $pH > 5$	Lim et al. (2009a)
	Calcium alginate encapsulated magnetic sorbent	Cu^{2+}	Aqueous solution	Maximum adsorption capacity was 0.99 m mol/g	Lim et al. (2008)
	Calcium alginate magnetic particles	AsO_4^{3-} Cu^{2+}	Aqueous solution	Higher pH enhances adsorption of Cu^{2+} and decreases adsorption of As^{5+}	Lim and Chen (2007)
Gum arabic	Adsorption of gum on the surface of magnetite nanoparticles	Cu^{2+}	Aqueous solution	Maximum adsorption capacity 38.5 mg/g; desorption in $pH \leq 2.0$	Banerjee and Chen (2007)
Gellan gum	Gel beads with magnetite nanoparticles	Pb^{2+} Cr^{3+} Mn^{2+}	Aqueous solution	Desorption with sodium citrate	Wang et al. (2009)

groups cause the binding of metal anions (*e.g.*, molybdate, vanadate, arsenate etc.) by electrostatic forces. However, the exact mechanism of the uptake of a certain metal ion may alter depending on the pH and the composition of the solution. These mechanisms suggest that the sorption efficiencies of metal ions are higher at acidic pH conditions and on chitosan with low degree of acetylation. For instance, it has been shown that chitosan may present four times higher capacity than chitin for the sorption of reactive red 222 and Pb(II) ions. Without modification, chitosan shows low sorptivity to alkaline and alkaline earth metal ions (Li et al. 2008).

Chitosan can also adsorb dyes over a wide pH range but with very different capacities. It is postulated that, under acidic conditions, chemical adsorption plays a major role with amino groups being the effective functional groups; under alkaline condition, however, physical adsorption dominates while the hydroxyl groups are the effective functional groups. Electrostatic attraction, hydrophobic interaction, and physical adsorption are among the proposed interactions between chitosan and dyes. The molecular size and anionicity of dyes are also the important parameters affecting the sorption efficiency (Li et al. 2008).

In addition to chitin and chitosan, other biopolymers like alginate and plant gums have been used for the preparation of magnetic biocomposites for xenobiotics removal (Tables 13.1 and 13.2).

In some cases magnetic biopolymer particles have served as a matrix to capture active adsorbents enabling their easy manipulation. For example, the hydrophobic octadecyl functionalized magnetite nanoparticles (Fe_3O_4@C-18) were caged into hydrophilic barium alginate particles to obtain a novel type of solid-phase extraction adsorbents, which were applied to the pre-concentration of polycyclic aromatic hydrocarbons and phthalate esters pollutants from environmental water samples. The hydrophilicity of the barium alginate cage enhances the dispersibility of adsorbents in water samples while the magnetic properties of the Fe_3O_4 core facilitate magnetic separation (Zhang et al. 2010).

13.4.2 Magnetic Plant Derivatives

There are numerous studies describing adsorption properties of naturally occurring low cost plant-based materials, such as agricultural and wood industry by-products and wastes (*e.g.*, sawdust, barley straw, tree bark, peanut skins, spent coffee grounds, spent grain etc.). Several studies have shown that these materials, especially sawdust, are very promising adsorbents for removing heavy metal ions, acid and basic dyes, oils and some other unwanted materials from waste water. Many agricultural by-products are very cheap, available in large quantities, and often there are problems with their disposal. The use of agricultural and wood wastes for removing pollutants would be beneficial both for the environment and the corresponding industries. Review papers describing the role of sawdust and agricultural waste materials in the removal of unwanted materials from water have been published recently (Farooq et al. 2010; Sud et al. 2008; Ahluwalia and Goyal 2007; Shukla et al. 2002).

Table 13.2 Magnetic biopolymer biocomposites for organic xenobiotics removal

Biopolymer	Type of composite	Adsorbed compounds	Matrix	Additional information	Reference
Chitosan	Magnetic chitosan/amino resin; magnetic chitosan/quarternary amine resin	Ractive Black 5	Aqueous solution	Elution with NH_4OH/NH_4Cl buffer (pH 10)	Elwakeel (2009)
	γ-Fe_2O_3/crosslinked chitosan	Methyl orange	Aqueous solution	Adsorption capacity decreased with increase of pH	Zhu et al. (2010)
	Magnetic chitosan gel particles with covalently immobilized copper phtalocyanine dye	Organic polycylic compounds	Water solution	Elution with methanol, methanol-conc. ammonia solution and acetic acid	Safarik (1995)
	Carboxymethylated chitosan covalently bound to magnetic nanoparticles	Acid dyes (crocein orange G, acid green 25)	Aqueous solution	Carbodiimide activation; desorption with NaCl and NaOH	Chang and Chen (2005a)
	Octadecyl functionalized magnetite nanoparticles caged into barium alginate particles	Polycyclic aromatic hydrocarbons, phthalate esters	Aqueous solution	Solid-phase extraction of analytes from large sample volumes	Zhang et al. (2010)
Chitin	Magnetite-immobilized chitin	Pentachloro-phenol	Aqueous solution	Adsorption ability of immobilized chitin decreased when pH and temperature increased	Pang et al. (2007)
Alginate	Magnetic calcium alginate beads	Methyl orange, methylene blue	Aqueous solution	Epichlorohydrin cross-linking	Rocher et al. (2010)
	Magnetic calcium alginate beads	Methyl orange, methylene blue	Aqueous solution	Activated carbon entrapped in alginate beads	Rocher et al. (2008)
	Barium alginate caged Fe_3O_4@C18 magnetic nanoparticles	Polycyclic aromatic hydrocarbons; phtalate esters	Environmental water samples	Used for magnetic solid phase extraction	Zhang et al. (2010)

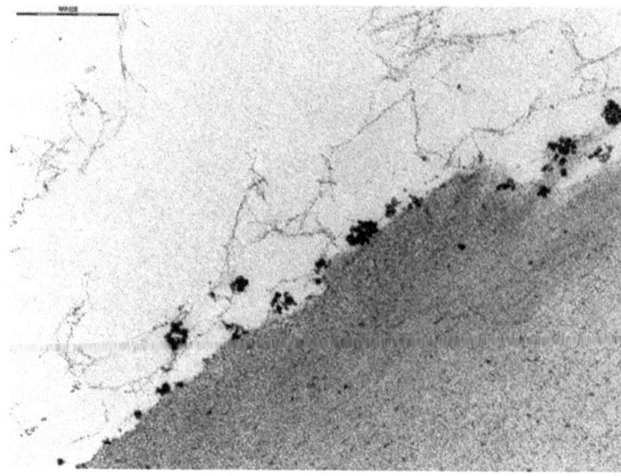

Fig. 13.2 Transmission electron microscope images of ultra thin sections of magnetic sawdust particles. The bar corresponds to 200 nm. (Reproduced, with permission, from Safarik et al. 2007a)

Currently there are only a few examples of magnetically responsive plant-based materials applicable for xenobiotics removal (see Table 13.3); currently especially adsorption of water soluble dyes has been studied. These materials have been applied in laboratory experiments. Water-based magnetic fluid stabilized with perchloric acid has been used for laboratory-scale magnetic modification of described plant-based materials (see Fig. 13.2). Maximum adsorption capacities reached values up to almost 100 mg/g (see Table 13.4).

13.4.3 Magnetically Modified Microbial and Algae Cells

Microbial cells, either in free or immobilized form, can be used for the preconcentration or removal of metal ions, organic and inorganic xenobiotics or biologically active compounds. The adsorption of xenobiotics on microbial and algae cells depends on the composition of their cell walls. For instance, the cell walls of yeasts contain large number of complex organic compounds and their polymers such as glucan (28%), mannan (31%), proteins (13%), lipids (8%), chitin and chitosan (2%). The cell walls of brown algae contain alginic acid (10–40%), fucoidan (5–20%) and cellulose (2–20%); red algae contain agar, carrageenan, xylans, pectin and cellulose, while the cell walls of green algae contain mainly pectic substances and cellulose. Such compounds possess numerous functional groups involved in xenobiotics binding process. The studies with fungal biomass and seaweed have suggested a dominant role for ion exchange metal binding to ionized carboxyl, phosphate and amine groups. Gold (III) is probably bound to the hydroxyl groups of polysaccharides and the carboxylate anions of amino acids from the peptidoglycan layer on the yeast cell walls. Positively charged amino groups are involved in the adsorption of Cr(VI) (anionic form $HCrO_4^-$), while imidazol groups of yeast histidine strongly interact with As(V) (in $HAsO_4^{2-}$ form). Different chemical and physical treatments

Table 13.3 Magnetic plant-based biocomposites for organic xenobiotics removal

Plant derivative	Type of composite	Adsorbed compounds	Matrix	Additional information	Reference
Spruce sawdust	Sawdust modified with water based magnetic fluid	Water soluble organic dyes	Aqueous solution	Change in pH can increases dyes adsorption	Safarik et al. (2007a)
Spruce sawdust	Sawdust modified with water based magnetic fluid	Water soluble organic dyes	Aqueous solution	Magnetic and microscopy characterization performed	Safarik et al. (2005)
Spruce sawdust	Sawdust modified with water based magnetic fluid			Detailed magnetic characterization of the prepared biocomposite	Mosiniewicz-Szablewska et al. (2007)
Peanut husks	Peanut husks modified with water based magnetic fluid	Water soluble organic dyes	Aqueous solution	Adsorption equilibrium reached in 60–90 min	Safarik and Safarikova (2010a)
Spent coffee grounds	Spent coffee grounds modified with water based magnetic fluid	Water soluble organic dyes	Aqueous solution	Magnetic solid phase extraction of triphenylmethane dyes performed	Safarik et al. (unpublished)
Spent grain	Spent grain modified with water based magnetic fluid	Water soluble organic dyes	Aqueous solution	Large differences in the adsorption of various dyes	Safarik et al. (2011)

Table 13.4 Comparison of maximum adsorption capacities Q_{max} (mg/g) of magnetically modified plant-based materials for tested dyes

Dyes	Colour index number	Maximum adsorption capacities of magnetically responsive nanobiocomposites (mg/g)			
		Spruce sawdust[a]	Peanut husks[b]	Spent coffee grounds[c]	Spent barley grains[d]
Acridine orange	46,005	24.1	71.4	49.3	
Aniline blue	42,755				44.7
Bismarck brown	21,000	52.1	95.3	97.8	72.4
Crystal violet	42,555	52.4	80.9	36.7	40.2
Malachite green	42,000			62.9	
Safranin O	50,240	25.0	86.1	34.3	

[a] Safarik et al. (2005); [b] Safarik and Safarikova (2010a); [c] Safarik et al. (unpublished); [d] Safarik et al. (2011)

of biomass can unmask or expose the metal binding groups through the disruption and permealization of membranes or through the modification of existing ones. Treatment with HCl, NaOH or organic solvents (DMSO, ethanol) leads to enhanced adsorption of some metal ions (Godlewska-Zylkiewicz 2006). Review papers describing the role of microbial biomass in the removal of unwanted materials from water have been published recently (Kaushik and Malik 2009; Vijayaraghavan and Yun 2008; Ahluwalia and Goyal 2007; Wang and Chen 2006; Fu and Viraraghavan 2001).

Magnetotactic bacteria are exceptional microorganisms having the ability to synthesize intracellular biogenic magnetic nanoparticles (based either on magnetite or greigite), which enable their magnetic separation. Other prokaryotic and eukaryotic microbial cells have to be magnetically modified, usually by forming complexes with magnetic particles. Generally microbial cells can be modified by the non-specific attachment of magnetic nanoparticles (*e.g.*, by the magnetic fluid treatment), by binding of maghemite or magnetite particles on the cell surface, by specific interactions with immunomagnetic nano- and microparticles, by the biologically driven precipitation of paramagnetic compounds on the cell surface, by covalent immobilization on magnetic carriers, by cross-linking of the cells or isolated cell walls with a bifunctional reagent in the presence of magnetic particles or by entrapment (together with magnetic particles) into biocompatible polymers. Alternatively the modification can be performed by binding paramagnetic cations on acid groups on the cell surface (Safarik and Safarikova 2007).

Recently a new and simple procedure for the preparation of magnetically modified cells, based on suspending the microbial cells in methanol and subsequent addition of perchloric acid stabilized ferrofluid has been developed (Safarik et al. 2007a). During the modification process, a specific precipitation of magnetic nanoparticles on the outer surface of treated microbial cells occurred (see Fig. 13.3).

Tables 13.5 and 13.6 show the possible applications of magnetically modified prokaryotic and eukaryotic microbial and algae cells for xenobiotics removal. In Table 13.7 maximum adsorption capacities of magnetically modified yeast and algae cells for dyes removal have been summarized and compared.

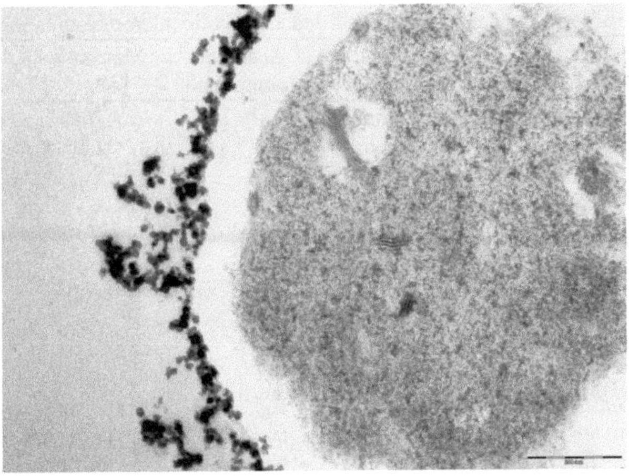

Fig. 13.3 Transmission electron microscopy picture of magnetically modified *Kluyveromyces fragilis* cells. The bar line corresponds to 200 nm. (Reproduced, with permission, from Safarik et al. 2007b)

Table 13.5 Magnetically modified prokaryotic microbial cells for xenobiotics removal

Prokaryotic cells	Type of composite	Adsorbed compounds	Matrix	Additional information	Reference
Enterobacter sp.	Magnetite-immobilized cells	Ni^{2+}	Aqueous solution	Batch procedure; elution with diluted citric acid	Wong and Fung (1997)
Pseudomonas putida	Magnetite-immobilized cells	Cu^{2+}	Aqueous solution	Batch type biosorption reactors; HCl pretreatment of cells	Chua et al. (1998)
Pseudomonas putida	Magnetite-immobilized cells	Cu^{2+}	Wastewater	HCl pretreatment of cells, desorption by acidic treatment	Lei et al. (2000)
Pseudomonas putida	Magnetite-immobilized cells	Cu^{2+}	electro-planting effluent	batch procedure; adsorption in bioreactor	Sze et al. (1996)
Pseudomonas putida	Magnetite-immobilized cells	Cu^{2+}	Industrial water effluent	Semi-continuous biosorption; HCl pretreatment of cells; desorption by acidic treatment	Wang et al. (2000)
Stenotrophomonas sp.	Magnetotactic bacterium	Au^{3+}	Aqueous solution	Desorption with thiourea	Song et al. (2008)
Rhodopseudomonas sphaeroides	Magnetite-immobilized cells	Chlorinated hydrocarbons	Wastewater	Six compounds tested	MacRae (1986)

Table 13.6 Magnetically modified eukaryotic microbial cells for xenobiotics removal

Eukaryotic cells	Type of composite	Adsorbed compounds	Matrix	Additional information	Reference
Saccharomyces cerevisiae	Cells immobilized on chitosan-coated magnetic nanoparticles	Cu^{2+}	Aqueous solution	Optimal pH=4.5; maximum adsorption capacity was 144.9 mg/g	Peng et al. (2010)
Saccharomyces cerevisiae cell walls	Biosorbent with covalently and non-covalently bound magnetite	Cu^{2+} Cd^{2+} Ag^{+}	Aqueous solution	Epichlorhydrin cross-linking	Patzak et al. (1997)
Saccharomyces cerevisiae subsp. *uvarum*	Cells modified with water based magnetic fluid	Hg^{2+} Cu^{2+} Ni^{2+} Zn^{2+}	Water solution; artificial wastewater	Regeneration with 0.1 M HNO_3	Yavuz et al. (2006)
Kluyveromyces fragilis	Cells modified with water based magnetic fluid	Sr^{2+}	Aqueous solution	Desorption with 0.1 M HNO_3	Ji et al. (2010)
Saccharomyces cerevisiae	Cells modified with water based magnetic fluid	Water-soluble dyes	Aqueous solution	Five dyes tested	Safarik et al. (2002)
Kluyveromyces fragilis	Cells modified with water based magnetic fluid	Water-soluble dyes	Aqueous solution	Seven dyes tested	Safarik et al. (2007b)
Saccharomyces cerevisiae subsp. *uvarum*	Cells modified with water based magnetic fluid	Water-soluble dyes	Aqueous solution	Five dyes tested	Safarikova et al. (2005)
Chlorella vulgaris	Algae cells modified with water based magnetic fluid	Water-soluble dyes	Aqueous solution	6 dyes tested	Safarikova et al. (2008)
Saccharomyces cerevisiae	Cells modified with water based magnetic fluid	Direct scarlet dye	Aqueous solution	99% of adsorbed dye eluted in 70% alcohol	Wu et al. (2009)

Table 13.7 Comparison of maximum adsorption capacities Q_{max} (mg/g) of magnetically modified yeast and algae cells for tested dyes

Dyes	Colour index number	Maximum adsorption capacities of magnetically responsive microbial cells (mg/g)			
		Saccharomyces cerevisiae[a]	*Saccharomyces cerevisiae* subsp. *uvarum*[b]	*Kluyveromyces fragilis*[c]	*Chlorella vulgaris*[d]
Acridine orange	46,005	82.8		62.2	
Amido black 10B	20,470		11.6	29.9	
Aniline blue	42,755	430.2	228.0		257.9
Bismarck brown	21,000			75.7	201.9
Congo red	22,120		93.1	49.7	156.7
Crystal violet	42,555	85.9	41.7	42.9	42.9
Malachite green	42,000	19.6			
Safranin O	50,240	90.3	46.6	138.2	115.7
Saturn blue LBRR	34,140			33.0	24.2

[a] Safarik et al. (2002); [b] Safarikova et al. (2005); [c] Safarik et al. (2007b); [d] Safarikova et al. (2008)

13.5 Concluding Remarks

Different types of biological materials, including agro-industrial by-products and microbial cells, can be successfully utilized as biosorbents for xenobiotics removal. Conversion of these biosorbents into "smart materials" exhibiting response to external magnetic field may be one of the possible ways how to improve biosorbents applicability, enabling their selective magnetic separation from waste water. Despite the fact, that currently magnetically responsive biocomposites have been tested only in laboratory experiments, there is a great potential for their large-scale applications in the near future.

References

Ahluwalia SS, Goyal D (2007) Microbial and plant derived biomass for removal of heavy metals from wastewater. Bioresour Technol 98:2243–2257

Aksu Z (2005) Application of biosorption for the removal of organic pollutants: a review. Process Biochem 40:997–1026

Bajpai SK, Armo MK (2009) Equilibrium sorption of hexavalent chromium from aqueous solution using iron(III)-loaded chitosan-magnetite nanocomposites as novel sorbent. J Macromol Sci A 46:510–520

Banerjee SS, Chen DH (2007) Fast removal of copper ions by gum arabic modified magnetic nano-adsorbent. J Hazard Mater 147:792–799

Chang YC, Chen DGH (2005a) Adsorption kinetics and thermodynamics of acid dyes on a carboxymethylated chitosan-conjugated magnetic nano-adsorbent. Macromol Biosci 5:254–261

Chang YC, Chen DH (2005b) Preparation and adsorption properties of monodisperse chitosan-bound Fe_3O_4 magnetic nanoparticles for removal of Cu(II) ions. J Coll Interf Sci 283:446–451

Chang YC, Chen DH (2006) Recovery of gold(III) ions by a chitosan-coated magnetic nano-adsorbent. Gold Bull 39:98–102

Chang YC, Chang SW, Chen DH (2006) Magnetic chitosan nanoparticles: studies on chitosan binding and adsorption of Co(II) ions. React Funct Polym 66:335–341

Chua H, Wong PK, Yu PHF, Li XZ (1998) The removal and recovery of copper (II) ions from wastewater by magnetite immobilized cells of *Pseudomonas putida* 5-X. Water Sci Technol 38:315–322

Donia AM, Atia AA, Elwakeel KZ (2007) Recovery of gold(III) and silver(I) on a chemically modified chitosan with magnetic properties. Hydrometallurgy 87:197–206

Donia AM, Atia AA, Elwakeel KZ (2008) Selective separation of mercury(II) using magnetic chitosan resin modified with Schiff's base derived from thiourea and glutaraldehyde. J Hazard Mater 151:372–379

Elwakeel KZ (2009) Removal of reactive black 5 from aqueous solutions using magnetic chitosan resins. J Hazard Mater 167:383–392

Elwakeel KZ (2010) Removal of Cr(VI) from alkaline aqueous solutions using chemically modified magnetic chitosan resins. Desalination 250:105–112

Elwakeel KZ, Atia AA, Donia AM (2009) Removal of Mo(VI) as oxoanions from aqueous solutions using chemically modified magnetic chitosan resins. Hydrometallurgy 97:21–28

Farooq U, Kozinski JA, Khan MA, Athar M (2010) Biosorption of heavy metal ions using wheat based biosorbents: a review of the recent literature. Bioresour Technol 101:5043–5053

Fu YZ, Viraraghavan T (2001) Fungal decolorization of dye wastewaters: a review. Bioresour Technol 79:251–262

Gadd GM (2009) Biosorption: critical review of scientific rationale, environmental importance and significance for pollution treatment. J Chem Technol Biotechnol 84:13–28

Godlewska-Zylkiewicz B (2006) Microorganisms in inorganic chemical analysis. Anal Bioanal Chem 384:114–123

Huang GL, Zhang HY, Shi JX, Langrish TAG (2009) Adsorption of chromium(VI) from aqueous solutions using cross-linked magnetic chitosan beads. Ind Eng Chem Res 48:2646–2651

Ji YQ, Hu YT, Tian Q, Shao XZ, Li JY, Safarikova M, Safarik I (2010) Biosorption of strontium ions by magnetically modified yeast cells. Sep Sci Technol 45:1499–1504

Kaushik P, Malik A (2009) Fungal dye decolourization: recent advances and future potential. Environ Int 35:127–141

Lei W, Chua H, Lo WH, Yu PHF, Zhao YG, Wong PK (2000) A novel magnetite-immobilized cell process for heavy metal removal from industrial effluent. Appl Biochem Biotechnol 84–86:1113–1126

Li CB, Hein S, Wang K (2008) Biosorption of chitin and chitosan. Mater Sci Technol 24:1088–1099

Lim SF, Chen JP (2007) Synthesis of an innovative calcium-alginate magnetic sorbent for removal of multiple contaminants. Appl Surf Sci 253:5772–5775

Lim SF, Zheng YM, Zou SW, Chen JP (2008) Characterization of copper adsorption onto an alginate encapsulated magnetic sorbent by a combined FT-IR, XPS and mathematical modeling study. Environ Sci Technol 42:2551–2556

Lim SF, Zheng YM, Zou SW, Chen JP (2009a) Removal of copper by calcium alginate encapsulated magnetic sorbent. Chem Eng J 152:509–513

Lim SF, Zheng YM, Zou SW, Chen JP (2009b) Uptake of arsenate by an alginate-encapsulated magnetic sorbent: process performance and characterization of adsorption chemistry. J Colloid Interface Sci 333:33–39

Liu XW, Hu QY, Fang Z, Zhang XJ, Zhang BB (2009) Magnetic chitosan nanocomposites: a useful recyclable tool for heavy metal ion removal. Langmuir 25:3–8

MacRae IC (1986) Removal of chlorinated hydrocarbons from water and waste-water by bacterial cells adsorbed to magnetite. Water Res 20:1149–1152

Monier M, Ayad DM, Wei Y, Sarhan AA (2010a) Adsorption of Cu(II), Co(II), and Ni(II) ions by modified magnetic chitosan chelating resin. J Hazard Mater 177:962–970

Monier M, Ayad DM, Wei Y, Sarhan AA (2010b) Preparation and characterization of magnetic chelating resin based on chitosan for adsorption of Cu(II), Co(II), and Ni(II) ions. React Funct Polym 70:257–266

Mosiniewicz-Szablewska E, Safarikova M, Safarik I (2007) Magnetic studies of ferrofluid-modified spruce sawdust. J Phys D 40:6490–6496

Namdeo M, Bajpai SK (2007) Chitosan-coated magnetite (CCM) nanoparticles as novel iron chelator for treatment of β-thalassemia by Tag and Drag (TAD) approach. J Bionanosci 1:131–133

Namdeo M, Bajpai SK (2008) Chitosan-magnetite nanocomposites (CMNs) as magnetic carrier particles for removal of Fe(III) from aqueous solutions. Colloid Surf A 320:161–168

Ngomsik AF, Bee A, Siaugue JM, Cabuil V, Cote G (2006) Nickel adsorption by magnetic alginate microcapsules containing an extractant. Water Res 40:1848–1856

Ngomsik AF, Bee A, Siaugue JM, Talbot D, Cabuil V, Cote G (2009) Co(II) removal by magnetic alginate beads containing Cyanex 272 (R). J Hazard Mater 166:1043–1049

Pang KM, Ng S, Chung WK, Wong PK (2007) Removal of pentachlorophenol by adsorption on magnetite-immobilized chitin. Water Air Soil Pollut 183:355–365

Patzak M, Dostalek P, Fogarty RV, Safarik I, Tobin, JM (1997) Development of magnetic biosorbents for metal uptake. Biotechnol Tech 11:483–487

Peng QQ, Liu YG, Zeng GM, Xu WH, Yang CP, Zhang JJ (2010) Biosorption of copper(II) by immobilizing *Saccharomyces cerevisiae* on the surface of chitosan-coated magnetic nanoparticles from aqueous solution. J Hazard Mater 177:676–682

Ren Y, Zhang M, Zhao D (2008) Synthesis and properties of magnetic Cu(II) ion imprinted composite adsorbent for selective removal of copper. Desalination 228:135–149

Rocher V, Siaugue J-M, Cabuil V, Bee A (2008) Removal of organic dyes by magnetic alginate beads. Water Res 42:1290–1298

Rocher V, Bee A, Siaugue JM, Cabuil V (2010) Dye removal from aqueous solution by magnetic alginate beads crosslinked with epichlorohydrin. J Hazard Mater 178:434–439

Rorrer GL, Hsien TY, Way JD (1993) Synthesis of porous magnetic chitosan beads for removal of cadmium ions from waste water. Ind Eng Chem Res 32:2170–2178

Safarik I (1995) Removal of organic polycyclic compounds from water solutions with a magnetic chitosan based sorbent bearing copper phthalocyanine dye. Water Res 29:101–105

Safarik I, Safarikova M (2004) Magnetic techniques for the isolation and purification of proteins and peptides. BioMagn Res Technol 2:7

Safarik I, Safarikova M (2007) Magnetically modified microbial cells: a new type of magnetic adsorbents. China Part 5:19–25

Safarik I, Safarikova M (2009) Magnetic nano- and microparticles in biotechnology. Chem Pap 63:497–505

Safarik I, Safarikova M (2010a) Magnetic fluid modified peanut husks as an adsorbent for organic dyes removal. Physics Procedia 9:274–278

Safarik I, Safarikova M (2010b) Magnetically responsive (nano)composites as perspective materials for environmental technology applications. ENT Mag (January–February) 2010:85–90

Safarik I, Horska K, Safarikova M (2011) Magnetically modified spent grain for dye removal. J Cereal Sci, in press

Safarik I, Ptackova L, Safarikova M (2002) Adsorption of dyes on magnetically labeled baker's yeast cells. Eur Cells Mater 3(Suppl 2):52–55

Safarik I, Safarikova M, Weyda F, Mosiniewicz-Szablewska E, Slawska-Waniewska A (2005) Ferrofluid-modified plant-based materials as adsorbents for batch separation of selected biologically active compounds and xenobiotics. J Magn Magn Mater 293:371–376

Safarik I, Lunackova P, Mosiniewicz-Szablewska E, Weyda F, Safarikova M (2007a) Adsorption of water-soluble organic dyes on ferrofluid-modified sawdust. Holzforschung 61:247–253

Safarik I, Rego LFT, Borovska M, Mosiniewicz-Szablewska E, Weyda F, Safarikova M (2007b) New magnetically responsive yeast-based biosorbent for the efficient removal of water-soluble dyes. Enzyme Microb Technol 40:1551–1556

Safarikova M, Ptackova L, Kibrikova I, Safarik I (2005) Biosorption of water-soluble dyes on magnetically modified *Saccharomyces cerevisiae* subsp. *uvarum* cells. Chemosphere 59:831–835

Safarikova M, Pona BMR, Mosiniewicz-Szablewska E, Weyda F, Safarik I (2008) Dye adsorption on magnetically modified *Chlorella vulgaris* cells. Fresenius Environ Bull 17:486–492
Shukla A, Zhang YH, Dubey P, Margrave JL, Shukla SS (2002) The role of sawdust in the removal of unwanted materials from water. J Hazard Mater 95:137–152
Song HP, Li XG, Sun JS, Xu SM, Han X (2008) Application of a magnetotactic bacterium, *Stenotrophomonas* sp to the removal of Au(III) from contaminated wastewater with a magnetic separator. Chemosphere 72:616–621
Srinivasan A, Viraraghavan T (2010) Decolorization of dye wastewaters by biosorbents: a review. J Environ Manage 91:1915–1929
Sud D, Mahajan G, Kaur MP (2008) Agricultural waste material as potential adsorbent for sequestering heavy metal ions from aqueous solutions: a review. Bioresour Technol 99:6017–6027
Sze KF, Lu YJ, Wong PK (1996) Removal and recovery of copper ion (Cu^{2+}) from electroplating effluent by a bioreactor containing magnetite-immobilized cells of *Pseudomonas putida* 5X. Resour Conserv Recycl 18:175–193
Vijayaraghavan K, Yun YS (2008) Bacterial biosorbents and biosorption. Biotechnol Adv 26:266–291
Wang J, Chen C (2006) Biosorption of heavy metals by *Saccharomyces cerevisiae:* a review. Biotechnol Adv 24:427–451
Wang JL, Chen C (2009) Biosorbents for heavy metals removal and their future. Biotechnol Adv 27:195–226
Wang L, Chua H, Wong PK, Lo WH, Yu PHF, Zhao YG (2000) An optimal magnetite immobilized *Pseudomonas putida* 5X cell system for Cu^{2+} removal from industrial waste effluent. Water Sci Technol 41:241–246
Wang X, Zhao CM, Zhao P, Dou PP, Ding Y, Xu P (2009) Gellan gel beads containing magnetic nanoparticles: an effective biosorbent for the removal of heavy metals from aqueous system. Bioresour Technol 100:2301–2304
Wong KS, Wong KH, Ng S, Chung WK, Wong PK (2007) Adsorption of copper ion on magnetite-immobilised chitin. Water Sci Technol 56:135–143
Wong PK, Fung KY (1997) Removal and recovery of nickel ion (Ni^{2+}) from aqueous solution by magnetite-immobilized cells of *Enterobacter* sp 4–2. Enzyme Microb Technol 20:116–121
Wu Q, Shan Z, Shen M, Li S, Chen H (2009) Biosorption of direct scarlet dye on magnetically modified *Saccharomyces cerevisiae* cells. Chin J Biotech 25:1477–1482
Yavuz H, Denizli A, Gungunes H, Safarikova M, Safarik I (2006) Biosorption of mercury on magnetically modified yeast cells. Sep Purif Technol 52:253–260
Zhang SX, Niu HY, Cai YQ, Shi YL (2010) Barium alginate caged Fe_3O_4@C-18 magnetic nanoparticles for the pre-concentration of polycyclic aromatic hydrocarbons and phthalate esters from environmental water samples. Anal Chim Acta 665:167–175
Zhou LM, Wang YP, Liu ZR, Huang QW (2006) Carboxymethyl chitosan-Fe_3O_4 nanoparticles: preparation and adsorption behaviors towards Zn^{2+} ions. Acta Physico-Chimica Sinica 22:1342–1346
Zhou LM, Wang YP, Huang QW, Liu ZR (2007) Adsorption properties of Cu^{2+}, Cd^{2+} and Ni^{2+} by modified magnetic chitosan microspheres. Acta Phys-Chim Sin 23:1979–1984
Zhou LM, Wang YP, Liu ZR, Huang QW (2009a) Characteristics of equilibrium, kinetics studies for adsorption of Hg(II), Cu(II), and Ni(II) ions by thiourea-modified magnetic chitosan microspheres. J Hazard Mater 161:995–1002
Zhou YT, Branford-White C, Nie HL, Zhu LM (2009b) Adsorption mechanism of Cu^{2+} from aqueous solution by chitosan-coated magnetic nanoparticles modified with α-ketoglutaric acid. Colloid Surf B 74:244–252
Zhou YT, Nie HL, Branford-White C, He ZY, Zhu LM (2009c) Removal of Cu^{2+} from aqueous solution by chitosan-coated magnetic nanoparticles modified with α-ketoglutaric acid. J Colloid Interface Sci 330:29–37
Zhu HY, Jiang R, Xiao L, Li W (2010) A novel magnetically separable gamma-Fe_2O_3/crosslinked chitosan adsorbent: preparation, characterization and adsorption application for removal of hazardous azo dye. J Hazard Mater 179:251–257

Index

P. Kotrba et al. (eds.), *Microbial Biosorption of Metals*,
DOI 10.1007/978-94-007-0443-5, © Springer Science+Business Media B.V. 2011

D

MIX
Papier | Fördert gute Waldnutzung
FSC® C083411

Kolibri 360 GmbH
Ferdinand-Jühlke-Straße 7
99095 Erfurt, Deutschland
produktsicherheit@kolibri360.de